Computer Performance Modeling

COMPUTER PERFORMANCE MODELING

Alexandre Brandwajn
University of California, Santa Cruz, USA

Thomas Begin
Université Claude Bernard Lyon 1, France

NEW JERSEY • LONDON • SINGAPORE • BEIJING • SHANGHAI • TAIPEI • CHENNAI

Published by

World Scientific Publishing Co. Pte. Ltd.
5 Toh Tuck Link, Singapore 596224
USA office: 27 Warren Street, Suite 401-402, Hackensack, NJ 07601
UK office: 57 Shelton Street, Covent Garden, London WC2H 9HE

Library of Congress Control Number: 2024041756

British Library Cataloguing-in-Publication Data
A catalogue record for this book is available from the British Library.

COMPUTER PERFORMANCE MODELING

ISBN 9789811292040 (hardcover)
ISBN 9789811292521 (paperback)
ISBN 9789811292057 (ebook for institutions)
ISBN 9789811292064 (ebook for individuals)

For any available supplementary material, please visit
https://www.worldscientific.com/worldscibooks/10.1142/13809#t=suppl

Desk Editors: Soundararajan Raghuraman/Steven Patt

Typeset by Stallion Press
Email: enquiries@stallionpress.com

Preface

This text incorporates many years of experience doing research, teaching, and practicing computer performance modeling. This book can be used as a textbook for a graduate-level course in computer performance modeling, as well as a reference for professionals working in this field.

The first chapter of this text defines the notion of a model, discusses in general terms different classes of approaches used to solve a computer model, and stresses the importance of prediction as the ultimate practical goal of computer modeling. This chapter is recommended for all users of this text.

In our second chapter, we use a simple high-level model of a transaction processing system to introduce essential terminology, as well as a common practical way to derive balance equations. We also cover the important Little's formula. Refinements of this basic model lead us to consider models with finite population and state-dependent service rates. We use this opportunity to develop some insight into the applicability of open models in practice. Additionally, we derive in this chapter the important Pollaczek–Khinchine formula for a model with Poisson arrivals and general service times, and touch on different approaches that can be used to obtain the steady-state distribution of the number of customers in such a system. This gives us the opportunity to introduce the quasi-general phase-type and Coxian distributions. We proceed to show that a simple recurrence can be used to solve a model with quasi-Poisson arrivals and such a service time distribution. An analogous solution is then developed for a model with quasi-general times between arrivals and memoryless service times. We use a model of systems where no queueing is allowed to have a first look at the computation of the state seen by an arriving customer from the standard steady-state probabilities. Unless one is using this text to teach only selected advanced topics in computer performance modeling, Chapter 2 is recommended for all users of this text.

Chapter 3 is devoted to fundamental results for models consisting of networks of interconnected queues. We discuss the classical central server model and the general Jacksonian networks. We study the state seen by an arriving customer in such product-form networks and the related mean value analysis as a solution approach. We then discuss the BCMP theorem as a generalization of the product-form solution to models with classes of customers and quasi-general service times for a set of service disciplines. Section 3.5, devoted to BCMP networks, is notation-heavy so an instructor might choose to cover its contents only in general terms. Unless one is using this text for specific advanced topics, this chapter is recommended for all users of this text.

Chapter 4 is entirely devoted to discrete-event simulation and the directly related analysis of simulation output. It touches on the strong points of this solution method, as well as on some of its difficulties. Additionally, we briefly mention some of the pros and cons of ready-to-use simulation environments. This chapter can be considered a background reading if no simulation language or environment is used in the class or student assignments.

In Chapter 5, we discuss selected aspects of numerical solutions of balance equations. If time is limited, we recommend covering the method of conditionals (Section 5.3) because of its connection to decomposition methods and its use in the recent approximate solutions of multi-server systems. The "more ambitious" variant of the method (discussed in the same Section 5.3) may be considered optional material. For most college students, we recommend reading Section 5.4 on numerical considerations in practice.

Chapter 6 presents selected approximation methods. Unless only specific approximations are of interest, we recommend covering Section 6.1 on equivalence and decomposition, as well as Section 6.2 since the latter discusses at some length the use of fixed-point iteration in conjunction with equivalence. We also recommend covering the material in Section 6.4 as an example of approximation using reduced state description. Depending on the focus of the class and the time available, the instructor may cover some or all of Sections 6.5–6.9. Section 6.10 gives an example of a hybrid solution of a model. This, and the fact that it pertains to cloud computing and energy considerations, warrant covering it or making it at least a background reading item.

In Chapter 7, devoted to multi-server queues, we recommend covering Sections 7.1, 7.4, 7.6, and 7.9. Other sections in this chapter each have their own merits and can be covered depending on specific interests and time constraints.

Chapter 8 is devoted to simple modeling techniques, such as the Operational Analysis, and is of interest to readers unfamiliar with such techniques. In particular, Section 8.3 deals with simple bottleneck analysis and should be covered. Section 8.4 can be viewed as advanced optional material.

The material in Chapter 9 related to model validation and robustness is highly recommended for inclusion in a general class on computer performance modeling.

About the Authors

Alexandre Brandwajn holds an Ingénieur Civil des Télécommunications degree from the Ecole Nationale Supérieure des Télécommunications in Paris and a Docteur d'Etat in Computer Science degree from the University of Paris VI. He worked as a Researcher at the Institut de Recherche en Informatique et Automatique (IRIA) France. Then, he served as faculty of the Ecole Nationale Superieure des Telecommunications in Paris where he directed a project in adaptive computer architecture. Later, he joined Amdahl Corporation in Sunnyvale, California, where he was Senior Computer Architect, and then Manager of Systems Analysis group. Since 1985 he has been a Professor of Computer Engineering at the University of California at Santa Cruz and President of PALLAS International Corporation through 2019 in San Jose. His current research interests include efficient solution of systems with large state space, application of conditional probability in the solution of performance models, models of virtualized systems, as well as efficient solution of priority systems.

Thomas Begin has been a Full Professor at the University Lyon 1 in France since 2022. He earned his Ph.D. in Computer Science from University Sorbonne in Paris. In 2009, he completed a post-doctoral fellowship at UC Santa Cruz, USA. From 2009 to 2022, he served as an Associate Professor at University Lyon 1, and during 2015–2016, he took a research leave at the University of Ottawa, Canada. His research focuses on performance evaluation, computer networks, cloud computing, and system modeling with primary applications in high-level modeling, wireless networks, resource allocation, and queueing systems. His work spans both theoretical and practical areas. He has authored over 100 papers in international peer-reviewed journals and conference proceedings and has received two best paper awards. In 2020, he was honored with the ACM MSWiM Rising Star Award.

Contents

Preface v

About the Authors ix

1. Introduction **1**

1.1 Importance of computer performance modeling . . . 1
1.2 The notion of a model 2
1.3 Types of computer performance models 2
1.4 Solution approaches 3
1.5 Prediction is the name of the game 4

2. Elementary Models **7**

2.1 A high-level model of transaction processing: Performance measures and model attributes 7
2.2 The M/M/1 model 9
2.3 A more realistic simple model with finite capacity: State dependent M/M/1/K model 15
2.4 Finite number of request sources 20
2.5 Relaxing distributional assumptions: The M/G/1 model 22
2.6 Using a Coxian distribution in our model: A simple recurrent solution 27
2.7 Phase-type distributions for times between arrivals — the Ph/M/1/K model 31

2.8 Phase-type distribution of service times: A recurrent solution of the M/Ph/1/K/model 36
2.9 Service with no queueing allowed: A loss system with finite sources 39

3. Networks of Queues 47

3.1 A model of thrashing in a system with virtual memory 47
3.2 Incorporating additional devices in our model: The central server model 51
3.3 Jacksonian networks of queues 54
3.4 State found on arrival and mean value analysis . . . 62
3.5 BCMP networks of queues: Models with classes of customers 68

4. Simulation 79

4.1 Basic structure of discrete-event simulation 79
4.2 Generation of random variates 80
4.3 Example of a simulation system 82
4.4 Simulation output analysis 83
4.5 Independent replications and batch means 88
4.6 More about output analysis and discrete-event simulation in practice 89
4.7 Example of variability issues 94
4.8 Ready-to-use simulation environments: Pros and cons 97

5. Numerical Methods 101

5.1 Characteristics of balance equations 101
5.2 The γ-method 102
5.3 The method of conditionals 110
5.4 Numerical considerations in practice 118

6. Selected Approximation Methods 123

6.1 Equivalence and decomposition for global dependencies 123
6.2 Fixed-point iteration and equivalence for approximate tandem network analysis 133
6.3 Approximate class aggregation to model a system with non-preemptive priorities 146

6.4 Partial state description to reduce complexity 153
6.5 Approximate mean value analysis 157
6.6 The X-model approach 160
6.7 Function of the average instead of the average of a function 162
6.8 Guided state sampling approximation to reduce complexity 165
6.9 Node-by-node analysis 167
6.10 Hybrid solution methods: A model of live VM migration in cloud computing 181

7. Multi-server Queues 193

7.1 M/M/C model and saturation patterns 193
7.2 Approximations for multiple servers and general service times 196
7.3 State upon arrival with general service times 202
7.4 Reduced-state approximation for the M/Ph/C/K model 208
7.5 Reduced-state approximation for the M/Ph/C model 227
7.6 Quasi-general time between arrivals and service times (Ph/Ph/C/K model) 235
7.7 Model with multiple servers, classes of customers, and FCFS service 241
7.7.1 Two classes of customers 244
7.7.2 More than two classes of customers 251
7.8 Model with multiple servers, quasi-general service times, and preemptive priorities: Level-by-level approximation 261
7.9 Model with multi-server jobs 278
7.10 Variance of the response time 289

8. Simple Techniques 297

8.1 Operational analysis 297
8.2 High-level modeling 306
8.3 Simple performance bounds: Bottleneck analysis 312
8.4 Sampling bounds with equivalence and decomposition 316

9. Model Validation and Robustness **325**

9.1 Basic notions . . . 325
9.2 Validation . . . 326
9.3 Calibration . . . 327
9.4 Iterative model development . . . 332
9.5 Robustness . . . 333
9.6 Art of computer modeling . . . 335

Index 337

Chapter 1

Introduction

1.1 Importance of computer performance modeling

Computer systems are so ubiquitous that their performance has a direct impact on many aspects of our lives. From our ability to access our bank accounts, healthcare or utilities to staying in touch with friends on social media, we depend on computer systems. Computer performance modeling provides tools to predict the performance of computer systems before they are built, as well as the effect of system modifications before they are released to the users. Ideally, outside emergency situations, system designs and modifications whose performance has not been assessed and is thus unknown, should not find their way into the marketplace (Denning, 1980; Denning & Martell, 2015). Computer performance modeling includes all aspects of computer systems, from their hardware and software components to their overall architecture. When dealing with existing systems, its goals include the determination of critical performance factors and the impact of projected changes. For hypothetical systems, performance modeling during the system design phase allows the designers to make critical choices and separate good solutions from poor ones. It has its place also in the process of competitive equipment sales if the prospective customers want to understand how a given vendor's equipment might perform in the customer's environment.

1.2 The notion of a model

Let's start by understanding the notion of a model. A model, in our view, is an abstraction of reality, taking into account only certain of its features. For a model to be useful, the features included in the model must be salient with respect to the phenomena being studied. If one is to rank models, a good model is the simplest model that reproduces the performance aspects of interest. As an example, if one is interested in the ability of a bank website to provide a reasonable response time to users, clearly, our model must correctly reproduce the evolution of the website response time as a function of the number of users trying to access the website per time unit. Unless such aspects are also of interest, it does not necessarily have to reproduce other aspects of the bank's website performance, and it certainly does not have to (in reality, should not) include irrelevant minutia of the website operation. As one of the co-authors learned early on during his work at a computer manufacturer in Silicon Valley in California, people have a tendency to substitute details for understanding but, at the end of the day, this only obscures the true performance issues. Thus, a realistic model is a model that correctly represents the phenomenon or feature being studied. By "correctly", we mean, of course, to a reasonable degree. It certainly makes little sense to insist on a model matching the system performance to less than 1% when the system workload is only vaguely known.

1.3 Types of computer performance models

Loosely speaking, there are two categories of computer performance models: deterministic and probabilistic. In the first category, we assume that all system parameters, such as job (or task) execution times and sequencing, are known exactly. Such deterministic models are mostly used in the area of real-time scheduling and control systems (Markenscoff, 1984; Adve & Vernon, 1998; Garland *et al.*, 2013) and are outside the scope of this book. In the second category, many workload and some system operation characteristics are represented through probability distributions. The resulting probabilistic model can usually be described by a set of mathematical equations. This book is devoted to selected aspects of probabilistic computer performance models.

1.4 Solution approaches

If we are lucky, there may be a formula (or a set of formulas) to represent the solution of our model's equations. This type of solution is usually referred to as analytical. If no formula is known or if it is very complex, it may be simpler to use a numerical method to solve the system of equations directly (e.g., one of many methods designed to solve a system of linear equations if our model is described by a system of linear equations). Appropriately, one refers to such a solution as numerical. The main advantage of an analytical solution over a numerical one, besides the intellectual satisfaction of dealing with a mathematical formula, is the potential ability to study in a symbolic form the influence of system parameters and possible limiting system behaviors. Keep in mind that an analytical solution still has to be evaluated numerically to obtain numerical results for specific parameter values. Depending on the complexity of the formula involved, this may be a challenging task in its own right. The potential difficulties for a numerical solution include the dimensionality, i.e., the number of equations to solve, which for some models may increase combinatorially with the size of the problem, as well as floating-point issues if the problem is "ill conditioned" as might be the case, for example, when the values of some model parameters differ by several orders of magnitude (Mitrani & Chakka, 1995; Woods *et al.*, 2023). Both analytical and numerical solution methods suffer from the frustrating issue that a seemingly minor change in model assumptions may result in a major difficulty, in the sense that the analytical solution may no longer exist or the system of equations may become much more complicated.

Simulation is an alternative solution technique, which does not require that specific equations be solved. Typically, we are talking about so-called discrete-event simulation, in which system operation is represented through a set of events, such as the arrival of a new transaction, the start or end of a CPU burst for the transaction, and simulation time advances in discrete increments from one event to the next one on the simulation schedule. The big advantage of simulation is that it is usually relatively easy to change model assumptions. Among the potential difficulties of the simulation solution method is making sure that the simulation does what one intends it to do, in particular, regarding event timing, and the fact that a simulation is in essence a statistical experiment whose results need to be interpreted accordingly. Hence, the issues of simulation duration and output analysis: how long should the simulation run for and how confident can we be that its results are sufficiently close to the quantities being estimated?

When the model incorporates events at several time scales, there may be very many events at a lower time scale for a single upper-scale event, which would result in a long model solution time. And sometimes, parts of the model, if considered separately, might have a very fast analytical or even numerical solution. Hence, the idea to solve the model by a combination of methods, for example, using an analytical solution for a part of the model, embedding it within a simulation of another part of the model. Such hybrid solution methods attempt to leverage the best features of each solution approach.

When the model considered turns out to be impossible or too difficult to solve exactly, it is oftentimes possible to find a good approximate solution. This is especially true for analytical and numerical solution methods. For example, when the exact analytical solution is not known or does not scale, it may be possible to find an approximate formula. Or, if the dimension of the problem is too big for standard computers, it may be possible to reduce the size of the problem through approximation and then solve it numerically, etc. Approximation is not a dirty word, and from a practical standpoint, a good approximate solution with a reasonable accuracy and fast execution time may well be all we need. We will have the opportunity to look closely at a few selected approximation methods.

1.5 Prediction is the name of the game

Whatever the nature of the model and its solution method, ultimately, the goal of computer performance modeling is to be able to predict. Whether it is the expected response time of an I/O controller for a specified attained I/O rate, or the maximum number of jobs a given system can process per time unit, or whatever other quantity of interest, the goal is to predict how the system will behave under some hypothetical conditions. Before leaving this chapter, we need to stress the important symbiotic relationship between measurements and models. Measurements provide values for model parameters and, conversely, a model of system behavior and structure is important to help us decide what quantities to measure. Without an underlying model, it is all too easy to measure a lot of quantities whose meaning and interpretation are not particularly useful when solving performance issues. And, of course, for existing systems, comparing measured values and model results is essential to properly calibrate and validate the model.

References

Adve, V. S., & Vernon, M. K. (1998). *A Deterministic Model for Parallel Program Performance Evaluation*. Technical Report (TR98-333), Center for Research on Parallel Computation, Rice University, Dec. 1998. http://www.cs.uiuc.edu/~vadve/Papers/detmodel,ps.gz.

Denning, P. J. (1980). ACM President's letter: What is experimental computer science? *Communications of the ACM*, 23(10), 543–544.

Denning, P. J., & Martell, C. H. (2015). *Great Principles of Computing*. MIT Press.

Garland, J., James, R., & Bradley, E. (2013). Determinism, complexity, and predictability in computer performance. *arXiv preprint arXiv: 1305.5408*.

Mitrani, I., & Chakka, R. (1995). Spectral expansion solution for a class of Markov models: Application and comparison with the matrix-geometric method. *Performance Evaluation*, 23(3), 241–260.

Markenscoff, P. (1984). A deterministic model for evaluating the performance of a multiple processor system with a shared bus. *IEEE Transactions on Computers*, 100(3), 281–285.

Woods, E. J., Kannan, D., Sharpe, D. J., Swinburne, T. D., & Wales, D. J. (2023). Analysing ill-conditioned Markov chains. *Philosophical Transactions of the Royal Society A*, 381(2250), 20220245.

Chapter 2

Elementary Models

2.1 A high-level model of transaction processing: Performance measures and model attributes

Let us start with a very simple high-level model of a transaction processing system. Assume that we know very little about the system other than that transactions arrive to the system with some rate λ, i.e., the mean time between two consecutive transaction arrivals is $1/\lambda$, they are processed with rate μ, i.e., the mean time to process a transaction is $1/\mu$, and once processed leave the system. In layman's terms, a new transaction arrives to the system on average every $1/\lambda$ time units, and the average processing time for a transaction is $1/\mu$ time units. A good example would be a bank ATM if we interpret each customer as a "transaction". Figure 2.1 shows our model. Arriving customers who find the ATM busy form a queue in front of the ATM (server) awaiting their turn. When discussing our model, we will use the terms "transaction" and "customer" interchangeably.

Before we go any further, we need to specify the performance measures (or indices) we would like our model to produce. This being our first attempt at computer performance modeling, we will use three performance measures. First, from the user quality of service side, the expected response time, i.e., how long on average a customer stays in the system, which we will denote by $E[W]$. Then, just because everybody seems to want to know this quantity, the expected number of transactions in the system, i.e., the expected number of customers at

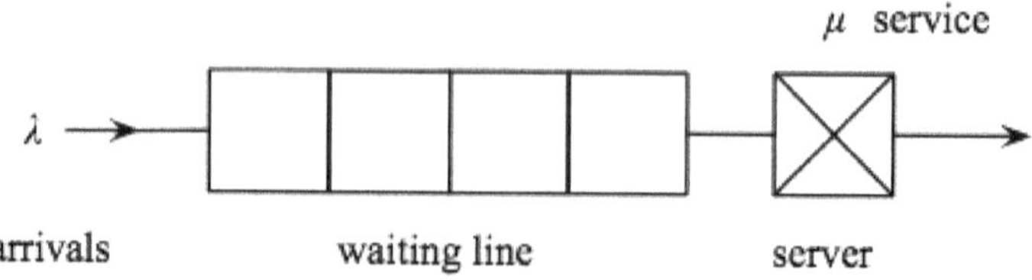

Figure 2.1. A simple model of a transaction processing system.

the ATM, in our case. We denote this quantity by $E[N]$. From the bank's perspective, an important measure is how much use the ATM gets, i.e., the utilization of the server in our model, denoted by U.

Our model is, in essence, a single server with its waiting line. Although in the case of a bank ATM things may seem obvious, in general, even for a single-server queue, several additional attributes need to be defined for our model. First, is there a single class or are there multiple classes of customers to distinguish? Next, how is the waiting line managed, i.e., do customers queue for service in the order of arrival or in some other order? This is referred to as the queueing discipline, and, among many others, the choices here include First-Come-First-Served (FCFS), Last-Come-First-Served Preemptive Resume (LCFS PR) where a newly arrived customer interrupts the customer currently in service and interrupted customers eventually resume their service from the point of interruption, Processor Sharing (PS) where customers present at any given time each get an infinitesimally small share of the server's time, etc. The PS queueing discipline is often presented as an abstraction of service by quanta, for example, CPU quanta, when the quanta are very small compared to the mean service time. Note that it is common to use the acronym FIFO (First-In-First-Out) instead of FCFS. This can be slightly misleading if there are multiple servers since customers don't necessarily complete service in the order in which they started it. This means that the first customer to enter the system may not be the first one to leave it, although the FIFO order remains true for the waiting line itself.

Next on our list of attributes to be specified is the issue of how many customers can there be in general. This is referred to as the population size, and it can be finite, meaning there is only a finite number of transaction sources, or this number is taken to be infinite. A closely related attribute is the maximum size of the waiting line.

This is referred to as a queueing room, which again can be finite or assumed infinite. The capacity of the queue is then the sum of the maximum size of the waiting line plus the number of servers. The term "queue" is used in this context to denote the whole model. Finally, two important attributes pertain to the distribution of the times between customer arrivals and the distribution of the customer service times.

Kendall's notation (Bolch *et al.*, 2006) is the standard system used to describe a queue of the kind of our model. The notation can be written as $A/S/c/K/N/\mathrm{D}$ where A denotes the distribution of the time between arrivals, S is the service time distribution, c is the number of servers, K is the capacity of the queue, N is the population size, and D is the queueing discipline. The final three (or two, or just one) parameters may not be specified, in which case it is assumed $K = \infty$, $N = \infty$, and D = FCFS (or $N = \infty$ and D = FCFS, or D = FCFS).

Some of the following codes are used to describe the probability distributions of the inter-arrival times or the service times:

D — for a constant (deterministic, hence the letter "D")
G — for general, meaning arbitrary distribution
GI — for general but independent of one another, e.g., service times are arbitrary but mutually independent
M — for exponential; the letter "M" comes from Markovian, but we like to think of it as M for "memoryless"
E_k — for Erlang of order k.

2.2 The M/M/1 model

Since we have little knowledge about our system besides the expected rate of transaction arrivals and the mean time to process a transaction, we will assume that times between consecutive arrivals are exponentially distributed with mean $1/\lambda$, and the service times are also exponentially distributed, with mean $1/\mu$. We are talking about the negative exponential probability distribution. In Kendall's notation, our model is then an M/M/1 queue. For transaction arrivals, this implies that, if one considers a very small interval of time of duration δt, $(t,\ t+\delta t]$, the probability that a new transaction arrives during such an interval is given by $\lambda\delta t + o(\delta t)$ regardless of the value of t, where $o(\delta t)/\delta t \to 0$ as $\delta t \to 0$. In other words, the probability of arrival during a very small interval of time is proportional to the duration of the interval and does not depend on the point in time this interval is located. This gives rise to the famous or infamous, depending on your point of view, memoryless property of the exponential distribution, by virtue of which the time that remains to wait has the same probability distribution regardless of how long one has already waited. Perhaps a good example might be what a wait for the inter-terminal shuttle feels like at some airports at times, where the probability of waiting an additional 10 minutes seems the same whether you just arrived at the shuttle stop or have been waiting there for 15 minutes already.

It may seem odd to choose a distribution with such a non-intuitive property, but, as we shall see soon, it is precisely this property that leads to the simplest mathematical description and elegant solution for our model. Before we proceed any further, let's state some consequences of our distributional assumptions. Referring to our very small interval of time $(t,\ t+\delta t]$, the probability that there is no customer arrival is given by $1-\lambda\delta t+o(\delta t)$, the probability that a service currently in progress ends is $\mu\delta t + o(\delta t)$, the probability that there is both an arrival and end of service is $\lambda\delta t\mu\delta t = o(\delta t)$, and, in general, the probability of two or more events occurring during the interval is $o(\delta t)$. Note that $o(\delta t)$ simply denotes a quantity that tends to 0 faster than linearly as $\delta t \to 0$.

Regarding customer arrivals, it can be shown that if the times between consecutive arrivals are exponentially distributed with mean $1/\lambda$, then the number of arrivals during any finite period of time x has a Poisson distribution with mean λx (Allen, 1990; Trivedi, 2008). This is why people sometimes refer to such arrivals as coming from a Poisson source or simply being Poisson.

An important step when dealing with a new model is to select an appropriate state description. In our case, we select (n, t), where $n = 0, 1, \ldots$ is the current number of transactions (customers) in the system, both queued and in service, to describe our system at time t. Our goal here is to derive the equations for the probability that there are n users in the system at time t, which we denote by $p(n, t)$. Bear with us. We do intend to make the material in our text as intuitive as possible, so this is the only time we show how the model equations actually arise. The technique is to consider system states one by one and, for each state, write the probability of the system being in the given state at time $t+\delta t$, expressing it in terms of the possible system states at time t, i.e., a very small time before. Let us illustrate it for the state $n = 0$. We have

$$\begin{aligned} p(0, t+\delta t) = {} & p(0,t)\text{Prob}\{\text{no arrival during } (t, t+\delta t]\} \\ & + p(1,t)\text{Prob}\{1 \text{ departure and no arrivals during} \\ & \times (t, t+\delta t]\} + o(\delta t). \end{aligned}$$

The term $o(\delta t)$ accounts for all possible terms involving multiple transitions during the interval $(t, t + o(\delta t)]$, e.g., $p(2,t)\text{Prob}\{2 \text{ departures and no arrival during } (t, t+\delta t]\}$.

This translates into

$$p(0, t+\delta t) = p(0,t)(1 - \lambda\delta t) + p(1,t)\mu\delta t + o(\delta t). \tag{2.1}$$

By an analogous reasoning, we get for $n = 1, 2, \ldots$

$$\begin{aligned} p(n, t+\delta t) = {} & p(n,t)(1 - \lambda\delta t - \mu\delta t) + p(n-1,t)\lambda\delta t \\ & + p(n+1,t)\mu\delta t + o(\delta t). \end{aligned} \tag{2.2}$$

It now suffices to take the terms $p(0,t)$ and $p(n,t)$ multiplied by 1 on the right-hand side in equations (2.1) and (2.2) to the left-hand side, divide both sides by δt and let $\delta t \to 0$ to obtain the differential-difference equations that describe how the probability that our model is in one of its states evolves over time:

$$\frac{\mathrm{d}}{\mathrm{d}t}p(0,t) = -\lambda p(0,t) + \mu p(1,t), \tag{2.3}$$

$$\frac{\mathrm{d}}{\mathrm{d}t}p(n,t) = -(\lambda+\mu)p(n,t) + \lambda p(n-1,t) + \mu p(n+1,t), \quad n = 1, 2, \ldots. \tag{2.4}$$

Of course, we must have

$$\sum_{n=0}^{\infty} p(n,t) = 1, \quad \forall t \geq 0. \tag{2.5}$$

Starting from a known system state at time $t_0 \geq 0$ (for example, system empty), the solution of this system of equations would give us the probability that there are n customers in the system for any time point $t > t_0$. Such a time-dependent solution is referred to as the transient solution. Although this solution is known for the M/M/1 model, it involves sums of modified Bessel functions, nothing too friendly for the common of mortals (Saaty, 1961). And, in general, for more complex models, explicit transient solutions are very difficult to obtain or simply unknown.

This is likely one of the reasons why computer performance modelers convinced themselves (and others) that one should be mostly interested in the long-run behavior of the system, a very long time after the known initial conditions have dissipated, i.e., technically, for $t \to \infty$. If the corresponding limit of $p(n,t)$ exists, we say that the system reaches a steady state, also sometimes referred to as statistical equilibrium. We denote by $p(n) = \lim_{t\to\infty} p(n,t)$, the steady-state probability that there are n users in the system. Although strictly speaking, there are some subtle differences between steady state and stationary probability distributions, we use the terms steady state, long run, stationary, and equilibrium probabilities interchangeably.

One could obtain equations for $p(n)$ from those for $p(n,t)$, but it is quite simple and instructive to obtain them directly from the fact that, if the system is in equilibrium, for each state, the rate of flow out of the state must be equal to the rate of flow into the state. For $n = 0$, the only way to exit the state is via an arrival of a customer. The corresponding rate of flow is $\lambda p(0)$. The state $n = 0$ can only be entered from the state $n = 1$ through an end of service. Thus, the rate of flow into the state is $\mu p(1)$. For a state $n > 0$, the rate of flow out of the state corresponds to an arrival or a departure (end of service): $(\lambda + \mu)p(n)$. The state $n > 0$ can be entered from the state $n-1$ through an arrival of a new customer or from the state $n+1$ through a completion of a customer's service. The corresponding rate of flow into the state is $\lambda p(n-1) + \mu p(n+1)$. Equating the rates of flow out and into each state, we get the following set of equations for the steady-state probabilities for our M/M/1 model:

$$\lambda p(0) = \mu p(1), \tag{2.6}$$

$$(\lambda + \mu)p(n) = \lambda p(n-1) + \mu p(n+1), \quad n = 1, 2, \ldots. \tag{2.7}$$

As always, we must have

$$\sum_{n=0}^{\infty} p(n) = 1. \tag{2.8}$$

Equations for the steady-state probabilities of a system are often called balance equations because they arise as a result of balancing the rate of flow out and into each state. The solution of the set of balance equations (2.6) and (2.7) together with the normalizing condition (2.8) is rather straightforward. Consider equation (2.7) for $n = 1$: $(\lambda + \mu)p(1) = \lambda p(0) + \mu p(2)$. If we use equation (2.6) in the latter, we get $\lambda p(1) = \mu p(2)$, which can in turn be used in the balance equation for $n = 2$, and so on, repeating the pattern $\lambda p(n) = \mu p(n + 1)$. This implies that, if there is a steady-state solution for our M/M/1 model, it must have the form $p(n) = p(0)(\lambda/\mu)^n$. The ratio of the rate of arrivals to the rate of service (or, equivalently, of the mean service time to the mean time between customer arrivals) is referred to as the traffic intensity. We will denote it by the Greek letter ρ : $\rho = \lambda/\mu$. With this notation, our long-run solution has the form

$$p(n) = p(0)\rho^n, \quad \text{for } n = 0, 1, \ldots. \tag{2.9}$$

It remains now to determine the value of the probability that the system is empty $p(0)$. This is where the normalizing condition (2.8) comes in handy. We must have $p(0)\sum_{n=0}^{\infty} \rho^n = 1$. This can only be true if $\rho < 1$, in which case we get $p(0)\frac{1}{1-\rho} = 1$, so that we have $p(0) = 1 - \rho$, and our steady-state solution is

$$p(n) = (1 - \rho)\rho^n, \quad \text{for } n = 0, 1, \ldots. \tag{2.10}$$

Clearly, this solution only exists if $\rho < 1$. This is the stability condition for our system. The stability condition is also referred to as the ergodicity condition in recognition of the link to the theory of Markov chains (Bolch *et al.*, 2006). The stability condition in our case simply states that, on average, the customer arrival rate must be smaller than the rate of service. Otherwise, the system is unstable in that it can never catch up, and the number of customers in the system grows without bound. It is remarkable that the steady-state probabilities $p(n)$ in the M/M/1 model depend only on the ratio λ/μ.

We are now ready to compute our performance indices, starting with the mean number of customers in the system.

$$E[N] = \sum_{n=1}^{\infty} np(n) = \rho/(1-\rho). \tag{2.11}$$

Our second customer quality of service performance index is the expected response time $E[W]$. We have concentrated our efforts on the probability distribution of the number of customers in the system, which may seem disconnected from the time a customer spends in the system. As it turns out, this is not the case, and there is a "magic formula" linking the mean number of customers in a system to the mean time customers spend in the system. This is the Little's formula (Stidham, 1974; Bertsimas & Nakazato, 1995), where $E[\Lambda]$ denotes the mean rate of customer arrivals to the system

$$E[N] = E[W]E[\Lambda]. \tag{2.12}$$

Quite remarkably, this result holds true regardless of the nature of the system under minimal assumptions. Applying it to the whole system, in our case, we have $E[\Lambda] = \lambda$ so that

$$E[W] = \frac{E[N]}{\lambda} = \frac{1}{\mu(1-\rho)}. \tag{2.13}$$

Our last performance index, the server utilization U, is simply the probability that the server is busy, i.e.,

$$U = 1 - p(0) = \rho. \tag{2.14}$$

It is interesting to look at some additional performance aspects of our system. The expected time a customer spends in the waiting line queueing for service, which we will denote by $E[W_q]$, can be expressed as the difference between the mean response time $E[W]$ and the mean service time $\frac{1}{\mu}$

$$E[W_q] = E[W] - \frac{1}{\mu} = \frac{\rho}{\mu(1-\rho)}. \tag{2.15}$$

We can apply Little's formula to the waiting line to obtain the mean number of transactions in the waiting line, $E[N_q]$:

$$E[N_q] = E[W_q]\lambda = \frac{\rho^2}{(1-\rho)}. \tag{2.16}$$

For our single-server system, $E[N_q]$ can also be computed as the mean number of transactions in the system $E[N]$ minus the server utilization U (and not minus 1, as one might be tempted to say hastily): $E[N_q] = E[N] - U$, yielding the same value as (2.16). Note that in a stable system with infinite capacity, the customer throughput, i.e., the number of customers processed by the system per time unit, always equals the arrival rate λ.

2.3 A more realistic simple model with finite capacity: State dependent M/M/1/K model

Examining formula (2.11), we notice that the factor $1/(1-\rho)$ plays the role of an "amplifier", leading to an increasingly fast growth in the mean number of customers in the system as the traffic intensity increases. In fact, $E[N]$ increases without bound as $\rho \to 1$. Of course, in a real transaction processing system, the capacity is not infinite. Imagine that the capacity of the system is limited to K customers. Additionally, the rate of customer arrivals might be a function of the current number of customers in the system, for example, if there is load balancing and fewer transactions are directed to a more congested system. Similarly, the rate of service might depend on the number of transactions in the system.

Let's denote by $\lambda(n)$ the rate of transaction arrivals and by $\mu(n)$ the rate with which transactions leave the system when there are n customers in the system. We assume that both the times between customer arrivals and the transaction service times follow memoryless distributions. In other words, considering a very small interval of time $(t, t+\delta t]$, the probability of an arrival during such an interval is given by $\lambda(n)\delta t + o(\delta t)$ if there are n customers at time t. Similarly, the probability of a transaction completing during such an interval is $\mu(n)\delta t + o(\delta t)$, regardless of the value of t. Clearly, we must have $\lambda(n) > 0$, for $n = 0, \ldots, K-1$, $\mu(n) > 0$ for $n = 1, \ldots, K$ and $\mu(0) = 0$ as there can be no customers leaving an empty system.

The number of transactions in the system is limited to K so that, if arrivals happen when the system is full, we assume that such arrivals are simply rejected or, as is customary to say, lost. State-dependent memoryless arrivals are sometimes referred to as quasi-Poisson.

We consider our new model in the steady state and we use the number of customers in the system as our state description: $n = 0, 1, \ldots, K$. We denote by $p(n)$ the corresponding steady-state probability. We will apply our "rate of flow out of a state = rate of flow into a state" principle to write the balance equations for $p(n)$:

$$\lambda(0)p(0) = \mu(1)p(1), \tag{2.17}$$

$$[\lambda(n) + \mu(n)]p(n) = \lambda(n-1)p(n-1) + \mu(n+1)p(n+1), \quad n = 1, \ldots, K-1, \tag{2.18}$$

$$\mu(K)p(K) = \lambda(K-1)p(K-1). \tag{2.19}$$

And, of course, we must have

$$\sum_{n=0}^{K} p(n) = 1. \tag{2.20}$$

As we did for the M/M/1 model, we can use equation (2.17) in equation (2.18) for $n = 1$, and so on, and we quickly note that the following pattern emerges: $\lambda(n)p(n) = \mu(n+1)p(n+1)$, for $n = 0, 1, \ldots, K-1$, which luckily for us happens to coincide with equation (2.19), the last boundary equation for $n = K$. From here, the form of $p(n)$ is obvious and rather elegant

$$p(n) = p(0) \prod_{i=1}^{n} \frac{\lambda(i-1)}{\mu(i)}, \quad \text{for } n = 0, 1, \ldots, K. \tag{2.21}$$

By convention, empty products are equal to 1 in (2.21) and similar expressions.

It remains to determine $p(0)$ from the normalizing condition (2.20)

$$p(0) = \frac{1}{\sum_{n=0}^{K} \prod_{i=1}^{n} \frac{\lambda(i-1)}{\mu(i)}}. \tag{2.22}$$

It is clear from the form of the solution that it always exists. We are dealing with a finite sum so that, unlike in the M/M/1 model, there are no stability issues. This is, in general, the case for models with finite population or finite capacity. Our model could be referred to as a state-dependent M/M/1/K queue. The reason we use the conditional in this statement is that a service rate that depends on the current number of customers can be, and often is, used to represent a system with multiple servers and memoryless service time distribution. For instance, if the mean service time for a transaction is $1/\mu$ and there are C servers to process the transactions (in our example, C operational ATM machines with a single waiting line), the service rate would be

$$\mu(n) = \begin{cases} n\mu, & n = 0, 1, \ldots, C-1 \\ C\mu, & n = C, C+1, \ldots, K \end{cases}.$$

Before we compute our performance measures, we need to find the expected rate of customer arrivals entering the waiting line $E[\Lambda]$. Since customers only enter the system if it is below capacity, we have

$$E[\Lambda] = \sum_{n=0}^{K-1} \lambda(n)p(n). \tag{2.23}$$

Now, we can obtain our customer-oriented performance measures

$$E[N] = \sum_{n=1}^{K} np(n) \tag{2.24}$$

and, from Little's formula,

$$E[W] = \frac{E[N]}{E[\Lambda]}. \tag{2.25}$$

When there are n customers in the system, the number of transactions queued awaiting service is $n - C$ if $n > C$ or 0 otherwise. This implies that we can obtain the expected number of transactions queued for service as

$$E[N_q] = \sum_{n=C+1}^{K} (n - C)p(n). \tag{2.26}$$

The expected time a customer spends waiting for service is then

$$E[W_q] = \frac{E[N_q]}{E[\Lambda]}, \tag{2.27}$$

again from Little's formula.

Since our model represents customers lost due to capacity constraints, we may add the loss probability to our performance indices. We denote this probability by p_{loss} and we evaluate it as the ratio of customers arriving when the system is at capacity to all arrivals, including the ones lost

$$p_{\text{loss}} = \frac{\lambda(K)p(K)}{\sum_{n=0}^{K} \lambda(n)p(n)}. \tag{2.28}$$

With possibly multiple servers, the notion of server utilization becomes a little tricky. We choose to define it as the expected number of busy servers, i.e.,

$$U = \sum_{n=1}^{K} \min(n, C)p(n). \tag{2.29}$$

Of course, with this definition, the utilization can be greater than 1. If one is particularly attached to the idea of utilization never exceeding 100%, it suffices to divide the quantity in (2.29) by the number of servers C.

Let us for a moment consider that we have only a single server and that the arrival and service rates in our model are not state-dependent: $\lambda(n) = \lambda$, $n = 0, 1, \ldots, K$, and $\mu(n) = \mu$, $n = 1, \ldots, K$. Letting $\rho = \frac{\lambda}{\mu}$, it is then possible to obtain the following closed-form formulas for the mean number of transactions in the system and for the server utilization in such an M/M/1/K model

$$E[N] = \begin{cases} \dfrac{\rho}{1-\rho} - \dfrac{(K+1)\rho^{K+1}}{1-\rho^{K+1}}, & \text{if } \rho \neq 1 \\ \dfrac{K}{2}, & \text{if } \rho = 1 \end{cases}, \tag{2.30}$$

$$U = \begin{cases} 1 - \dfrac{1-\rho}{1-\rho^{K+1}}, & \text{if } \rho \neq 1 \\ 1 - \dfrac{1}{K+1}, & \text{if } \rho = 1 \end{cases}. \tag{2.31}$$

Clearly, since the system has a finite capacity, ρ can be equal to or even exceed 1. For $\rho < 1$, formulas (2.30) and (2.31) tend to their corresponding M/M/1 expressions (2.11) and (2.14) when $K \to \infty$. As we saw when discussing the properties of the M/M/1 model, the mean number of transactions in the system in this model grows without bound when the traffic intensity approaches unity and thus diverges from the values produced by the seemingly more realistic M/M/1/K model with a finite capacity. This leads to an important question from a computer performance modeling perspective: up to what values of the traffic intensity is the unrestricted M/M/1 queue an acceptable model?

To get some elements of answer to this question, let's compare the values for the expected number of customers in the system $E[N]$ and the expected time in system $E[W]$ obtained from both models. On purpose, we start by fixing the capacity of the system at $K = 8$. As we see in Table 2.1, with this system capacity, both models produce essentially equivalent results up to a server utilization of about 0.6. After that, the results diverge, and, for a server utilization $U = 0.9$, the M/M/1 model results make no sense: obviously, the mean number in the system cannot exceed the system capacity. Since the M/M/1 model can be viewed as a limit of the M/M/1/K model for $K \to \infty$, we can expect that the point at which and the amount

Table 2.1. Comparison of mean numbers and mean times in system in M/M/1/K and M/M/M/1 models.

	M/M/1/K with $K = 8$		M/M/1	
U	$E[N]$	$E[W]$	$E[N]$	$E[W]$
0.1	0.1111	1.1111	0.1111	1.1111
0.2	0.2500	1.2500	0.2500	1.2500
0.3	0.4284	1.4281	0.4286	1.4286
0.4	0.6647	1.6618	0.6667	1.6667
0.5	0.9860	1.9720	1.0000	2.0000
0.6	1.4294	2.3823	1.5000	2.5000
0.7	2.0440	2.9200	2.3333	3.3333
0.8	2.9044	3.6305	4.0000	5.0000
0.9	4.1717	4.6353	9.0000	10.0000
0.95	5.1188	5.3882	19.0000	20.0000

Table 2.2. Mean number and mean time in system in an M/M/1/K model as a function of queue capacity for a server utilization $U = 0.7$.

	M/M/1/K with $U = 0.7$		M/M/1 with $U = 0.7$	
K	$E[N]$	$E[W]$	$E[N]$	$E[W]$
5	1.7211	2.4587	2.3333	3.3333
8	2.0440	2.3200	2.3333	3.3333
10	2.1615	3.0875	2.3333	3.3333
15	2.2901	3.2716	2.3333	3.3333
20	2.3234	3.3192	2.3333	3.3333

by which the results diverge depends on the system capacity. Indeed, as we observe in Table 2.2, for a server utilization $U = 0.7$, the results from both models become practically identical for $K = 20$. If we were to create a rule of thumb, we might say that the value produced by the unrestricted M/M/1 model is probably acceptable (compared to M/M/1/K results) as long as the mean number of customers in the system predicted by the model does not exceed about 25% of the system capacity. Of course, rules of thumb are just that, but it is clear that a model with infinite capacity stops being relevant as the server utilization approaches 1.

2.4 Finite number of request sources

State-dependent arrival rates are useful to represent many forms of load control and also situations where we have a well-defined set of sources of transaction requests, such as set of terminals, intelligent or not. As an example, consider a system with N_t terminals. We denote $1/\alpha$ the mean time between a system response to the last transaction originated from a given terminal and the next transaction request from the same terminal. In the context of human users, dating back to early days of interactive computer systems, this time is commonly referred to as "think time". In other contexts, this time may be referred to as the idle time of the source. Let us assume that the think time has an exponential distribution. Then, if n out of the N_t terminals are awaiting response to their last transaction request, the

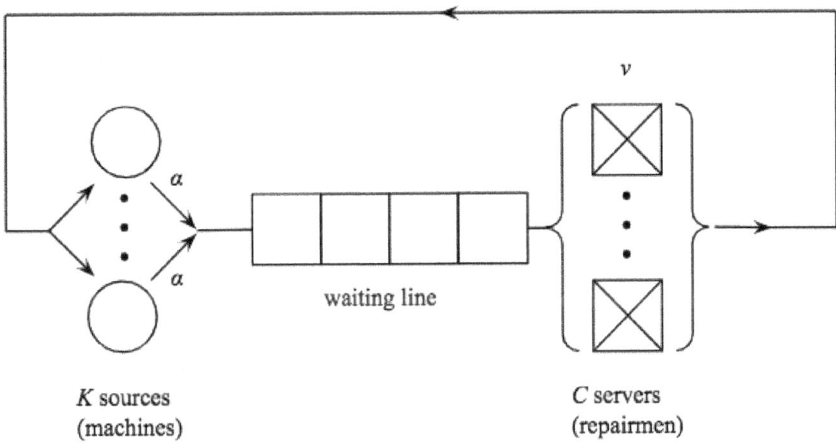

Figure 2.2. Machine repair model.

rate with which new transactions are generated can be written as $\lambda(n) = (N_t - n)\alpha$, for $n = 0, \ldots, N_t$.

Figure 2.2 shows an example of a system with K sources of requests and a total of C servers. The mean service time to complete a request issued by a source is $1/\nu$, and the mean time a source is idle after the completion of its last request is $1/\alpha$. The idle time ("think time") of a source and the service times are exponentially distributed. If at any time there are more requests than there are servers, requests are queued in FCFS order. For notation fans, this model is an M/M/C/K/K system. Interestingly, besides its many other applications, including one of the first, if not the first, successful model of an interactive computer system (Scherr, 1965), this model has been used to represent machines subject to failures. The set of sources represents the machines and the mean source idle time is the mean time to failure for a machine. Servers represent repairmen, and the mean service time is the expected machine repair time, exclusive of any time waiting for repairman availability. This gave the model its name machine repair model.

In such an application, the performance indices of interest would be the expected number of down machines, the expected time a machine has to wait for service, and the expected number of repairmen busy. In our notation, these quantities correspond to $E[N]$ given by formula (2.24), $E[W_q]$ from (2.27), and U given by formula (2.29), respectively. The steady-state probability that there are n machines in repair is given by formulas (2.21) and (2.22) with $\lambda(n) = (K-n)\alpha$ and $\mu(n) = \min(n, C)\nu$. The expected rate of machine failures can be written as $E[\Lambda] = \sum_{n=0}^{N} \alpha(K-n)p(n) = \alpha(K - E[N])$.

2.5 Relaxing distributional assumptions: The M/G/1 model

Using memoryless distributions, we have been able to obtain simple solutions for our models. What happens if we try to relax this distributional assumption? Consider again our initial model of a transaction processing system with a single server, but this time, let's assume that the service time is general, not necessarily exponentially distributed. The mean service time is, as before, $1/\mu$. Not to make things any more complicated, we keep the assumption that arrivals come from a Poisson source, i.e., the times between arrivals are exponentially distributed with rate λ. If we try to use n, the number of customers in this M/G/1 model, as our state description, we immediately run into a problem. We don't know how to express the rate at which a customer completes service for the simple reason that this rate depends on how long the current service has been in progress, except, of course, if the service times are exponentially distributed! So, we have no way to write directly the balance equations for the steady-state probabilities $p(n)$.

Before we briefly discuss a few ways to address this issue and solve our model, we wish to focus on the expected time a customer spends in the waiting line queueing for service. Denote by $E_A[N_q]$ the expected number of customers already waiting for service found by an arriving customer. Clearly, because of FCFS service discipline, the new arrival will have to wait for these customers to be served. And before these customers get the use of the server, any customer already in service at the instant of arrival must complete its remaining service time. We denote by $E[S_{\text{res}}]$ the expected remaining (or residual) service time, and by $\text{Prob}_A\{\text{server busy}\}$ the probability that the new arrival finds the server busy. From our reasoning, we have

$$E[W_q] = E_A[N_q]\frac{1}{\mu} + \text{Prob}_A\{\text{server busy}\}E[S_{\text{res}}]. \qquad (2.32)$$

Our arrivals come from a Poisson source, and what such arrivals find in steady state is the "time average" state of the system. This is the PASTA property of Poisson arrivals: "Poisson Arrivals See Time Averages" (Wolff, 1982). The PASTA property implies that $E_A[N_q] = E[N_q]$ and $\text{Prob}_A\{\text{server busy}\} = U = \rho$, where $\rho = \lambda/\mu$ is the traffic intensity, from Little's law. We also have $E[W_q] = E[N_q]/\lambda$ again from Little's law so that we get

$$E[N_q] = \frac{1}{(1-\rho)}\rho\lambda E[S_{\text{res}}]. \tag{2.33}$$

Since Poisson arrivals happen "at random" with respect to the progress of the service, the expected residual service time can be expressed in terms of the mean and of the coefficient of variation of the service time distribution, denoted by c_s (Kleinrock, 1975), as

$$E[S_{\text{res}}] = \frac{1}{\mu}\left(\frac{1+c_s^2}{2}\right). \tag{2.34}$$

The coefficient of variation is defined as the ratio of the standard deviation of the distribution to its mean. If we use formula (2.34) in our result for $E[N_q]$, we obtain

$$E[N_q] = \frac{\rho^2}{(1-\rho)}\left(\frac{1+c_s^2}{2}\right). \tag{2.35}$$

This important result is one of the forms of the Pollaczek–Khintchine formula (Kleinrock, 1975) for the M/G/1 model. It states that the mean number of customers waiting for service in an M/G/1 queue (and, therefore, also the mean queueing time) depends on the service time distribution only through its first two moments. For exponentially distributed service times, we have $c_s = 1$ and formula (2.35) gives us exactly the same result we obtained for the M/M/1 queue in formula (2.16). For the expected number of customers in the M/G/1 system, we have simply

$$E[N] = \frac{\rho^2}{(1-\rho)}\left(\frac{1+c_s^2}{2}\right) + \rho. \tag{2.36}$$

Sometimes, it is this last result that is referred to as the Pollaczek–Khintchine formula. Both formulas (2.35) and (2.36) show the effect of the service variability on the mean number of customers: the more variable the service time (as measured by the squared coefficient of variation), the higher the expected number of customers in the queue. The factor $(1 + c_s^2)/2$ captures the distributional dependence of the mean number of customers in the M/G/1 model, and, indeed, it has inspired a number of approximations over the years (Allen, 1990). It is clear from formula (2.33) or (2.35) that for the M/G/1 queue to possess a steady state, the traffic intensity must be less than 1, i.e., we must have $\rho < 1$.

If all we are interested in are the mean numbers of customers in the queue and in the system, or the mean queueing time and the mean time in the system (response time), formulas (2.35) and (2.36) give us the answer. If, however, we need quantities such as the variance of the number of transactions in the system or the probability that the queue length exceeds or not a specific value, then it becomes necessary to know the probability distribution for the number of customers in the system. We saw earlier that it is not directly possible to write balance equations for our system because the rate at which the server completes its service depends on how long the current service has been in progress. One of the possible approaches is to consider the state of the system at selected instants of time, specifically, just after a service completion. Indeed, at these instants, the elapsed service time does not enter the picture.

With this in mind, we denote by $P(n)$, $n = 0, 1, \ldots$ the probability that there n transactions in the system immediately following a customer departure. We also denote by b_k the probability that there are exactly k arrivals during a single service period. If $f(x)$ is the density of the service time distribution, since arrivals come from a Poisson source with rate λ, we have

$$b_k = \int_0^\infty \frac{(\lambda x)^k}{k!} e^{-\lambda x} f(x) dx. \tag{2.37}$$

We can write the equations for $P(n)$ by considering only instants after a customer departure. We get

$$\begin{aligned} P(0) &= P(0)b_0 + P(1)b_0, \\ P(n) &= P(0)b_n + \sum_{k=0}^{n} P(n+1-k)b_k, \quad \text{for } n = 1, 2, \ldots. \end{aligned} \tag{2.38}$$

To understand the underlying reasoning, consider the equation for $n = 0$. We are saying that in order to have 0 customers right after a departure, one possibility is that a preceding departure left the system empty, then at some point, a customer arrived (with probability 1) and there were no new arrivals during the service of this customer. Another possibility is that the preceding departure left one customer in the system, and there were no arrivals during that customer's service. An analogous reasoning applies to other values of n. We obtain a relatively simple system of linear equations, albeit an infinite one. As always, we must have $\sum_{n=0}^{\infty} P(n) = 1$.

Assume that we can find a solution for this system of equations and thus obtain the probabilities $P(n)$. An important question then is how these probabilities at instants just after an end of service relate to the continuous–time steady-state probabilities $p(n)$. As luck would have it because the arrival process is Poisson, these two probability distributions are the same (Saaty, 1961; Bertsimas & Nakazato, 1991). This approach to a system with non-memoryless distributions is called the Embedded Markov Chain method in recognition of the fact that a memoryless behavior is found within the model. For all its elegance, it is difficult or simply impossible to generalize. For instance, even keeping Poisson arrivals, we would be hard-pressed to apply this method to a system with multiple general servers.

As we saw, the issue in the M/G/1 system is that the probability of a service completion depends on how long the current service has been in progress. A possible approach is to simply add a variable to keep track of the progress of the service. This is the underlying idea of the method of supplementary variable (Cox, 1955; Cox & Miller, 1977). For $n > 1$, we define the system state by the couple (n, u) where n is the current number of customers in the system and u measures how long the current service has been in progress. The service time distribution is then defined through its instantaneous departure rate given u, $\gamma(u)$. We note in passing that $\gamma(u)$ is closely related to the notion of failure rate in reliability (Trivedi, 2008). For $n = 0$, the supplementary variable u is superfluous. The steady-state equations describing our M/G/1 model have the following form:

$$0 = -\lambda p(0) + \int_0^{\infty} p(1, u)\gamma(u)\mathrm{d}u,$$

$$\frac{\partial p(n,u)}{\partial u} = -[\lambda + \gamma(u)]p(n,u) + \lambda p(n-1,u),$$
$$\text{for } n > 1 \quad \text{and} \quad u > 0,$$
$$p(n,u=0) = \int_0^\infty p(n+1,u)\gamma(u)\mathrm{d}u, \quad \text{for } n > 1.$$

This is a system of integro-differential difference equations whose solution, interesting as it may be, is beyond the reach of the common of mortals. The difficulty of solving the system equations related to the mixture of discrete and continuous variables seems to be the bane of the supplementary variables method.

Interestingly, a related approach, but in which all state variables are discrete, has found a much wider use. Indeed, it has been shown (O'Cinneide, 1990; Neuts, 1994; Bolch *et al.*, 2006) that any distribution can be approximated arbitrarily closely by a finite set of exponential phases. Appropriately, we then refer to such distributions as phase-type distributions. Figure 2.3(a) shows a general representation of a phase-type distribution. There is a given number of exponential phases with their individual intensities, a set of probabilities

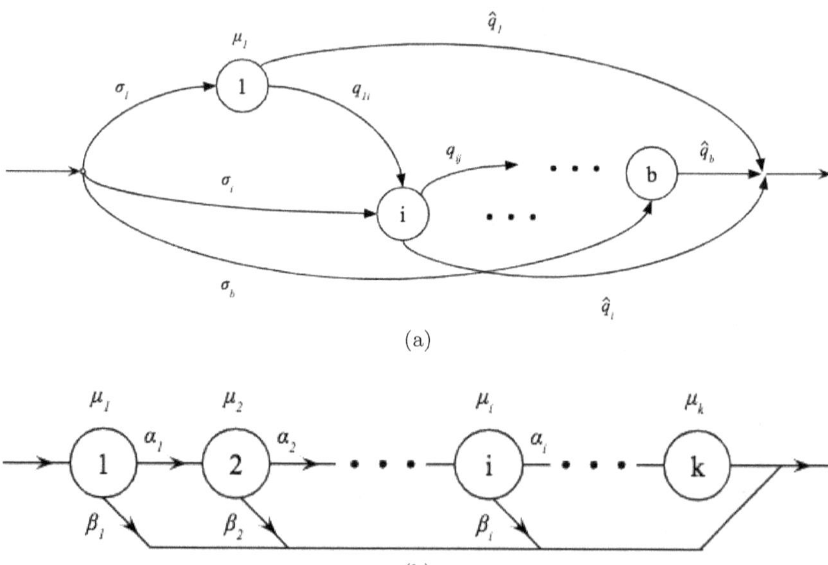

Figure 2.3. (a) Phase type distribution and (b) Coxian distribution.

that the service starts in any of the phases, for each phase, a set of probabilities that after completing this phase, the service moves on to another phase, as well as probabilities that the service ends. By intensity of a phase, we mean here the parameter of the exponential distribution of the phase, i.e., the inverse of its mean.

A subset of phase-type distributions with a well-defined structure are so-called Coxian distributions represented in Figure 2.3(b). Here, we have a set of k exponential phases, referred to as stages, each with its own intensity μ_i, $i = 1, \ldots, k$. We denote by α_i the probability that the service moves on to the next stage after completing stage i, $i = 1, \ldots, k-1$. With probability $\beta_i = 1 - \alpha_i$ the service completes after stage i. The service always completes after stage k. Remarkably, this subset of phase-type distributions can also be used to approximate any distribution with arbitrary accuracy. Of course, the number of phases (or stages) needed depends on the nature of the distribution being represented, and algorithms have been developed to match various distributions, including empirical distributions (Bobbio *et al.*, 2005; Osogami & Harchol-Balter, 2006; Khayari *et al.*, 2003; Thummler *et al.*, 2006; Feldmann & Whitt, 1998). In some areas of computer performance, there have been extensive load characterization studies (e.g., internet traffic (Jiang & Dovrolis, 2005; Crovella & Bestavros, 1997; Hsu & Smith, 2003; McNutt, 2005)), but, oftentimes, we may not know much more than the first two moments (or, equivalently, the mean and the coefficient of variation) of an empirical distribution. How many exponential phases are needed to match the first two moments of a distribution depends on the coefficient of variation of the distribution. For coefficients of variation no smaller than $1/\sqrt{2}$, two phases (or stages) suffice. For smaller coefficients of variation, more phases are needed.

2.6 Using a Coxian distribution in our model: A simple recurrent solution

We are now ready to apply the fact that a Coxian distribution can represent (at least approximately) an arbitrary distribution to the simple model with non-memoryless service times shown in Figure 2.4.

We denote by K the system capacity, and we let $n = 0, \ldots, K$ be the current number of customers in our system. We assume that

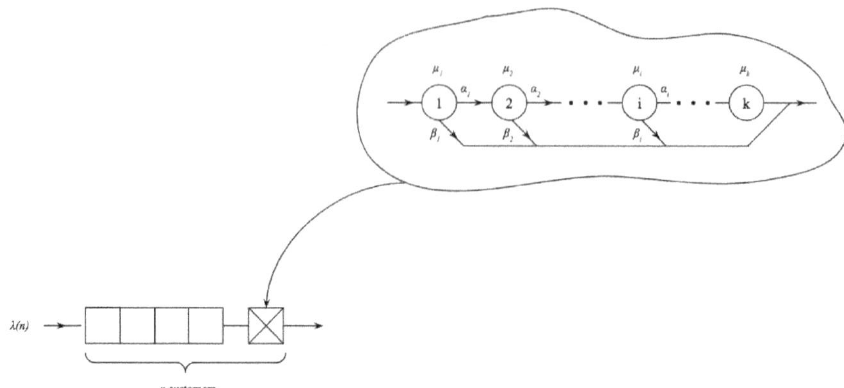

Figure 2.4. Single server queue with Coxian service times.

the times between customer arrivals follow a memoryless distribution with a state-dependent rate $\lambda(n)$. For $n > 0$, the state of the system can be fully described by the couple (n, i), where $i = 1, \ldots, k$ refers to the current stage of the customer being served. Let's consider the system in its steady state and denote by $p(n, i)$ the corresponding steady-state probability. Using our "rate of flow out = rate of flow in" principle, we have no trouble writing the balance equations for our model. Let us start with $n = 1$. We get

$$p(1,1)[\lambda(1) + \mu_1] = \sum_{i=1}^{k} p(2,i)\mu_i\beta_i + p(0)\lambda(0),$$
$$\text{for } i = 1 \text{ (the first stage)}, \tag{2.39}$$
$$p(1,i)[\lambda(1) + \mu_i] = p(1,i-1)\mu_{i-1}\alpha_{i-1},$$
$$\text{for } i > 1 \text{ (other stages)}. \tag{2.40}$$

For values of $n > 1$, we have

$$p(n,1)[\lambda(n) + \mu_1] = \sum_{i=1}^{k} p(n+1,i)\mu_i\beta_i$$
$$+ p(n-1,1)\lambda(n-1), \text{ for the first stage,} \tag{2.41}$$

$$p(n,i)[\lambda(n)+\mu_i] = p(n,i-1)\mu_{i-1}\alpha_{i-1} + p(n-1,i)\lambda(n-1), \text{ for stages } i = 2,\ldots,k. \tag{2.42}$$

In the above equations for $n = K$, we set the rate of arrivals to 0 and impossible probabilities are assumed to vanish.

That wasn't particularly complicated. So how could we solve this system of equations subject to the normalizing condition $p(0) + \sum_{n=1}^{K}\sum_{i=1}^{k} p(n,i) = 1$? As it turns out, these balance equations can be transformed into equations for the conditional probabilities of the service stage number given the current number of customers and then solved as a simple recurrence.

We denote by $p(n)$ the marginal probability that there are n customers in our system and by $p(i|n)$ the conditional probability that the current service is in stage i given that there are n customers. Directly from the definition of conditional probability, we know that

$$p(n,i) = p(i|n)p(n). \tag{2.43}$$

The conditional probabilities $p(i|n)$ must sum to 1 for each value of $n > 0$, i.e., we must have $\sum_{i=1}^{k} p(i|n) = 1$ for all $n = 1,\ldots,K$. We also have $\sum_{n=0}^{K} p(n) = 1$.

If we look only at the probability $p(n)$, the rate with which n increases is obviously $\lambda(n)$, i.e., the rate of customer arrivals for $n = 0,1,\ldots,K-1$. The number of customers in the system decreases through service completions, which happen with a rate that we will denote by $u(n)$, for $n = 1,\ldots,K$. If the customer in service is in stage i, the rate of completion is $\mu_i\beta_i$. The probability that the customer is in stage i when there are n customers in the system is given by $p(i|n)$. Hence,

$$u(n) = \sum_{i=1}^{k} p(i|n)\mu_i\beta_i. \tag{2.44}$$

All this implies that, with respect to the probability $p(n)$, our system behaves like an M/M/1/K queue with a state-dependent arrival rate $\lambda(n)$ and a state-dependent service rate $u(n)$. We know from formula (2.21) that $p(n)$ can be expressed as

$$p(n) = p(0)\prod_{\ell=1}^{n}\frac{\lambda(\ell-1)}{u(\ell)}, \quad \text{for } n = 0,1,\ldots,K,$$

where

$$p(0) = \frac{1}{\sum_{n=0}^{K} \prod_{\ell=1}^{n} \frac{\lambda(\ell-1)}{u(\ell)}}.$$

From the form of the marginal probabilities $p(n)$, it follows that $p(n-1)/p(n) = u(n)/\lambda(n-1)$. We now use formula (2.43) together with this relationship and the definition of $u(n)$ (2.44) in the balance equations for $p(n)$. We get for $n = 1$

$$p(1|1)[\lambda(1) + \mu_1] = \lambda(1) + u(1), \quad \text{for } i = 1, \tag{2.45}$$

$$p(i|1)[\lambda(1) + \mu_i] = p(i-1|1)\mu_{i-1}\alpha_{i-1}, \quad \text{for } i = 2, \ldots, k. \tag{2.46}$$

For $n > 1$, we obtain

$$p(1|n)[\lambda(n) + \mu_1] = \lambda(n) + p(1|n-1)u(n), \quad \text{for } i = 1, \tag{2.47}$$

$$p(i|n)[\lambda(n) + \mu_i] = p(i-1|n)\mu_{i-1}\alpha_{i-1} + p(i|n-1)u(n), \quad \text{for } i = 2, \ldots, k. \tag{2.48}$$

Let's consider these equations in the order of increasing values of n, starting with the equation for $n = 1$. Equation (2.46) implies a straightforward product-form solution for $p(i|1)$:

$$p(i|1) = p(1|1) \prod_{j=2}^{i} \frac{\mu_{j-1}\alpha_{j-1}}{\lambda(1) + \mu_j}, \quad i = 1, 2, \ldots, k. \tag{2.49}$$

Since we must have $\sum_{i=1}^{k} p(i|1) = 1$, we readily determine $p(1|1)$ as

$$p(1|1) = \frac{1}{\sum_{i=1}^{k} \prod_{j=2}^{i} \frac{\mu_{j-1}\alpha_{j-1}}{\lambda(1)+\mu_j}}. \tag{2.50}$$

As always, empty products are set to 1 by convention.

Now consider equations (2.47) and (2.48) for increasing values of n and, for each n, for increasing values of i. The probabilities $p(i|n-1)$ are known from the preceding step so that we can express the conditional probabilities $p(i|n)$ in the form $p(i|n) = b_i(n) + c_i(n)u(n)$.

For $i = 1$, from (2.47), we have

$$b_1(n) = \lambda(n)/[\lambda(n) + \mu_1] \quad \text{and} \quad c_1(n) = p(1|n-1)/[\lambda(n) + \mu_1]. \tag{2.51}$$

Equation (2.48) gives us the following recurrence for $i = 2, \ldots, k$:

$$\begin{aligned} b_i(n) &= b_{i-1}(n)\mu_{i-1}\alpha_{i-1}/[\lambda(n) + \mu_i], \\ c_i(n) &= [c_{i-1}(n)\mu_{i-1}\alpha_{i-1} + p(i|n-1)]/[\lambda(n) + \mu_i]. \end{aligned} \tag{2.52}$$

Once we have computed all the coefficients $b_i(n)$ and $c_i(n)$ for the current value of n, we can determine the completion rate $u(n)$ from the normalizing condition $\sum_{i=1}^{k} p(i|n) = 1$:

$$u(n) = \frac{1 - \sum_{i=1}^{k} b_i(n)}{\sum_{i=1}^{k} c_i(n)}. \tag{2.53}$$

The knowledge of $u(n)$ allows us to easily compute the steady-state probability that there are n customers in our system, $p(n)$, and, hence any performance measure that can be derived from this distribution. This approach transforms the solution of the balance equations for our model into a simple recurrent computation. The recurrence in formulas (2.51) and (2.52) involves no subtraction, and it has been shown to be numerically stable (Brandwajn & Wang, 2008). As we show in Section 2.8, one can develop an analogous solution for general non-cyclical phase-type distributions, i.e., for the M/Ph/1/K model. It is interesting that using such phase-type distributions (Coxian, in our example) we don't need to assume infinite system capacity nor rely explicitly on specific properties of Poisson arrivals.

2.7 Phase-type distributions for times between arrivals — the Ph/M/1/K model

A largely analogous approach can be used for a Ph/M/1/K queue in which the times between arrivals are given by a phase-type distribution, and the service times have a memoryless distribution with rate $\mu(n)$ when there are n customers in the system (Brandwajn & Begin, 2012). Here, we assume that the phase-type distribution of the times between arrivals is acyclic with a memoryless phases.

Given n customers currently in system, we denote by $\tau_j(n)$ the probability that the arrival process starts in phase j, $j = 1, \ldots, a$, by $\lambda_j(n)$ the completion rate of phase j, by $r_{jl}(n)$ $(l > j)$ probability that phase j is followed by phase l, and by $\hat{r}_j(n)$ the probability that the arrival process terminates after phase j. As you probably noted, we use a generalized formulation in which the components of the phase distribution may depend on the current number of customers in the system. This way, we can represent, among other features, situations where the rate of arrivals varies according to how busy the system is. This is the case, for example, with load balancing. To keep things simple, we assume that customers arriving to find the buffer full, i.e., $n = K$, are simply lost and the arrival process continues unperturbed. The original paper (Brandwajn & Begin, 2012) also considers the case where the arrival process stops until there is a departure from the system.

As usual, we consider this queue in steady state, and we use (n, j), the current number of customers in the system, and the current phase of the arrival process as its state description. We denote by $p(n, j)$ the corresponding steady-state probability. We use the "rate of flow out = rate of flow in" principle for each state to obtain the balance equations for $p(n, j)$:

$$p(n,j)[\lambda_j(n) + \mu(n)] = p(n+1,j)\mu(n+1) + \sum_{l=1}^{j-1} \lambda_l(n) r_{lj}(n) p(n,l) + \tau_j(n) \sum_{l=1}^{a} \lambda_l(n-1)\hat{r}_l(n-1)p(n-1,l),$$
$$j = 1, \ldots, a, \quad \text{for } n = 0, 1, \ldots, K-1, \tag{2.54}$$

$$p(K,j)[\lambda_j(K) + \mu(K)] = \sum_{l=1}^{j-1} \lambda_l(K) r_{lj}(K) p(K,l) + \tau_j(K) \times \sum_{l=1}^{a} \lambda_l(K-1)\hat{r}_l(K-1)p(K-1,l) + \tau_j(K) \sum_{l=1}^{a} \lambda_l(K)\hat{r}_l(K)p(K,l),$$
$$j = 1, \ldots, a, \quad \text{for } n = K. \tag{2.55}$$

In the equation for $n = 0$, the terms for $p(n-1, l)$ are assumed to vanish. We denote by $p(j|n)$ the conditional probability that the arrival process is in phase j given the number of customers in the system, and by $p(n)$ the steady-state probability that the current number of customers is n, $n = 0, \ldots, K$. From the definition of conditional probability, we have

$$p(n, j) = p(j|n)p(n). \tag{2.56}$$

Let's concentrate for a moment on the current number of customers in the system. Obviously, the rate with which n decreases is $\mu(n)$, and $\mu(0) = 0$. But what is the rate with which n increases? If the arrival process is in its phase j, it is $\lambda_j(n)\hat{r}_j(n)$. The probability that the current phase is j given the current number of customers is $p(j|n)$, so that the rate with which n increases, denoted by $\alpha(n)$, can be expressed as

$$\alpha(n) = \sum_{j=1}^{a} \lambda_j(n)\hat{r}_j(n)p(j|n), \quad n = 0, \ldots, K-1. \tag{2.57}$$

This implies that, with respect to the probability $p(n)$, our Ph/M/1/K queue behaves like an M/M/1/K queue with a state-dependent arrival rate $\alpha(n)$ and service rate $\mu(n)$. Hence, the probability that there are n customers in the system is given by

$$p(n) = \frac{1}{\mathrm{G}} \prod_{i=1}^{n} \frac{\alpha(i-1)}{\mu(i)}, \quad n = 0, 1, \ldots, K, \tag{2.58}$$

where G is a normalizing constant and empty products are set to 1. We could reach the same conclusion if we used the relationship (2.56) in the balance equations (2.54) and (2.56) and summed the equations for a given value of n over all values of $j = 1, \ldots, a$.

We now use the relationship (2.56) together with the fact that $p(n-1)/p(n) = \mu(n)/\alpha(n-1)$ and $p(n+1)/p(n) = \alpha(n)/\mu(n+1)$ from formula (2.58), in the balance equations (2.54) and (2.56) to obtain the following equations for the conditional probabilities $p(j|n)$:

$$\begin{aligned} p(j|n)[\lambda_j(n) + \mu(n)] = p(j|n+1)\alpha(n) &+ \sum_{l=1}^{j-1} \lambda_l(n) r_{lj}(n) p(l|n) \\ &+ \tau_j(n)\mu(n), \quad j = 1, \ldots, a, \\ &\text{for } n = 0, \ldots, K-1, \end{aligned} \tag{2.59}$$

$$p(j|K)[\lambda_j(K)+\mu(K)] = \sum_{l=1}^{j-1} \lambda_l(K) r_{lj}(K) p(l|K) + \tau_j(K)\mu(K)$$
$$+\tau_j(K)\alpha(K), \quad j=1,\ldots,a, \quad \text{for } n=K. \tag{2.60}$$

Additionally, the probabilities $p(j|n)$ are normalized for each value of n

$$\sum_{j=1}^{a} p(j|n) = 1, \quad \forall n = 0,1,\ldots. \tag{2.61}$$

The form of equations (2.60) and (2.59) suggests that we can express the probabilities $p(j|n)$ in the form

$$p(j|n) = \varphi_j(n)\alpha(n) + \xi_j(n)\mu(n). \tag{2.62}$$

Of course, for $n = 0$, $\mu(n) = 0$ and all $\xi_j(0) = 0$. We start with $n = K$. We use relationship (2.62) in equation (2.60) to obtain the coefficients $\varphi_j(K)$ and $\xi_j(K)$ in the order $j = 1,\ldots,a$. We then consider consecutively decreasing values of $n = K-1,\ldots,0$. Equation (2.59) implies

$$\varphi_j(n) = \frac{1}{[\lambda_j(n)+\mu(n)]}\left\{p(j|n+1) + \sum_{l=1}^{i-1} \lambda_l(n) r_{lj}(n)\varphi_l(n)\right\},$$
$$j = 1,\ldots,a, \tag{2.63}$$

and

$$\xi_j(n) = \frac{1}{[\lambda_j(n)+\mu(n)]}\left\{\sum_{l=1}^{i-1} \lambda_l(n) r_{lj}(n)\xi_l(n) + \tau_j(n)\right\},$$
$$j = 1,\ldots,a. \tag{2.64}$$

The unknown conditional rate of arrivals $\alpha(n)$ is determined from the normalizing condition (2.61)

$$\alpha(n) = \frac{1-\mu(n)\sum_{j=1}^{a}\xi_j(n)}{\sum_{j=1}^{a}\varphi_j(n)}. \tag{2.65}$$

Recurrence relations (2.63) and (2.64) can, in fact, be used also for $n = K$ if we let $p(j|K+1) = \tau_j(K)$, $\forall j$.

To summarize, we start with $n = K$, compute the coefficients $\varphi_j(K)$ and $\xi_j(K)$, find the value of $\alpha(K)$ from formula (2.65), and then compute $p(j|K)$ using formula (2.62). We then consider consecutive decreasing values of $n = K - 1, \ldots, 0$. For each n, the values of $p(j|n+1)$ are known, and we determine the coefficients $\varphi_j(n)$ and $\xi_j(n)$ in the order $j = 1, \ldots, a$ using the recurrence formulas (2.63) and (2.64). Once these coefficients have been determined, we obtain $\alpha(n)$ from formula (2.65) and $p(j|n)$ from formula (2.62), and so on until all values of n have been enumerated. The simple recurrence (2.63) and (2.64) requires that the phase-type distribution of the time between arrivals be acyclic (only moving forward in phase transitions). The recurrence used can be shown to be numerically stable (Brandwajn & Begin, 2012), which is an important point for any recurrent computation.

Clearly, once we have the values of $\alpha(n)$ for $n = 0, 1, \ldots, K - 1$, we are ready to compute the steady-state probabilities $p(n)$ from formula (2.58). In general, $\alpha(n)$, the conditional rate of customer arrivals exhibits a strong dependence on n, except in the case of Poisson arrivals, in which case $\alpha(n)$ is constant.

You probably noted that since our queue can have service rates that depend on the current number of customers in the system, it can represent a system with multiple servers. Indeed, if there are C servers and the intrinsic service rate of a single server when there are n users in the system is $\nu(n)$, we set $\mu(n) = \min(n, C)\nu(n)$. Thus, our simple recurrence applies in fact to the Ph/M/C/K model as well.

Measurements indicate that arrival processes in I/O systems (McNutt, 2005; Hsu & Smith 2003), as well as in computer networks and Internet traffic (Jiang & Dovrolis, 2005; Crovella & Bestavros, 1997), can be bursty and behave quite differently from a Poisson process. In particular, these distributions have been found to exhibit heavy-tail characteristics (Willinger *et al.*, 2004). Such high-variability distributions can be successfully represented as phase-type distributions (Brandwajn & Begin, 2012). Figure 2.5 shows an example of a Pareto-like distribution obtained using the PhFit software (Horváth & Telek, 2002). It contains 16 phases, 6 of which are used for the heavy-tail part of the distribution.

As we will have the opportunity to see later in this book, the use of phase-type distributions turns out to be a useful angle of attack for a number of more involved models.

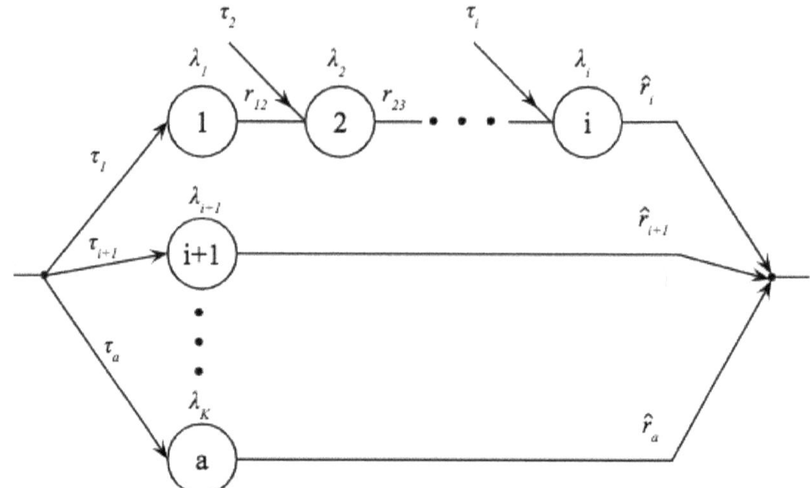

Figure 2.5. Example of a high-variability phase-time distribution.

2.8 Phase-type distribution of service times: A recurrent solution of the M/Ph/1/K/model

In the preceding section, we saw that it was possible to obtain a simple recurrent solution to a queue with phase-type times between arrivals and memoryless service times. It turns out that we can obtain an analogous recurrent solution of a queue with memoryless times between arrivals and a phase type, i.e., quasi-general, service time distribution shown in Figure 2.3(a). We denote by $\alpha(n)$ the rate of arrivals when the current number of customers in the system (queued and in service) is n, $n = 0, 1, \ldots, K$. K is the maximum number of customers that can be present in the system, and, as mentioned before, the times between arrivals are assumed to be memoryless.

The balance equations for a single-server queue with arrival rate $\alpha(n)$ and our phase-type service time distribution are as follows:

For $n = 1$,

$$p(1,i)[\alpha(1) + \mu_i] = p(0)\alpha(0)\sigma_i + \sum_{l=1}^{i-1} \mu_l q_{li} p(1,l) + \sum_{l=1}^{b} p(2,l)\mu_l \hat{q}_l \sigma_i, i = 1, \ldots, b; \quad (2.66)$$

for $n = 1, \ldots, K-1$,

$$p(n,i)[\alpha(n)+\mu_i] = p(n-1,i)\alpha(n-1) + \sum_{l=1}^{i-1} \mu_l q_{li} p(n,l)$$

$$+\sum_{l=1}^{b} p(n+1,l)\mu_l \hat{q}_l \sigma_i, \quad i=1,\ldots,b; \quad (2.67)$$

and for $n = K$,

$$p(K,i)\mu_i = p(K-1,i)\alpha(n-1) + \sum_{l=1}^{i-1} \mu_l q_{li} p(K,l),$$

$$i = 1,\ldots,b. \quad (2.68)$$

Let us consider only the steady-state probability that there are n users in this M/Ph/1/K system, denoted by $p(n)$. The instantaneous rate of increase in n is obviously $\alpha(n)$, and the instantaneous rate with which n decreases can be expressed as $u(n) = \sum_{i=1}^{b} \mu_i \hat{q}_i p(i|n)$, $n = 1,\ldots,K$, where $p(i|n)$ is the conditional probability that the service phase is i given that there are n customers in the system. This implies that the probability $p(n)$ is given by

$$p(n) = \frac{1}{\mathrm{G}} \prod_{k=1}^{n} \alpha(k-1)/u(k), \quad n = 0,\ldots,K. \quad (2.69)$$

G is a normalizing constant such that $\sum_{n=0}^{K} p(n) = 1$, and, as usual, empty products are set to 1. As a consequence of formula (2.69), we must have $\frac{p(n-1)}{p(n)} = \frac{u(n)}{\alpha(n-1)}$ and $\frac{p(n+1)}{p(n)} = \frac{\alpha(n)}{u(n+1)}$. From the definition of conditional probability, we have $p(i,n) = p(i|n)p(n)$. We use these relationships in the balance equations for our M/Ph/1/K queue to obtain the following equations for the conditional probabilities $p(i|n)$:

For $n = 1$,

$$p(i|1)[\alpha(1)+\mu_i] = u(1)\sigma_i + \sum_{l=1}^{i-1} \mu_l q_{li} p(l|1) + \alpha(1)\sigma_i,$$

$$i = 1,\ldots,b; \quad (2.70)$$

for $n = 1,\ldots,K-1$,

$$p(i|n)[\alpha(n)+\mu_i] = p(i|n-1)u(n) + \sum_{l=1}^{i-1} \mu_l q_{li} p(l|n) + \alpha(n)\sigma_i,$$

$$i = 1,\ldots,b; \quad (2.71)$$

and for $n = K$,

$$p(i|K)\mu_i = p(i|K-1)u(n) + \sum_{l=1}^{i-1} \mu_l q_{li} p(l|K), \quad i = 1, \ldots, b. \tag{2.72}$$

The form of these equations suggests a solution of the form

$$p(i|n) = d_i(n) + e_i(n)u(n), \quad i = 1, \ldots, b, \quad n = 1, \ldots, K. \tag{2.73}$$

The coefficients $d_i(n)$ and $e_i(n)$ can be determined as a recurrence from the equations (2.71), (2.72) and (2.72). For $n = 1$, we have

$$d_1(1) = \frac{1}{[\alpha(1) + \mu_1]}\alpha(1)\sigma_1;$$

$$e_1(1) = \frac{1}{[\alpha(1) + \mu_1]}\sigma_1, \quad \text{and,} \quad \text{for } i > 1,$$

$$d_i(1) = \frac{1}{[\alpha(1) + \mu_i]}\left\{\sum_{l=1}^{i-1} \mu_l q_{li} d_l(1) + \alpha(1)\sigma_i\right\};$$

$$e_i(1) = \frac{1}{[\alpha(1) + \mu_i]}\left\{\sum_{l=1}^{i-1} \mu_l q_{li} e_l(1) + \sigma_i\right\}. \tag{2.74}$$

Then, for $n = 1, \ldots, K-1$, we get

$$d_i(n) = \frac{1}{[\alpha(n) + \mu_i]}\left\{\sum_{l=1}^{i-1} \mu_l q_{li} d_l(n) + \alpha(n)\sigma_i\right\} \quad \text{and}$$

$$e_i(n) = \frac{1}{[\alpha(n) + \mu_i]}\left\{\sum_{l=1}^{i-1} \mu_l q_{li} e_l(n) + p(i|n-1)\right\}, \quad i = 1, \ldots, b. \tag{2.75}$$

Finally, for $= K$, we have

$$d_i(K) = \frac{1}{\mu_i}\left\{\sum_{l=1}^{i-1} \mu_l q_{li} d_l(K)\right\};$$

$$e_i(K) = \frac{1}{\mu_i}\left\{\sum_{l=1}^{i-1} \mu_l q_{li} e_l(K) + p(i|K-1)\right\}, \quad i = 1, \ldots, b. \tag{2.76}$$

Thus, we can compute in a recurrent fashion the coefficients $d_i(n)$ and $e_i(n)$ in the order $i = 1, \ldots, b$ for consecutive values of $n = 1, 2, \ldots, K$. For each value of n, once these coefficients have been obtained, we can compute the conditional rate of departures $u(n)$ from the normalizing condition $\sum_{i=1}^{b} p(i|n) = 1$, i.e., from $\sum_{i=1}^{b} d_i(n) + u(n) \sum_{i=1}^{b} e_i(n) = 1$

$$u(n) = \frac{1 - \sum_{i=1}^{b} d_i(n)}{\sum_{i=1}^{b} e_i(n)}, \quad n = 1, \ldots, K. \tag{2.77}$$

Knowing $u(n)$ and the coefficients $d_i(n)$ and $e_i(n)$, we obtain $p(i|n)$ from (2.73) and move on to the next value of n, until $n = K$. Thus, we have a way to solve the M/Ph/1/K queue using a simple recurrence, which can be shown to be numerically stable.

2.9 Service with no queueing allowed: A loss system with finite sources

We have considered so far systems in which customers (transactions) have some room to queue if all servers are busy. There are systems in which no queueing is possible so that a transaction either finds an available server or has to be rejected (lost). One clear example is cell phone tower where a new call or a handover must be rejected if all trunks are busy.

Figure 2.6 shows such a loss system with C servers and N_s sources of requests (transactions, customers). The term "request" finds its justification in the applications of this model to represent access to resources or devices. We denote by $1/\mu$ the mean service time of a customer and by $1/\alpha$ the mean time a source spends generating a new request following the completion of its preceding request. To keep things simple, we assume that both the service time and the time for a source to generate a new request are exponentially distributed. Consider this system in steady state and denote by m, $m = 0, .1, \ldots, C$ the current number of busy servers (we take $N_s > C$ so that it is possible to have all servers busy). We denote by $p(m)$ the steady-state probability that m servers are busy.

We let $\lambda(m) = (N_s - m)\alpha$, and, as usual, we employ the "rate of flow out=rate of flow in" principle to derive the balance equations

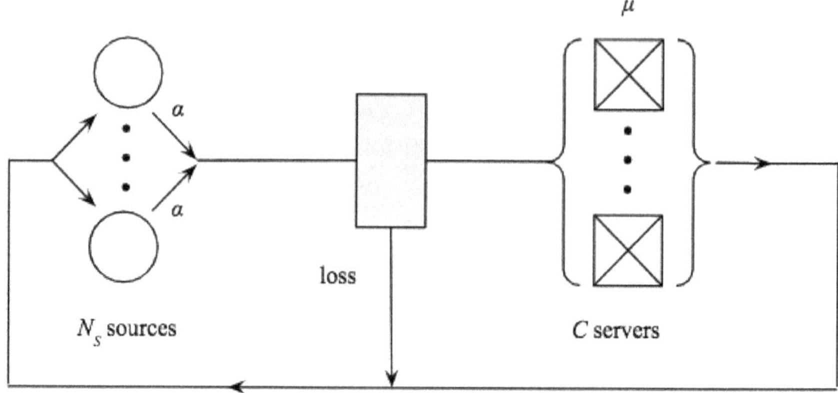

Figure 2.6. Loss system with multiple servers and discrete sources of requests.

for $p(m)$:

$$p(0)\lambda(0) = p(1)\mu, \quad \text{for } m = 0, \tag{2.78}$$

$$p(m)[\lambda(m) + m\mu] = p(m-1)\lambda(m-1) + p(m+1)(m+1)\mu,$$
$$\text{for } m = 1, \ldots, C-1, \tag{2.79}$$

$$p(C)C\mu = p(C-1)\lambda(C-1). \tag{2.80}$$

Let's consider equation (2.79) for $m = 1$, and let's use (2.78) in it. We get $p(1)\lambda(1) = p(2)2\mu$. Continuing this process for other values of m, we obtain $p(m)\lambda(m) = p(m+1)(m+1)\mu$, for $m = 0, \ldots, C-1$. Interestingly, for $m = C-1$ this is exactly equation (2.80). In essence, the general balance equation (2.79) happens to separate into two sub-equations: $p(m)\lambda(m) = p(m+1)(m+1)\mu$ and $p(m)m\mu = p(m-1)\lambda(m-1)$. These sub-equations balance one arrival with one departure and one departure with one arrival. And, as luck would have it, they are compatible with the limit equations (2.78) and (2.80). When the balance equations for a system separate into such balancing sub-equations, the system is said to possess a local balance. Local balance in turn means that it is possible (or easier) to obtain an explicit solution. In our case, we readily get

$$p(m) = \frac{1}{\mathrm{G}} \prod_{j=1}^{m} \frac{\lambda(j-1)}{j\mu}, \quad \text{for } m = 0, 1, \ldots, C. \tag{2.81}$$

G is a normalizing constant such that $\sum_{m=0}^{C} p(m) = 1$, and we have $p(0) = 1/G$.

It is important to stress that only the global balance equations, i.e., the balance equations derived from the principle "rate of flow out=rate of flow in" for each state, have to hold. Depending on the model and the state description used, local balance may or may not exist. A good example of a system in which there is no local balance is the system of balance equations (2.39–2.42) for a single-server queue with Coxian service times. Try as we might, we are unable to find balancing sub-equations for the probabilities $p(n,i)$. On the other hand, an M/M/1/K queue clearly exhibits local balance for the probability $p(n)$, where n is the current number of transactions in the system. As does our single-server queue with Coxian or phase-type service times when one considers the marginal distribution $p(n)$.

In a system where customers are subject to rejection, an important performance metric is the loss probability, i.e., the probability that an arriving customer will be rejected, which in our system happens when the arrival finds all servers busy. So, what does an arriving request see in the system of Figure 2.6? It is easy to convince oneself that an arriving request cannot see the steady-state distribution $p(m)$. Indeed, if there were as many sources as there are servers, an arriving request could never find all servers busy, regardless of the value of $p(C)$. This is quite unlike what happens with Poisson arrivals, which see the steady state of the system. Let us use an informal reasoning to find out. Consider a long period of time T. The number of arrivals that happen when there are m servers busy is simply $Tp(m)\lambda(m)$. The total number of arrivals, including the ones that are rejected, during the same period of time, can be expressed as $\sum_{k=0}^{N_s} Tp(k)\lambda(k)$. Then the probability that an arriving request finds m servers busy, denoted by $P_A(m)$, is simply the fraction of requests that happen when there are m busy servers:

$$P_A(m) = \frac{p(m)\lambda(m)}{\sum_{k=0}^{C} p(k)\lambda(k)}. \tag{2.82}$$

From here, we get the loss probability as $P_A(C)$. (Incidentally, formula (2.28) for the M/M/1/K model was derived using an analogous reasoning.)

Imagine that we are dealing with requests subject to loss but in the context of two different engineering designs. One design works just like our loss system described above, i.e., any available server can be used by an arriving request, but in the other requests have an affinity to a specific server. By affinity, we mean that a request must select with equal probability one of the servers in the system, and needs this specific server to be available not to suffer a rejection. We wish to use the loss probability as a performance metric to compare the two designs.

How can we modify our loss model of Figure 2.6 to represent the second design? We note that with affinity, there is a non-zero probability that an available server might not be usable for an arriving request. So, we can simply introduce an activation function, $\gamma(m)$, to represent the probability that a request that arrives when there are m servers busy can activate an additional server. Of course, in the original loss model, $\gamma(m) = 1$ for all $m = 0, \ldots, C-1$. In the design with server affinity, we have $\gamma(m) = (C-m)/C$. As an example, with a total of $C = 4$ servers, when there are $m = 3$ servers already busy, the probability that the required server is free would be $\gamma(3) = 1/4$. The balance equations (2.78–2.80) can be readily modified to include the activation function. Indeed, it suffices to replace $\lambda(m)$ by $\lambda(m)\gamma(m)$ in the balance equations. The resulting steady-state probability that m servers are busy becomes

$$p(m) = \frac{1}{G} \prod_{j=1}^{m} \frac{\lambda(j-1)\gamma(j-1)}{j\mu}, \quad \text{for } m = 0, 1, \ldots, C. \tag{2.83}$$

As usual, G is a normalizing constant such that $\sum_{m=0}^{C} p(m) = 1$, and we have $p(0) = 1/\text{G}$. The probability that an arriving customer finds m servers busy is still given by formula (2.71), but $P_A(C)$ no longer is the loss probability. Instead, let's compute first the probability that an arriving customer successfully acquires a server, which can be expressed as

$$p_{\text{success}} = \frac{\sum_{m=0}^{C-1} p(m)\lambda(m)\gamma(m)}{\sum_{k=0}^{C} p(k)\lambda(k)}. \tag{2.84}$$

The loss probability for this design with server affinity is then $p_{\text{loss}} = 1 - p_{\text{success}}$.

Before leaving the subject of loss systems, in which no queueing is allowed, let us note that such systems tend to be insensitive to the service time distribution (Burman *et al.*, 1984; Adan & Resing, 2001). In other words, with respect to the steady-state distribution $p(m)$, our assumption of exponential service times is not restrictive. The same tends to be true for a number of distributions of the time a source remains idle before it generates a new request (Burman *et al.*, 1984; Adan & Resing, 2001).

References

Adan, I., & Resing, J. (2001). *Queueing theory: Ivo adan and jacques resing.* Eindhoven University of Technology. Department of Mathematics and Computing Science.

Allen, A. O. (1990). *Probability, Statistics, and Queueing Theory.* Gulf Professional Publishing.

Bertsimas, D. J., & Nakazato, D. (1995). The general distributional Little's law and its applications. *Operations Research*, 43(2), 298–310.

Bobbio, A., Horváth, A., & Telek, M. (2005). Matching three moments with minimal acyclic phase type distributions. *Stochastic Models*, 21(2–3), 303–326.

Bolch, G., Greiner, S., De Meer, H., & Trivedi, K. S. (2006). *Queueing Networks and Markov Chains: Modeling and Performance Evaluation with Computer Science Applications.* John Wiley & Sons.

Brandwajn, A., & Begin, T. (2012). A recurrent solution of Ph/M/c/N-like and Ph/M/c-like queues. *Journal of Applied Probability*, 49(1), 84–99.

Brandwajn, A., & Wang, H. (2008). A conditional probability approach to M/G/1-like queues. *Performance Evaluation*, 65(5), 366–381.

Burman, D. Y., Lehoczky, J. P., & Lim, Y. (1984). Insensitivity of blocking probabilities in a circuit-switching network. *Journal of Applied Probability*, 21(4), 850–859.

Cox, D. R. (1955, July). The analysis of non-Markovian stochastic processes by the inclusion of supplementary variables. In *Mathematical Proceedings of the Cambridge Philosophical Society* (Vol. 51, No. 3, pp. 433–441). Cambridge University Press.

Cox, D. R., & Miller, H. D. (1977). *The Theory of Stochastic Processes* (Vol. 134). CRC Press.

Crovella, M. E., & Bestavros, A. (1997). Self-similarity in world wide web traffic: Evidence and possible causes. *IEEE/ACM Transactions on Networking*, 5(6), 835–846.

Feldmann, A., & Whitt, W. (1998). Fitting mixtures of exponentials to long-tail distributions to analyze network performance models. *Performance Evaluation*, 31(3–4), 245–279.

Horváth, A., & Telek, M. (2002, April). PhFit: A general phase-type fitting tool. In *International Conference on Modelling Techniques and Tools for Computer Performance Evaluation* (pp. 82–91). Springer Berlin Heidelberg.

Hsu, W. W., & Smith, A. J. (2003). Characteristics of I/O traffic in personal computer and server workloads. *IBM Systems Journal*, 42(2), 347–372.

Jiang, H., & Dovrolis, C. (2005, June). Why is the internet traffic bursty in short time scales? In *Proceedings of the 2005 ACM SIGMETRICS International Conference on Measurement and Modeling of Computer Systems* (pp. 241–252). Association for Computing Machinery, New York.

Khayari, R. E. A., Sadre, R., & Haverkort, B. R. (2003). Fitting world-wide web request traces with the EM-algorithm. *Performance Evaluation*, 52(2–3), 175–191.

Kleinrock, L. (1975). *Queueing Theory* (Vol. I). John Wiley & Sons.

McNutt, B. (2005). *The Fractal Structure of Data Reference: Applications to the Memory Hierarchy* (Vol. 22). Springer Science & Business Media.

Neuts, M. F. (1994). *Matrix-Geometric Solutions in Stochastic Models: An Algorithmic Approach.* Courier Corporation.

O'Cinneide, C. A. (1990). Characterization of phase-type distributions. *Stochastic Models*, 6(1), 1–57.

Osogami, T., & Harchol-Balter, M. (2006). Closed form solutions for mapping general distributions to quasi-minimal PH distributions. *Performance Evaluation*, 63(6), 524–552.

Saaty, T. L. (1961). *Elements of Queueing Theory: With Applications* (Vol. 34203). McGraw-Hill.

Scherr, A. L. (1965). *An Analysis of Time-Shared Computer Systems* (Vol. 535). Massachusetts Institute of Technology.

Stidham Jr., S. (1974). A last word on $L = \lambda W$. *Operations Research*, 22(2), 417–421.

Thummler, A., Buchholz, P., & Telek, M. (2006). A novel approach for phase-type fitting with the EM algorithm. *IEEE Transactions on Dependable and Secure Computing*, 3(3), 245–258.

Trivedi, K. S. (2008). *Probability & Statistics with Reliability, Queuing and Computer Science Applications.* John Wiley & Sons.

Willinger, W., Alderson, D., & Li, L. (2004, October). A pragmatic approach to dealing with high-variability in network measurements. In *Proceedings of the 4th ACM SIGCOMM Conference on Internet Measurement* (pp. 88–100). Association for Computing Machinery, New York.

Wolff, R. W. (1982). Poisson arrivals see time averages. *Operations Research*, 30(2), 223–231.

Chapter 3

Networks of Queues

3.1 A model of thrashing in a system with virtual memory

Our models considered so far viewed the system as a server (or a set of servers) but included no details of the computer system itself. We will now explore selected classical models that can be used to represent parts of the computer system. We start with a very simple model of a virtual memory system. Our goal is to develop a model that can successfully reproduce the phenomenon of thrashing. This term is used to describe the empirically observed fact that, as the number of concurrently running jobs (or applications) in a virtual memory system increases, the job throughput first increases only to decrease precipitously at a sufficiently high level of multiprogramming. Using classical terminology, we refer here to the number of jobs concurrently sharing the computer memory as the level (or degree) of multiprogramming.

The model shown in Figure 3.1 has a CPU with its queue of jobs and a secondary memory device (SM), such as a Solid-State Drive or disk drive, with its own queue. In our model, once a job gets the control of the CPU, it keeps it until it generates a page fault or it completes. To study the system for different values of the multiprogramming degree, we assume that as soon as a job completes, it is replaced by a new one. This is the meaning of the "new job" loop in Figure 3.1. We let n be the total number of jobs in the system, i.e., the degree of multiprogramming considered. We denote by v_0

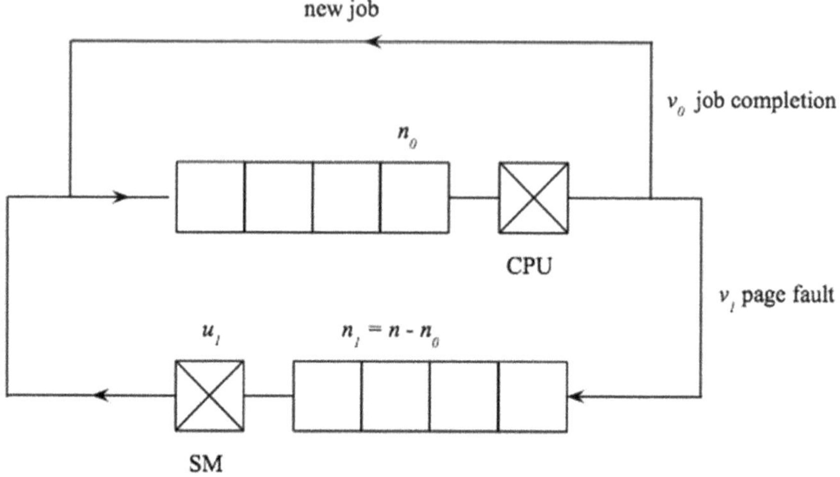

Figure 3.1. Simple mode of a virtual memory system.

the rate with which a job currently executing on the CPU completes and by v_1 the rate with which a job executing on the CPU generates a page fault. The mean time for the SM to complete the service of page fault is denoted by $1/u_1$. To keep things simple, we assume that all service times are exponentially distributed.

Since thrashing is the result of excessive page fault rates caused by the sharing of main memory by jobs, we need a model of page fault rates, so that we can link the rate v_1 to the degree of multiprogramming. Measurements done at IBM during the early days of virtual memory come in handy (O'Neill, 1976). The so-called lifetime function $e(m)$ links the mean virtual (CPU) time between page faults to the memory space available to a job, which we denote by m. Several approximations have been proposed for the observed lifetime function. The one proposed by Belady & Kuehner (1969) has the form $e(m) = am^{\kappa}$, where κ reflects program locality (κ is in the range of $1-2$) and a is a coefficient that depends on the speed of the processor. Chamberlin *et al.* (1973) proposed another two-parameter approximation of the form $e(m) = \frac{2b}{1+(\frac{d}{m})^2}$, where b and d are the two parameters of the approximation. Both approximations have their week points (Denning, 1980), but they are a good starting point and sufficient for our purpose here.

Let's now return to our model of thrashing. We assume that all jobs are statistically identical and that they share the available memory space equally, so that, if M denotes the total memory available to jobs, $m = M/n$. From here, we get for the rate of page faults in virtual time $v_1 = 1/e(M/n)$ when there are n jobs sharing main memory. Denote by U_0 the CPU utilization and by Θ the job throughput, i.e., the number of jobs completed per time unit in our system. Clearly, we must have

$$\Theta = U_0 v_0. \tag{3.1}$$

The rate v_0 is the inverse of the mean CPU time per job, and we view it as independent of the degree of multiprogramming n. This means that the job throughput is proportional to the CPU utilization, and we will concentrate on the latter for our study of thrashing. Denote by $p(n_0)$ the steady-state probability that there are n_0, $n_0 = 0, \dots, n$, jobs at the CPU, queued and in service, and thus, $n_1 = n - n_0$ jobs at the SM. The following balance equations for $p(n_0)$ are derived using our customary "rate of flow out = rate of flow in" technique:

$$p(0)u_1 = p(1)v_1, \tag{3.2}$$

$$p(n_0)(v_1 + u_1) = p(n_0 - 1)u_1 + p(n_0 + 1)v_1,$$

$$\text{for } n_0 = 1, \dots, n-1, \quad (\text{i.e., } n_1 > 0), \tag{3.3}$$

$$p(n)v_1 = p(n-1)u_1. \tag{3.4}$$

As usual, the state probabilities must sum to one, i.e., we must have $\sum_{n_0=1}^{n} p(n_0) = 1$.

You may have noticed that the job completion rate v_0 does not appear in the balance equations (3.2)–(3.4). This is because the "new job" transition does not change the number of jobs at the CPU and is therefore "invisible" with respect to the state description used.

It is easy to convince oneself that the solution to the balance equations for our system must have the form

$$p(n_0) = p(0)\left(\frac{u_1}{v_1}\right)^{n_0}. \tag{3.5}$$

From the normalizing condition, we obtain $p(0) = \frac{1}{\sum_{n_0=0}^{n}(\frac{u_1}{v_1})^{n_0}}$. We readily obtain the CPU utilization as

$$U_0 = \sum_{n_0=1}^{n} p(n_0). \tag{3.6}$$

Figure 3.2 illustrates the results our model produces for the CPU utilization as a function of the degree of multiprogramming. We clearly see the sudden collapse of CPU utilization as the number of jobs sharing main memory increases.

In our model, jobs circulate between the CPU and the SM, and we can obtain the SM utilization, which we denote by U_1 through a simple informal reasoning based on job conservation. Consider a long period of time T. During this period, the CPU was busy for TU_0 and there were TU_0v_1 page faults generated. By the same reasoning, during this same period, there were TU_1u_1 page faults processed by the SM. As T tends to infinity, the two numbers must become equal since all page faults generated are eventually served by the SM:

$$U_0v_1 = U_1u_1. \tag{3.7}$$

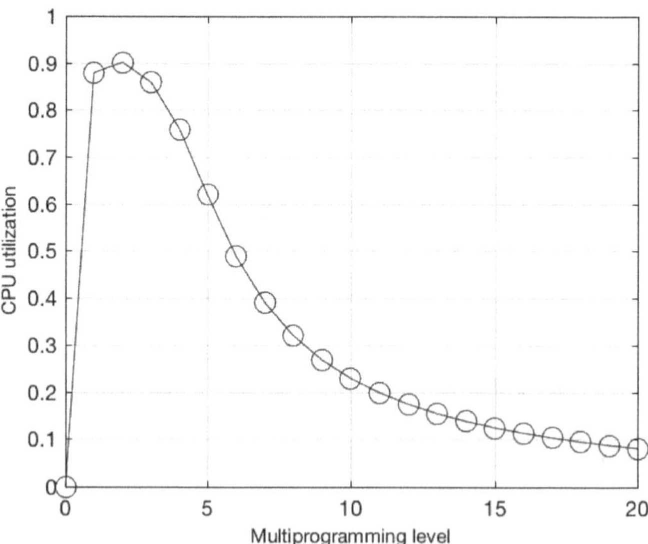

Figure 3.2. Example of results illustrating thrashing.

This means that we must have $U_1 = U_0 \frac{v_1}{u_1}$. We can easily check that a direct computation of the SM utilization as $\sum_{n_0=0}^{n-1} p(n_0)$, i.e., a sum over all states for which the SM is busy, produces exactly the same result. As we mentioned earlier, $1/v_1$ is the mean CPU time between page faults and $1/u_1$ is the mean service time at the SM. Examining the solution given by formula (3.5), we note that only the ratio of these times matters in the steady-state probability $p(n_0)$, or as one could put it, only the total work put on the server counts for the state probabilities.

3.2 Incorporating additional devices in our model: The central server model

Our model of Figure 3.1, successful as it may be in reproducing thrashing, lacks an important element found in larger systems: other I/O devices. Figure 3.3 shows a model with a separate device to support non-paging I/O. We denote by $1/u_2$ the mean service time

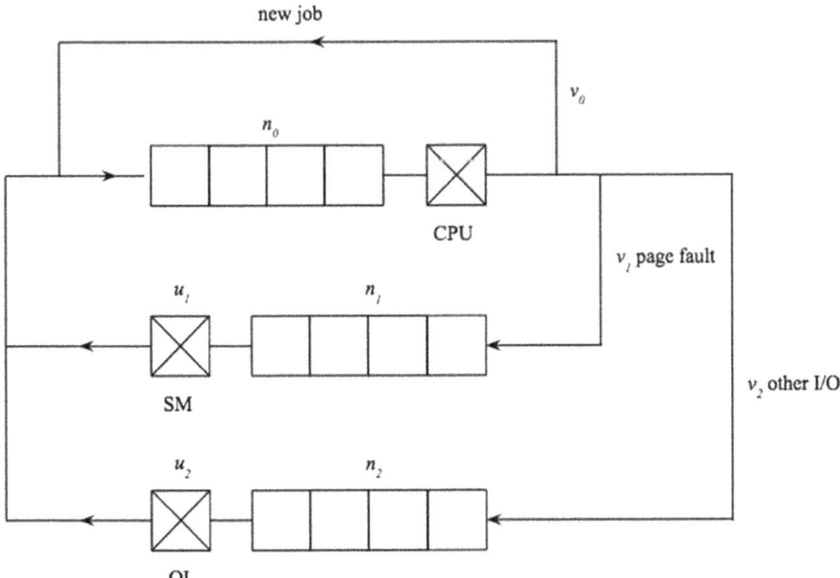

Figure 3.3. Model with an additional I/O device.

for this device and by $1/v_2$ the mean CPU time between such I/O requests. As before, there is a total of n jobs in the system.

Another way of formulating our model is to say that the rate with which a job completes a CPU service burst is $u_0 = v_0 + v_1 + v_2$. If we let $v_0 = u_0 p_0$, $v_1 = u_0 p_1$, $v_2 = u_0 p_2$, and $\sum_{i=0}^{2} p_i = 1$, we can interpret $1/u_0$ as the mean duration of a CPU burst, p_0 the probability that a job completes at the end of the burst, p_1 the probability that a page fault ends the burst, and p_2 the probability that the CPU burst ends in an I/O request.

Regardless of the formulation adopted, our goal is to obtain the steady-state probability that there are n_0 jobs at the CPU, n_1 jobs at the SM device and n_2 jobs at the other I/O device. Since the total number of jobs is kept constant, any two of these quantities fully define the state of the system if one assumes, as we do here, exponentially distributed service times. If we can obtain the probability $p(n_0, n_1)$, for example, we can easily compute the CPU utilization for different values of the degree of multiprogramming n. There is no particular difficulty in writing the balance equations for this state description. Solving them is another issue since we are now dealing with linear difference equations in two independent variables, and, unlike for a single variable, there is no general methodology to find a solution. Luckily for us, it turns out that our model is a special case of the so-called central server model, perhaps best known to the computer performance modeling community from the work of Buzen (1971) although the network can be traced back to the work of Gordon & Newell (1967).

The model, shown in Figure 3.4, includes a central server and a set of L peripheral servers. The mean service times of the peripheral servers are denoted by $1/u_1, \ldots, 1/u_L$ for peripheral servers $1, \ldots, L$. The mean service time at the central server is denoted by $1/u_0$. All service times are assumed to be exponentially distributed. A total of n jobs circulate in the system. Following the end of a service period at the central server with probability p_0 the job returns to the central server queue and with probabilities $p_1, p_2, \ldots, p_L$, the job is directed to peripheral server $1, 2, \ldots, L$, respectively. The probabilities p_j are referred to as routing or branching probabilities. Of course, we have $\sum_{j=0}^{L} p_j = 1$. All jobs are statistically identical and all queueing is in the FCFS order. We denote by n_0 the current number of jobs (queued and in service) at the central server and by n_i the current number of

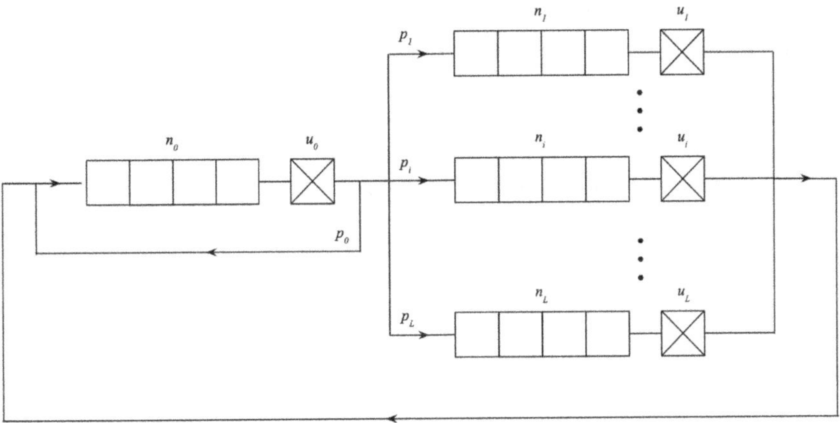

Figure 3.4. Central server model.

jobs at peripheral server i, $i = 1, \ldots, L$. The corresponding steady-state probability $p(n_0, n_1, \ldots, n_L)$, in which one of the variables is "captive" since $\sum_{j=0}^{L} n_j = n$, has the following surprisingly simple form:

$$p(n_0, n_1, \ldots, n_L) = \frac{1}{G(n)} \prod_{i=1}^{L} \left(\frac{p_i u_0}{u_i} \right)^{n_i}. \tag{3.8}$$

We use the notation $G(n)$ for the normalizing constant to emphasize the fact that it depends on the number of jobs in the system. If we denote by $\hat{s}$ a system state $(n_0, n_1, \ldots, n_L)$ and let $S(n) = \left\{ (n_0, n_1, \ldots, n_L) : \sum_{i=1}^{L} n_i \leq n) \right\}$, we can write

$$G(n) = \sum_{\hat{s} \in S(n)} \prod_{i=1}^{L} \left(\frac{p_i u_o}{u_i} \right)^{n_i}. \tag{3.9}$$

Formula (3.8) is quite remarkable. The solution has the form of a product of factors, each factor pertaining to the state of a single peripheral server. You have probably noticed that $p_i u_0$ is the rate with which the central server, when active, generates requests for service by peripheral i. By the same token, u_i is the rate with which the latter peripheral server services such requests. Such a product-form solution is directly tied to the existence of local balance in the global balance equations for the central server model. These local balance equations balance an end of service at each peripheral server with a request for service at the same peripheral. Formula (3.9) simply states that the normalizing constant is a sum over all feasible states of the simple products in formula (3.8).

If our goal is to compute the central server utilization, denoted by U_0, it is interesting to note that this utilization can be expressed in terms of the normalizing constant as follows:

$$U_0 = \frac{G(n-1)}{G(n)}. \tag{3.10}$$

We invite the reader to derive formula (3.10) as an exercise. This formula is particularly important if we have a way of computing the normalizing constant with n jobs through a recurrence involving normalizing constants in the same model with smaller numbers of users (Buzen, 1973). Once we know the central server utilization, we can easily derive the utilization of any peripheral server. We let U_i denote the utilization of peripheral server i, $i = 1, \ldots, L$. Just like we did for our simple two-server model of thrashing, we can use conservation laws in the central server model to obtain

$$U_i = \frac{p_i u_o}{u_i} U_o, \quad i = 1, \ldots, L. \tag{3.11}$$

3.3 Jacksonian networks of queues

The central server model proved to be quite successful as a base for models of centralized computer systems. However, with its rigid topology, it may not be suitable to represent different architectures of computer systems, their components or networks. As it turns out, a paper published by Jackson (1963) several years before the central server model got widely adopted, proposed a more general model, with not only arbitrary topology but also one that could be either open, i.e., with customers arriving from an outside source and departing the network, or closed, i.e., with a constant number of customers in the network. We use the term "customers" here to refer to transactions, jobs, requests, packets, or any other entities using the servers.

We refer to the class of networks of queues solved in Jackson's paper as Jacksonian networks, and we start by the case of open

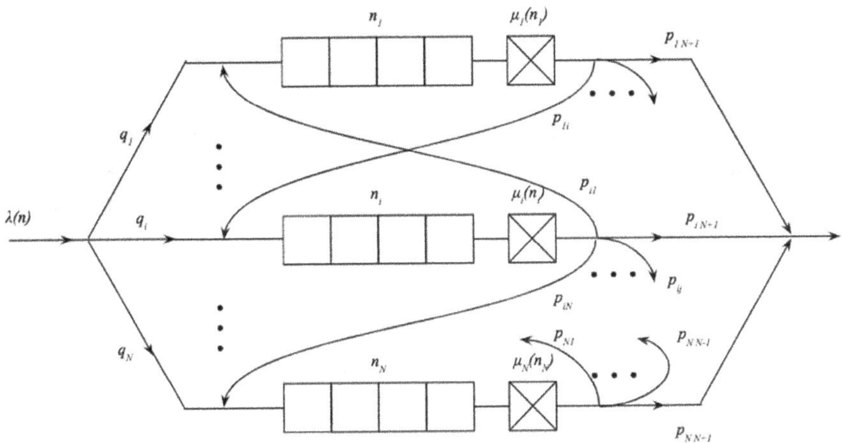

Figure 3.5. Open Jacksonian network.

networks. An open Jacksonian queueing network is represented in Figure 3.5.

The network consists of N arbitrarily interconnected service centers, numbered $1 - N$, each with its unrestricted queueing room. We denote by n_i the current number of customers at center i and by $\mu_i(n_i)$ the rate of service at this center when there are n_i customers. All service times are assumed to be memoryless, i.e., the probability that a service completes at center i during a small interval of time $(t, t + \delta t]$ when there are n_i customers at time t is given by $\mu_i(n_i)\delta t + o(\delta t)$. After completion of service at center i, $i = 1, \ldots, N$, with probability p_{ij}, the customer proceeds to center j, $j = 1, \ldots, N$, or with probability p_{iN+1} leaves the system. Customers arrive to the system from an outside memoryless source with rate $\lambda(n)$, where $n = \sum_{i=1}^{N} n_i$ denotes the total current number of customers in the system. In other words, the probability of an arrival during a small interval of time of length δt is $\lambda(n)\delta t + o(\delta t)$. The probability that an arriving customer proceeds to center i is denoted by q_i. Routing probabilities must sum to one, so that we must have $\sum_{i=1}^{N} q_i = 1$, and, for each center, $\sum_{j=1}^{N+1} p_{ij} = 1$, $i = 1, \ldots, N$. All customers are assumed to be statistically identical, and the queueing discipline at the service centers is FCFS.

The system state is described by the vector $\hat{n} = (n_1, \ldots, n_N)$, and our goal is to obtain the steady-state probability $p(\hat{n})$ if the steady state exists. Indeed, since the system is open, the existence of a steady state is not guaranteed. The solution proceeds in five steps:

- **Compute the frequency of visits:** In this step, we compute how often customers visit each center during their stay in the network. The frequencies of visits for each center, denoted by e_i, $i = 1, \ldots, N$, are determined from the following set of equations:

$$e_i = q_i + \sum_{j=1}^{N} e_j p_{ji}, \quad i = 1, \ldots, N. \tag{3.12}$$

- **Compute the factor related to service:** The factor related to service is the following product of factors, each pertaining to a single service center:

$$v(\hat{n}) = \prod_{i=1}^{N} \left[\prod_{m=1}^{n_i} \frac{e_i}{\mu_i(m)} \right]. \tag{3.13}$$

Since $\hat{n} = (n_1, \ldots, n_N)$, $v(\hat{n})$ is simply a factorized function of the numbers of customers at each center.

- **Compute the factor related to arrivals:** The factor related to arrivals is a simple product

$$w(n) = \prod_{m=0}^{n-1} \lambda(m). \tag{3.14}$$

- **Compute the normalizing constant if possible:** Denote by $V(k)$ the sum of all factors $v(\hat{n})$ such that $\sum_{i=1}^{N} n_i = k$, i.e., $V(k) = \sum_{\hat{n}:n=k} v(\hat{n})$. The normalizing constant, if it exists, is given by $1/G$ with

$$G = \sum_{k=0}^{\infty} w(k)V(k). \tag{3.15}$$

- **Obtain the solution:** If the solution to the set of equations (3.12) is unique with all frequencies of visit $e_i \geq 0$, and if the sum G exists, i.e., converges to a finite non-zero value, then the steady-state probability $p(\hat{n})$ is given by

$$p(\hat{n}) = \frac{1}{G} w(n) v(\hat{n}). \tag{3.16}$$

Jackson's (1963) paper introducing this solution is remarkable for at least two reasons. First, while it proposed a very general solution, it remained unnoticed for many years, with hardly anyone realizing its contribution. Second, the paper has an uncanny degree of sincerity in that the author did not try to cast his solution in some "higher" general framework; he just tried and found that this solution would work. One should keep those things in mind when judging papers as well as when presenting one's results.

Let us now try to apply Jackson's solution to the simple two-server model shown in Figure 3.6. Jobs arrive to the system from a Poisson source with rate λ. Arriving jobs join the queue, if any, at server 1. The mean service time at this server is $1/u_1$, and, upon completion of service, the jobs either exit the system with probability p_1 or with probability $p_2 = 1 - p_1$ require service at server number 2. The mean service time for that server is denoted by $1/u_2$. All service times are exponentially distributed. The current numbers of users at servers 1 and 2 are n_1 and n_2, respectively. The steady-state probability, if it exists, is denoted by $p(n_1, n_2)$.

It doesn't take long to convince ourselves that the model is a Jacksonian network. The equations for the frequencies of visit in our case are

$$e_1 = 1 + e_2, \tag{3.17}$$

$$e_2 = 0 + e_1 p_2. \tag{3.18}$$

We readily obtain the solution to this system of equations as $e_1 = 1/p_1$ and $e_2 = p_2/p_1$. We note that $1/p_1$ is the expected number of times a job visits server 1 and e_2 is the expected number of times a job visits server 2. Consequently, the factor related to servers

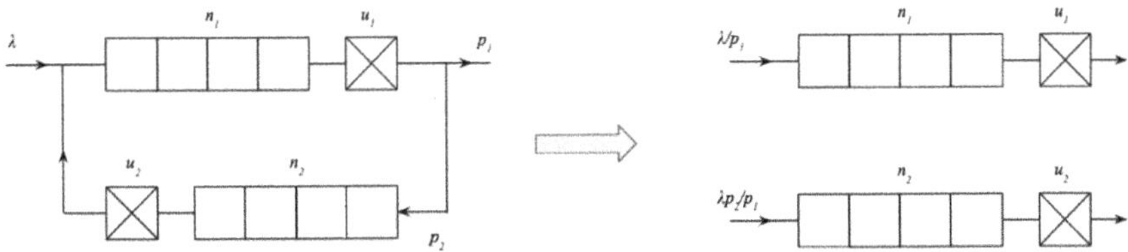

Figure 3.6. Simple two-server open queueing network model.

is $v(n_1, n_2) = [\frac{1}{(p_1u_1)^{n_1}}][(\frac{p_2}{p_1u_2})^{n_2}]$. The factor related to arrivals is simply $w(n) = \lambda^{n_1+n_2}$. If we let

$$\rho_1 = \frac{\lambda}{p_1u_1} \quad \text{and} \quad \rho_2 = \frac{\lambda p_2}{p_1u_2}, \tag{3.19}$$

the steady-state solution becomes

$$p(n_1, n_2) = \frac{1}{G}{\rho_1}^{n_1}{\rho_2}^{n_2}. \tag{3.20}$$

The normalizing constant G can be written as $G = \sum_{n_1=0}^{\infty} \sum_{n_2=1}^{\infty} {\rho_1}^{n_1}{\rho_2}^{n_2}$. For G to converge, we must have $\rho_1 < 1$ and $\rho_2 < 1$. Under these conditions, we have

$$p(n_1, n_2) = (1-\rho_1)(1-\rho_2){\rho_1}^{n_1}{\rho_2}^{n_2}. \tag{3.21}$$

You may have noticed that formula (3.21) is what we would get if we had two independent M/M/1 queues, one with arrival rate λ/p_1 and service rate u_1 and another one with arrival rate $\lambda p_2/p_1$ and service rate u_2. It is, in fact, a general property of open Jacksonian networks with state-independent arrival rates that, in the steady state, each center behaves (with respect to the distribution of the number of customers) as an independent queue. This property has been used for many years in the industry, whereby people would focus on a single facility or device, such as an I/O device in a large system and treat it as independent of other devices except, of course, for the correctly adjusted rates of customer arrivals.

In a closed Jacksonian network, like the one shown in Figure 3.7, there are N arbitrarily interconnected service centers, numbered $1-N$. The current number of customers at center i is denoted by n_i. There are no outside arrivals, and customers never leave the network so that the total number of customers in the network $n = \sum_{i=1}^{N} n_i$ remains constant. As for the open Jacksonian networks, we denote by $\mu_i(n_i)$ the rate of service at center i when the current number of customers at this center is n_i. All service times are assumed to be memoryless. After completion of service at center i, $i = 1, \ldots, N$, with probability p_{ij}, the customer proceeds to center j, $j = 1, \ldots, N$. As always, routing probabilities must sum to one, which means that we must have $\sum_{j=1}^{N} p_{ij} = 1$ for each station $i = 1, \ldots, N$. All customers are assumed to be statistically identical, and the queueing discipline at the service stations is FCFS.

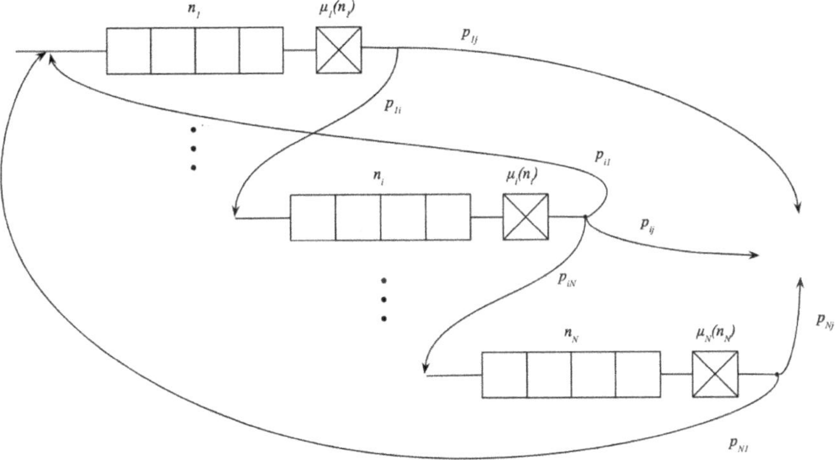

Figure 3.7. Closed Jacksonian network.

The state of the system is described by the vector $\hat{n} = (n_1, \ldots, n_N)$. We must have $n_i \geq 0$ and, since the network is closed, $\sum_{i=1}^{N} n_i = n$, where n is the constant number of customers in the system. In other words, one of the state variables is "captive". The computation of the steady-state probability $p(\hat{n})$ proceeds in steps analogous to the solution of open systems, except, of course, that there is no term related to arrivals and there are no steady-state existence issues:

- **Compute the relative frequency of visits:** In a closed network, in steady state, every center is visited an infinite number of times, so it only makes sense to talk about the frequency of visits relative to some other center. We arbitrarily select center1 and compute the frequencies of visits relative to this center from the following system of equations:

$$e_1 = 1, \tag{3.22}$$

$$e_i = \sum_{j=1}^{N} e_j p_{ji}. \tag{3.23}$$

- **Compute the factor related to service:** The factor related to service is the following product of factors, each pertaining to a single service center:

$$v(\hat{n}) = \prod_{i=1}^{N} \left[\prod_{m=1}^{n_i} \frac{e_i}{\mu_i(m)} \right]. \tag{3.24}$$

This factor is by convention equal to 0 if the state $\hat{n}$ is not feasible.

- **Compute the normalizing constant:**

$$G = \sum_{\hat{n}:\sum_{i=1}^{N} n_i = n} v(\hat{n}). \tag{3.25}$$

The normalizing constant is simply the sum over all feasible states of the corresponding factors related to service. Since the number of such states is finite, so is the sum, provided all service rates are greater than 0.

- **Obtain the steady-state solution:** If the relative frequencies of visits are such that $e_i > 0$ and thus the normalizing constant G is greater than 0, we have

$$p(\hat{n}) = \frac{1}{G} v(\hat{n}). \tag{3.26}$$

The solution given by formula (3.26) for closed Jacksonian networks is truly straightforward. As an exercise, let us apply it to compute the steady-state solution of the central server model in Figure 3.4. To keep the server numbering used in Figure 3.4, we number servers $0-L$ and adapt the formulas (3.22)–(3.26) accordingly. We start by the relative frequencies of visits, and we set $e_0 = 1$ for the central server. Then, it is immediately obvious that the relative frequencies of visits for the peripheral servers can be obtained as $e_i = p_i$, $i = 1, \ldots, L$. We continue with the factor related to service. For the state description $\hat{s} = (n_0, n_1, \ldots, n_L)$, since $n_o = n - \sum_{i=1}^{L} n_i$, we get

$$v(\hat{s}) = \frac{1}{u_0{}^{n_0}} \prod_{i=1}^{L} \left(\frac{p_i}{u_i} \right)^{n_i} = \frac{1}{u_o{}^{n}} \prod_{i=1}^{L} \left(\frac{p_i u_0}{u_i} \right)^{n_i}. \tag{3.27}$$

For a given value of n, the term $1/u_0{}^n$ is a constant; it appears for all feasible state vectors $\hat{s}$, which means that it will cancel out in the normalization process. This implies that we can write the solution as

$$p(n_0, n_1, \ldots, n_L) = \frac{1}{G(n)} \prod_{i=1}^{L} \left(\frac{p_i u_0}{u_i} \right)^{n_i}. \tag{3.28}$$

The normalizing constant is then $G(n) = \sum_{\hat{s} \in S(n)} \prod_{i=1}^{L} \left(\frac{p_i u_o}{u_i} \right)^{n_i}$, where $S(n)$ denotes the set of feasible state vectors $\hat{s}$. This is exactly the solution given by formulas (3.8) and (3.9).

You probably noticed that we have been using the term "service center" and not "server" when speaking of Jacksonian networks. The reason for this choice of vocabulary is that the service rate in a Jacksonian network can depend on the current number of customers at the center, and, as noted in Chapter 2, such a state-dependent service rate can represent multiple memoryless servers. In other words, in a Jacksonian network, we can have multiple servers at some or all of the centers.

Sometimes, when dealing with a model, we may not immediately realize that the resulting queueing network is a Jacksonian network. Take, for example, the model of an interactive computer system represented in Figure 3.8.

There is a set of N_t terminals generating transactions. The idle time between transaction requests at a terminal ("think time") is exponentially distributed with mean $1/\alpha$. There is no queueing at the terminals, and they are viewed as a "delay server", i.e., a pure delay. The system consists of a CPU and two peripherals. The mean service times at the peripherals, numbered 1 and 2, are $1/u_1$ and $1/u_2$, respectively. A service period at the CPU ends because the transaction is complete (with rate v_o), or a transaction needs to

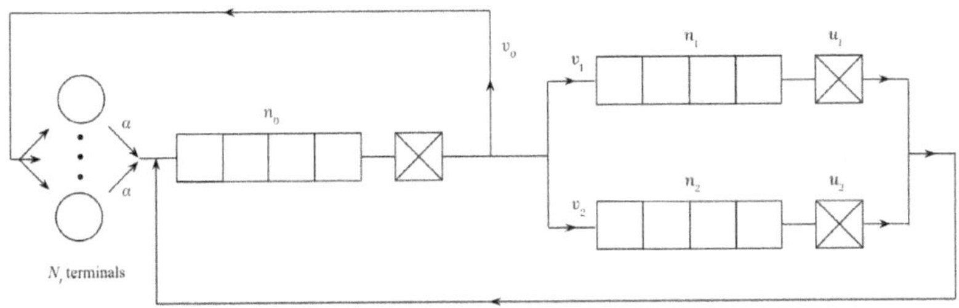

Figure 3.8. Model of an interactive system.

access peripherals 1 or 2, with respective rates v_1 and v_2. This formulation of the CPU service is equivalent to saying that the rate of service at the CPU is $u_0 = v_0 + v_1 + v_2$ and that the probability that a transaction is complete after a CPU service is $p_0 = v_0/u_0$, the probability that the transaction needs access to peripheral 1 or 2 is given by v_1/u_0 and v_2/u_0, respectively. Service times at the CPU and at the peripheral devices are assumed to be exponentially distributed. Strictly speaking, our description of Jacksonian networks doesn't mention delay servers, but nothing prevents us from replacing the set of terminals by a single server with a state-dependent service rate $u(n_t) = n_t\alpha$, where $n_t = N_t - \sum_{i=0}^{2} n_i$ denotes the current number of idle ("thinking") terminals. This transforms our model into a closed Jacksonian network with four servers. As an exercise, we invite the reader to derive the steady-state solution of this network, i.e., steady-state probability $p(n_t, n_0, n_1, n_2)$ for the numbers of transactions at each center.

3.4 State found on arrival and mean value analysis

As we have seen, Jacksonian networks have a simple product-form solution. It can be shown that the product form of the solution is directly tied to the existence of local balance in the balance equations for these models. From a practical standpoint, the main difficulty lies in the computation of the normalizing constant. If one follows the formulas literally, the computation of the normalizing constant requires the enumeration of all system states. For open networks, this can be avoided if the arrival rate does not depend on the total number of customers in the network since such networks behave like a set of independent M/M/1 queues. For closed networks, however, the size of the state space grows combinatorically with the number of customers and of service stations. This realization led to the development of several algorithms to compute the normalizing constant in closed networks without enumerating all the states (Buzen, 1973; Chandy & Sauer, 1980). The enumeration of states is particularly frustrating if one is actually only interested in mean numbers of customers at devices or in their utilization, which may oftentimes be the case.

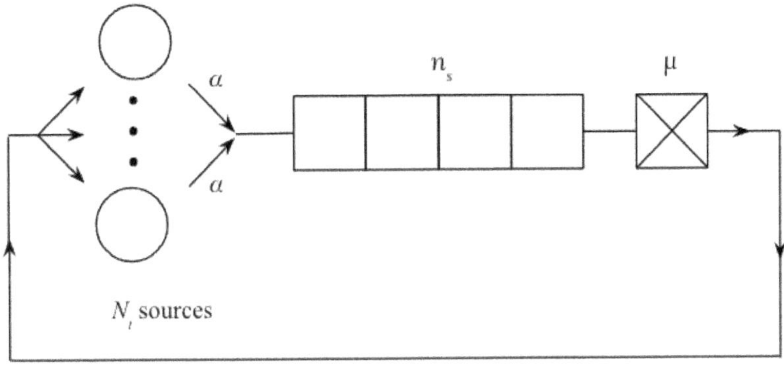

Figure 3.9. Simple closed network: Scherr's model.

Let us digress for a moment and consider the very simple closed network shown in Figure 3.9. In this model, transactions are generated by a set of N_t exponential sources. The mean idle time of a source is $1/\alpha$. There is single server with a mean service time $1/\mu$ to process the transactions in FCFS order. The service time is exponentially distributed. This model can be viewed as a very simple closed Jacksonian model. We denote by n_s the current number of transactions at the server (queued and in service), and by $p(n_s)$, the corresponding steady-state probability. Let $\rho = \alpha/\mu$. We readily get

$$p(n_s) = \frac{1}{G}\frac{\rho^{n_s}}{(N_t - n_s)!}. \tag{3.29}$$

The normalizing constant G is simply $\sum_{n_s=0}^{N_t} \frac{\rho^{n_s}}{(N_t-n_s)!}$. Formula (3.29) gives us the steady-state probability of the number of transactions at the server, but this is not what a new request arriving from a source finds. Denote by $P_A(n_s)$ the probability that an arriving request finds n_s transactions at the server. Clearly, we can only have $n_s = 0, \ldots, N_t - 1$, since no transactions can be generated when all sources are waiting for a transaction completion. To derive $P_A(n_s)$, we use a simple informal reasoning. Consider a long period of time T. During this time, the number of new transactions that arrived when there were n_s transactions already in the system is given by $Tp(n_s)(N_t - n_s)\alpha$. The total number of new transactions during the same period of time is $T\sum_{i_s=0}^{N_t-1} p(i_s)(N_t - i_s)\alpha$. We compute $P_A(n_s)$

as the fraction of new transactions that arrive when there are n_s transactions already at the server:

$$P_A(n_s) = \frac{p(n_s)(N_t - n_s)}{\sum_{i_s=0}^{N_t-1} p(i_s)(N_t - i_s)}, \quad \text{for } n_s = 0, \ldots, N_t - 1. \tag{3.30}$$

If we take into account the form of $p(n_s)$, we can transform formula (3.30) into the following expression:

$$P_A(n_s) = \frac{\frac{\rho^{n_s}}{(N_t-1-n_s)!}}{\sum_{i_s=0}^{N_t-1} \frac{\rho^{i_s}}{(N_t-1-i_s)!}}.$$

Comparing this expression with formula (3.29), we quickly note that the probability $P_A(n_s)$ is in fact the same as the steady-state probability that there are n_s transactions in a system with $N_t - 1$ sources, i.e., with one less customer than in the original network. This result turns out to be true, in general, for closed Jacksonian networks (Jackson, 1963) and is perhaps most strikingly stated as "a customer arriving at a subsystem sees the steady state of the subsystem with the arriving customer removed". It is a remarkable result. Intuitively, one would certainly expect that the arriving customer cannot "see itself" in the subnetwork it is about to join, but finding exactly the steady state of the subnetwork without the arriving customer is far from obvious.

As mentioned above, this property holds for closed Jacksonian networks. It is directly tied to the product form of the steady-state solution in these networks. It is also an essential element of a method for computing the mean numbers of customers, the mean time at each server, as well as the throughputs through each service station without enumerating the states of the system or even computing the normalizing constant. The method in question, known as the mean value analysis, is the most straightforward when the service times are not state-dependent.

To explain how the method works, we start by the simple two-server model shown in Figure 3.10. The system consists of two servers, numbered 1 and 2. The service times at these servers are assumed to be exponentially distributed with respective mean values t_1 and t_2. The total number of customers in the system is denoted by n. The mean value analysis leverages the state seen upon arrival,

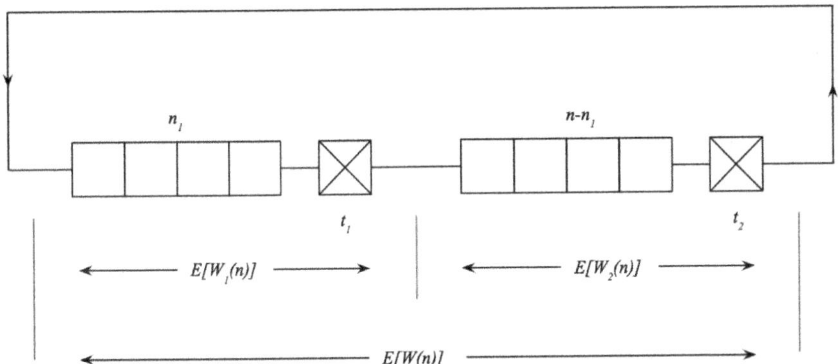

Figure 3.10. Closed tandem network with two service centers.

exponential service times, as well as Little's law applied to the network as a whole and to each server. We denote by $E[N_i(j)]$ the mean number of customers and by $E[W_i(j)]$ the mean sojourn time for a customer at server i, $i = 1, 2$ in our network with a total of j customers, $j = 0, \ldots, n$. We denote by $E[W(j)]$ the mean turnaround time, i.e., the mean time for a single pass through the network. The throughput of customers through the system is denoted by $\Theta(j)$ and the throughput of customers through server i with j customers in the whole system is $\theta_i(j)$. We also let $\tilde{n}_i(j)$ be the mean number of customers found by an arriving request at server i in our network with j customers. Due to the network topology, we have $E[W(j)] = E[W_1(j)] + E[W_2(j)]$ and $\theta_i(j) = \Theta(j)$, $i = 1, 2$.

For any population level (i.e., total number of customers) $j \geq 1$, the time a customer spends at a service station during a single pass is the time a customer has to wait for the customers found at the service station upon arrival plus its own service time. Due to the memoryless property of the exponential distribution, the mean residual service time for the customer already in service, if any, is the same as the mean service time, so that

$$E[W_i(j)] = \tilde{n}_i(j)t_i + t_i. \tag{3.31}$$

By the property of the state seen upon arrival in a closed Jacksonian network, we have $\tilde{n}_i(j) = E[N_i(j-1)]$, and formula (3.31) becomes

$$E[W_i(j)] = E[N_i(j-1)]t_i + t_i. \tag{3.32}$$

In other words, the mean time a customer spends at a server (including queueing) at population level j can be determined if we know the mean number of customers at the server at population level $j-1$. We can apply Little's law to the system as a whole to compute the overall throughput

$$\Theta(j) = \frac{j}{E[W(j)]}. \tag{3.33}$$

Knowing $\Theta(j)$, and hence $\theta_i(j)$, we can use Little's law to compute the mean numbers of customers at each server at population level j.

$$E[N_i(j)] = E[W_i(j)]\theta_i(j). \tag{3.34}$$

Clearly, we have $E[N_i(0)] = 0$ for $i = 1, 2$. So, for $j = 1$, we can write

$$E[W_i(1)] = t_i, \quad i = 1, 2,$$

$$\Theta(1) = \frac{1}{E[W(1)]}, \quad \text{where } E[W(1)] = E[W_1(1)] + E[W_2(1)],$$

$$E[N_i(1)] = E[W_i(1)]\theta_i(1), \quad \text{where } \theta_i(1) = \Theta(1), \quad i = 1, 2.$$

From here, we can move on to population level $j = 2$, and so on, until we reach the desired population level $j = n$. What emerges is a recurrent computation scheme whose complexity grows linearly with population level or the number of servers.

As another example, let us use the mean value analysis in the central server model of Figure 3.4. We use similar notation as in our introductory example, and we let $t_i = 1/u_i$ denote the mean service time at server i, $i = 0, 1, \ldots, L$. We denote by $\Theta(j)$ the throughput of the peripherals in our model with j jobs and by φ_i the frequency of visits to server i relative to the set of peripherals. From the topology of the network, we readily obtain $\varphi_0 = 1/(1 - p_0)$ for the central server and $\varphi_i = p_i//(1 - p_0)$ for the peripheral servers $i = 1, \ldots, L$. We note in passing that, as we would expect, the ratios φ_i/φ_0 are equal to the relative frequencies of visits we used in formula (3.27). From the same topology, we get

$$E[W(j)] = E[W_0(j)]\varphi_0 + \sum_{i=1}^{L} \varphi_i E[W_i(j)], \tag{3.35}$$

$$\theta_i(j) = \Theta(j)\varphi_i, \quad i = 0, 1, \ldots, L. \tag{3.36}$$

For a given population level j, we have also the following relationships:

$$E[W_i(j)] = E[N_i(j-1)]t_i + t_i, \quad i = 0, 1, \ldots, L, \tag{3.37}$$

$$\Theta(j) = \frac{j}{E[W(j)]}, \tag{3.38}$$

$$E[N_i(j)] = E[W_i(j)]\theta_i(j), \quad i = 0, 1, \ldots, L. \tag{3.39}$$

Formula (3.37) comes from the state upon arrival in Jacksonian networks, formula (3.38) comes from Little's law applied to the whole system, and formula (3.39) comes from Little's law applied to each server.

We start with population level $j = 1$. Obviously, $E[N_i(0)] = 0$ for all servers, so that we can immediately obtain the set of mean sojourn times $E[W_i(1)]$ for all servers $i = 0, 1, \ldots, L$. We then use formula (3.35) to compute the turnaround time $E[W(1)]$. Next, formula (3.38) gives us the system throughput and from there the set of throughputs $\theta_i(1)$ via formula (3.36). Finally, we compute the mean numbers $E[N_i(1)]$ with the help of formula (3.39). This process continues recurrently for increasing population levels until we reach the desired value $j = n$. At the end of the process, we have the mean sojourn times and the mean numbers of customers at all stations in the network. We also have the throughputs of customers through each server, $\theta_i(n)$. To compute the utilization of each server, it suffices then to multiply the throughput by the mean service time (from Little's law)

$$U_i = \theta_i(n)t_i, \quad i = 0, 1, \ldots, L. \tag{3.40}$$

The throughput through the "new job" loop is given by $\theta_0(n)p_0 = \Theta(n)p_0/(1-p_o)$, and we can obtain the mean total time per job from Little's formula by dividing the number of jobs in the model, n, by this throughput.

The mean value analysis can be extended to Jacksonian networks with state-dependent service rates (Sauer, 1983) although the method loses a little of its simplicity.

3.5 BCMP networks of queues: Models with classes of customers

Jacksonian networks combine a general topology and state-dependent service rates with a simple product-form solution, which provides the basis to model a large number of systems. However, they miss several potentially important features one may want to incorporate in a computer system model. Among them, the ability to distinguish different classes of customers (jobs, users, transactions). Also, it would be nice to be able break away from memoryless service times assumed in Jacksonian networks.

The so-called BCMP theorem, named after the four authors who developed and published the solution (Baskett *et al.*, 1975), gives the steady-state joint distribution of the numbers of customers in a network with several classes of customers. In our presentation of the BCMP theorem, we use largely the notation of the original article. The network of queues considered consists of N service centers and R classes of customers. We denote by n_{ir} the number of customers of class r at service center i. The vector $y_i = (n_{i1}, \ldots, n_{iR})$ gives the numbers of customers of each class at service center i, $i = 1, \ldots, N$. The state of the whole network is defined by the vector $S = (y_1, \ldots, y_i, \ldots, y_N)$. The steady-state solution for the network, denoted by $P(S)$, has the following form:

$$P(S) = Cd(S)g_1(y_1) \ldots \; g_N(y_N). \tag{3.41}$$

In the above expression, $d(S)$ is a function of the state of the system related to arrivals, $g_i(.)$ is a function depending on the type of service center i, and C is a normalizing constant.

There are four possible types of service centers according to the queueing discipline: FCFS, Processor Sharing (PS), pure delay with no queueing also called Infinite Server (IS), and Last Come First Served Preemptive Resume (LCFS PR). A PS service center has a single server whose service rate is inversely proportional to the number of customers at the center. The PS service discipline is an abstraction of quanta-based CPU service with infinitesimal quanta.

At any given time, each customer belongs to a single class at a given service center. Upon completion of a service request, the customer has a fixed probability of moving to another service center, where it may also change its class. We denote by $p_{i,r;j,s}$ the probability that a customer of class r upon completion of service at center i will next proceed to center j and be there of class s. In a closed system, we have $\sum_{(j,s)} p_{i,r;j,s} = 1$ for all pairs (i, r). In an open network after completion of service at center i, a customer of class r may leave the system with probability $1 - \sum_{(j,s)} p_{i,r;j,s}$. We assume that the routing and class change probabilities $p_{i,r;j,s}$ define m routing chains (also referred to as ergodic subchains), which partition the pairs (service center, class). We denote these routing chains E_k, $k = 1, \ldots, m$.

For open networks, times between arrivals are assumed to be memoryless with potentially state-dependent arrival rates. Two types of state dependencies are allowed for arrivals. In the first type, there is a single arrival stream whose instantaneous rate $\lambda(.)$ may be a function of the total current number of customers in the system. This is the same as for open Jacksonian networks. In the second type, there may be m memoryless arrival streams corresponding to the m routing chains. The instantaneous arrival rate for such a chain, denoted by $\lambda_k(.)$, may depend on the current number of customers in the given routing chain (chain number k). We denote by q_{js} the probability that an outside arrival starts at service center j and is there of class s.

At service centers where non-memoryless service time distributions are allowed, such distributions are assumed to possess a rational Laplace transform. This is the case for the phase-type distributions we considered when discussing elementary models in Chapter 2, and, in particular, for Coxian distributions. This means that a distribution with a rational Laplace transform can be used to approximate arbitrarily closely any distribution. The queueing room at each service center is assumed to be arbitrarily large. As mentioned earlier, BCMP networks can have four types of service centers. The assumptions on the service times depend on the type of the service center:

- **Type 1:** The queueing discipline is FCFS, the service time distribution is memoryless with rate $\mu_i(n_i)$ if there are currently n_i customers at this center. Note that all customers, regardless of their class, must have the same rate of service at FCFS service stations.

- **Type 2:** There is a single server and the queueing discipline is PS. Each class may have a distinct service time distribution with a rational Laplace transform. The mean service time for a customer of class r at center i is denoted by $1/\mu_{ir}$.
- **Type 3:** This is a pure delay (IS) with no queueing and each class may have its own delay time with a rational Laplace transform. The mean service time for a customer of class r at center i is denoted by $1/\mu_{ir}$.
- **Type 4:** There is a single server and the queueing discipline is LCFS-PR. Each class may have a distinct service time distribution with a rational Laplace transform. With the LCFS-PR discipline, a newly arrived customer interrupts the customer currently in service, if any. An interrupted customer eventually resumes service at the stage the customer was in when the service was interrupted. Not counting service interruptions, the mean service time for a customer of class r at center i is $1/\mu_{ir}$.

The product form of the solution is tied to the existence of local balance in BCMP networks. This local balance (also called independent balance by Schweitzer (1977)) can be stated as follows: "the rate of flow into a state by a customer entering a stage of service equates the rate of flow out of that state due to a customer leaving that stage of service". For a customer queued at a service center, the stage of service refers to the stage the customer will enter when given service.

The solution proceeds in steps broadly analogous to those of Jacksonian networks:

- **Compute the frequencies of visits:** We denote by e_{js} the frequency (or relative frequency for closed networks) with which a customer of class s visits service center j. For each routing chain E_k, $k = 1, \ldots, m$, we solve the following set of equations:

$$e_{js} = q_{js} + \sum_{(i,r)\epsilon E_k} e_{ir} p_{ir;js}, \quad (j, s)\epsilon E_k. \tag{3.42}$$

Recall that q_{js} is the initial selection probability that a customer arriving from outside the network is of class s and starts at service center j. A BCMP model can be open for some and closed for other classes of customers. If the system is closed for a class, the corresponding initial selection probabilities are equal to zero for all centers.

- **Compute factors related to service:** We let $n_i = \sum_{r=1}^{R} n_{ir}$ denote the total number of customers at service center i. The factors related to service depend on the type of the service center. If center i is of type 1 (FCFS), the corresponding factor related to service is

$$g_i(y_i) = n_i! \left\{ \prod_{r=1}^{R} \frac{e_{ir}^{n_{ir}}}{n_{ir}!} \right\} \prod_{j=1}^{n_i} \frac{1}{\mu_i(j)}. \tag{3.43}$$

If center i is of type 2 (PS) or of type 4 (LCFS-PR), we have the following. If there is no state dependency of the service rates, the factor related to service is

$$g_i(y_i) = n_i! \prod_{r=1}^{R} \left(\frac{1}{n_{ir}!} \right) \left[\frac{e_{ir}}{\mu_{ir}} \right]^{n_{ir}}. \tag{3.44}$$

If the class service rate is a function of the number of users of that class at the given center, $\mu_{ir}(n_{ir})$, the factor related to service is given by

$$g_i(y_i) = n_i! \prod_{r=1}^{R} \left(\frac{1}{n_{ir}!} \right) [e_{ir}]^{n_{ir}} \prod_{j=1}^{n_{ir}} \mu_{ir}(j). \tag{3.45}$$

If the service center is of type 3 (IS), in the state-independent case, the factor related to service is given by

$$g_i(y_i) = \prod_{r=1}^{R} \left(\frac{1}{n_{ir}!} \right) \left[\frac{e_{ir}}{\mu_{ir}} \right]^{n_{ir}}. \tag{3.46}$$

If the delay for each customer class may depend on the number of users of that class at the given server, we have

$$g_i(y_i) = \prod_{r=1}^{R} \left(\frac{1}{n_{ir}!} \right) [e_{ir}]^{n_{ir}} \prod_{j=1}^{n_{ir}} \frac{1}{\mu_{ir}(j)}. \tag{3.47}$$

- **Compute factor related to arrivals:** We denote by $M(S)$ the total number of customers in the system with state S. If there is a single arrival stream, we have

$$d(S) = \prod_{i=0}^{M(s)-1} \lambda(i). \tag{3.48}$$

If there are distinct arrival streams to each routing chain (ergodic subchain), we let $M(S|E_k)$ denote the total number of customers in chain k corresponding to system state S. We then have

$$d(S) = \prod_{k=1}^{m} \prod_{i=0}^{M(S|E_k)} \lambda_k(i). \tag{3.49}$$

If the system is closed, i.e., there are no outside arrivals,

$$d(S) = 1.$$

- **Compute the normalizing constant:** The normalizing constant C must be such that the $P(S)$ sum to one when all system states are considered. Of course, for an open network, this is only possible if the values of the model parameters are such that the network is stable, i.e., if the network possesses a steady-state solution.

  Although the notation with multiple indices used above may be a bit overwhelming at times, you may have noticed that the steady-state probability factors $g_i(y_i)$ for service centers that allow non-exponential service time distributions don't in fact depend on the nature of the distribution. All that matters is the mean service time or the mean service rate. So, for centers with PS, pure delay (IS), and LCFS-PR service disciplines, we may as well assume memoryless service times! Keep in mind that this is not true for the "simple" First Come First Served queueing discipline. At a FCFS service center in a BCMP network, the service times must be exponentially distributed (in fact, memoryless) and cannot be class-dependent.

  Let us now consider the model with four service centers and two classes of jobs shown in Figure 3.11. The model is an extension of our first model of thrashing in a virtual memory system.

The system consists of a CPU, a SM and two additional devices, labeled DV2 and DV3. Each class of jobs has its own mean service

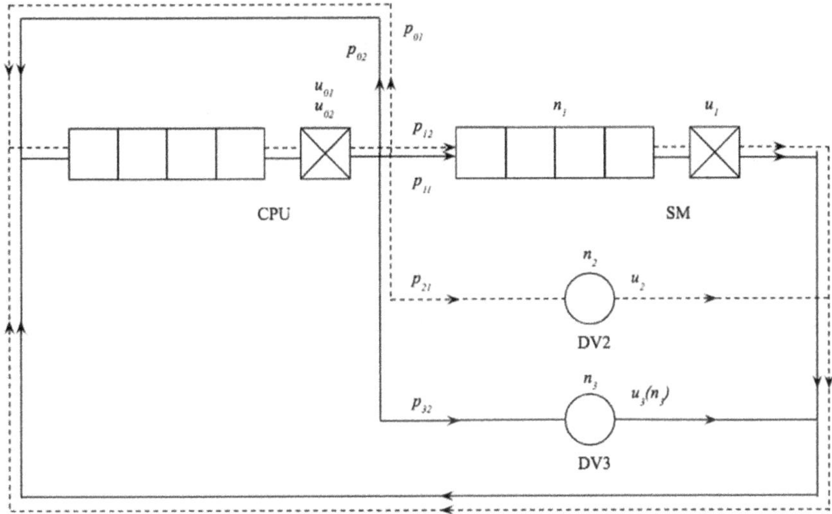

Figure 3.11. Model with two classes of processes (class 1 is represented by dashed lines and class 2 by solid lines).

time at the CPU (server 0), denoted by $1/u_{o1}$ and $1/u_{o2}$, for jobs of class 1 and 2, respectively. The service discipline at the CPU is PS. After completion of service at the CPU, a job of class r completes with probability p_{0r}, in which case it is replaced by a new job of the same class. With probability p_{1r}, the job requires service from the SM device. The service time at this device is exponentially distributed with mean $1/u_1$ regardless of the class of the process. The service discipline at the SM is FCFS. With probability p_{21}, jobs of class 1 access device DV2 but never access device DV3. Jobs of class 2, on the other hand, never access device DV2 but have a probability p_{32} of being routed to device DV3 after completion of service at the CPU. Service at device DV2 is represented as pure delay of mean duration $1/u_2$. Service at device DV3 is also a pure delay, but its rate of completion $u_3(n_3)$ is a function of the current number of jobs at the device, n_3. Pure delays are useful to represent at a high level various system devices or activities (e.g., intranet access, time spent at a software lock, wait for record retrieval from near-online storage, etc.). The total number of class r jobs in the system is denoted by M_r, $r = 1, 2$. Our goal is to compute the CPU utilization for processes of each class.

With the assumption of PS queueing discipline at the CPU and exponentially distributed service times identical for both classes of jobs at the SM with FCFS service discipline, the network clearly satisfies the assumptions for a BCMP network. We will, therefore, apply the solution procedure described above. The state of the model is described using the vectors $y_i = (n_{i1}, n_{i2})$, $i = 0, 1, 2, 3$. For $i = 2$, we must have $n_{22} = 0$, and for $i = 3$, we must have $n_{31} = 0$ because of how the job classes use devices DV2 and DV3. We let $n_i = n_{i1}+n_{i2}$ for all service centers. There are two routing chains (ergodic subchains) in our model corresponding to the two job classes. We start by the relative frequency of visits. For class 1, we get

$$e_{01} = 1, \quad e_{11} = p_{11}, \quad e_{21} = p_{21}, \quad e_{31} = 0. \tag{3.50}$$

For class 2, we get

$$e_{02} = 1, \quad e_{12} = p_{12}, \quad e_{22} = 0, \quad e_{32} = p_{32}. \tag{3.51}$$

The solution factors related to service are as follows:

$$g_0(n_{01}, n_{02}) = n_0! \prod_{r=1}^{2} \left(\frac{1}{n_{0r}!}\right) \left[\frac{1}{u_{0r}}\right]^{n_{0r}},$$

$$n_{01} = 0, \ldots, M_1, \quad n_{02} = 0, \ldots, M_2, \tag{3.52}$$

$$g_1(n_{11}, n_{12}) = n_1! \left\{ \prod_{r=1}^{2} \frac{p_{1r}^{n_{1r}}}{n_{1r}!} \right\} \frac{1}{u_1^{n_1}},$$

$$n_{01} = 0, \ldots, M_1, \quad n_{02} = 0, \ldots, M_2, \tag{3.53}$$

$$g_2(n_{21}, 0) = \left(\frac{1}{n_{21}!}\right) \left[\frac{p_{21}}{u_2}\right]^{n_{21}}, \quad n_{21} = 0, \ldots, M_1, \tag{3.54}$$

$$g_3(0, n_{32}) = \left(\frac{1}{n_{32}!}\right) [p_{32}]^{n_{32}} \prod_{j=1}^{n_{32}} \frac{1}{u_3(j)}, \quad n_{32} = 0, \ldots, M_2. \tag{3.55}$$

Since the network is closed, there is no factor related to arrivals to compute, and the steady-state solution becomes

$$P(S) = \frac{1}{G} \prod_{i=0}^{3} g_i(n_{i1}, n_{i2}).$$

Of course, only states $S = (y_o, y_1, y_2, y_3)$ such that $\sum_{i=0}^{3} n_{ir} = M_r$, $r = 1, 2$, are possible. The normalizing constant is computed as the sum over all possible system states of the terms $\prod_{i=0}^{3} g_i(n_{i1}, n_{i2})$. The CPU utilization for class r is then the sum of the probabilities $P(S)$ over all states such that $n_{0r} > 0$.

BCMP networks may be seen as an important generalization of Jacksonian networks. The presence of several classes of customers increases significantly the number of possible states, and, on the surface of it, we would have to enumerate all system states to compute the normalizing constant. The issue of having to enumerate system states to compute the normalizing constant is eliminated in an open BCMP network if there is a single stream of Poisson arrivals (arrivals of the first type with state-independent rate). In this case, with respect to the steady-state distribution of the number of customers, service centers of types 1, 2, and 4 behave like independent M/M/1 queues with appropriately chosen traffic intensities. Similarly, with respect to the steady-state distribution of the number of customers, a station of type 3 (pure delay) behaves like an M/M/∞ queue (Kleinrock, 1975). We are not implying that arrivals to a center within the network are necessarily Poisson; just that the steady-state distribution of the number of customers at a service center in an open network with overall Poisson arrivals is the same as if they were.

For closed BCMP networks, things are not as simple, although algorithms have been developed to compute the normalizing constant (Bruell & Balbo, 1980; Chandy & Sauer, 1980; Conway & Georganas, 1986; Casale, 2006). Also, like for the single-class closed Jacksonian networks, if one is only interested in expected numbers of customers, expected sojourn times and throughputs of customers, it is possible to extend the mean value analysis. The extension uses the fact that in a closed BCMP network, a customer arriving at a subnetwork "sees" the steady-state of the subnetwork with the arriving customer removed (Reiser & Lavenberg, 1980).

Jacksonian and BCMP networks both have a product-form solution in steady state. These networks are also called separable precisely because of the product form of their solution: it separates into a product of factors. The product-form solution, directly tied to the existence of local balance (independent balance), makes these networks easier to analyze. Unfortunately, many features found in computer systems and networks destroy local balance. Among them are many forms of blocking, FCFS service with different service requirements for different customer classes, priorities among customers (other than LCFS-PR), and simultaneous resource possession where a customer holds more than one resource at a time, to name a few. To handle such features, we resort to discrete-event simulation, numerical solutions, and approximation methods.

References

Baskett, F., Chandy, K. M., Muntz, R. R., & Palacios, F. G. (1975). Open, closed, and mixed networks of queues with different classes of customers. *Journal of the ACM*, 22(2), 248–260.

Belady, L. A., & Kuehner, C. J. (1969). Dynamic space-sharing in computer systems. *Communications of the ACM*, 12(5), 282–288.

Bruell, S. C., & Balbo, G. (1980). *Computational Algorithms for Closed Queueing Networks* (pp. 1–206). Elsevier North-Holland.

Buzen, J. P. (1971). *Queueing Network Models of Multiprogramming* (Doctoral dissertation, Harvard University).

Buzen, J. P. (1973). Computational algorithms for closed queueing networks with exponential servers. *Communications of the ACM*, 16(9), 527–531.

Casale, G. (2006). An efficient algorithm for the exact analysis of multiclass queueing networks with large population sizes. *ACM SIGMETRICS Performance Evaluation Review*, 34(1), 169–180.

Chamberlin, D. D., Fuller, S. H., & Liu, L. Y. (1973). An analysis of page allocation strategies for multiprogramming systems with virtual memory. *IBM Journal of Research and Development*, 17(5), 404–412.

Chandy, K. M., & Sauer, C. H. (1980). Computational algorithms for product form queueing networks. *Communications of the ACM*, 23(10), 573–583.

Conway, A. E., & Georganas, N. D. (1986). RECAL — a new efficient algorithm for the exact analysis of multiple-chain closed queuing networks. *Journal of the ACM (JACM)*, 33(4), 768–791.

Denning, P. J. (1980). Working sets past and present. *IEEE Transactions on Software Engineering*, 6(1), 64–84.

Gordon, W. J., & Newell, G. F. (1967). Closed queuing systems with exponential servers. *Operations Research*, 15(2), 254–265.

Jackson, J. R. (1963). Jobshop-like queueing systems. *Management Science*, 10(1), 131–142.

Kleinrock, L. (1975). *Queueing Theory* (Vol. 1). John Wiley & Sons.

O'Neill, R. W. (1967, April). Experience using a time-shared multiprogramming system with dynamic address relocation hardware. In *Proceedings of the April 18–20, 1967, Spring Joint Computer Conference* (pp. 611–621).

Reiser, M., & Lavenberg, S. S. (1980). Mean-value analysis of closed multichain queuing networks. *Journal of the ACM*, 27(2), 313–322.

Sauer, C. H. (1983). Computational algorithms for state-dependent queueing networks. *ACM Transactions on Computer Systems*, 1(1), 67–92.

Schweitzer, P. J. (1977). Maximum throughput in finite-capacity open queueing networks with product-form solutions. *Management Science*, 24(2), 217–223.

Chapter 4

Simulation

4.1 Basic structure of discrete-event simulation

Discrete-event simulation is a solution technique which allows, in theory, the analysis of arbitrarily complex models (Heidelberger & Lavenberg, 1984) (the interested reader may refer to this text for a discussion of merits and drawbacks of different solution methods). It derives its name from the fact that the time in the simulation, the simulation time, advances in discrete increments from one event to the next on the simulation time schedule. Many people tend to refer to discrete-event simulation as a modeling technique, but we prefer to view it as a model solution approach. The ability of the simulation to handle arbitrarily complex models is certainly one of its appealing features. On the flip side, simulations may be orders of magnitude slower than other solution methods, their results need some amount of statistical sophistication to be interpreted, and, importantly, their development may be lengthy as well as fraught with subtle and frustrating errors.

With classical computers, the simulation is really a computer program whose general basic structure is loosely as follows.

Step 1. Generate the model: In this step, we initialize simulation data structures, including statistics collection data, i.e., simulation instrumentation, and put in place the initial event of the simulation.

Step 2. Execute the simulation model: This step is in reality a loop that comprises the following general sub-steps:

a. Select the next event on the simulation time schedule.
b. Update simulation time to the time of the event selected.
c. Change the state of the simulation to reflect the event.
d. Collect statistics.
e. If the end of the simulation has been reached, proceed to Step 3, otherwise return to phase a in Step 2.

Step 3. Generate results: Here, we compute the general statistics and prepare the simulation results.

In practice, as we will see shortly, this general outline may be altered to suit the needs of the analysis of the simulation results.

4.2 Generation of random variates

A fundamental characteristic of discrete-event simulation is its repeatability. For reasons ranging from the desire to be able to reproduce the results of a simulation experiment to the need to debug a simulation, we must be able to reproduce exactly the sequence of simulated events. Thus, embedded in all simulation systems (or languages) is a pseudo-random number generator, typically generating a series of numbers in the interval $(0, 1)$ that appear to be random to statistical randomness tests. Since these series of numbers can be repeated exactly, they are not truly random, hence the name *pseudo-random*. As an example, denoting by x_i the ith consecutive pseudo-random number generated, a congruential linear generator has the general form

$$x_{i+1} = (ax_i + b)\text{modulo } m, \tag{4.1}$$

where a and b are appropriately selected constants and m is chosen to make the "*modulo*" operation simple on a computer, e.g., $m = 2^{31}$ (Knuth, 1981). By selecting a different starting number x_0 (the "seed" of the generator), we can generate different streams of consecutive random numbers x_i. A significant amount of research effort has been put over the years into the selection of "good" values for a and b, and into the generation of pseudo-random numbers, so that this area is well established (L'Ecuyer, 2017; Steele & Vigna,

2022). It is important to keep in mind that a given generator may in general generate a long but finite sequence of numbers before recycling and repeating the sequence of numbers.

The pseudo-random number generator is then used as a base to generate variates, as they are called, from different probability distributions, such as the exponential, Erlang, and uniform. Usually, a more or less extensive set of generation routines, all based on the pseudo-random numbers in the interval $(0, 1)$, is included in the simulation system. The specific method used to generate a variate depends on the distribution. For the negative exponential distribution with parameter λ, a simple transformation suffices. If we denote by $\{\nu_i\}$ a string of (pseudo-) random numbers in the interval $(0, 1)$ (we use the notation $\{\nu_i\}$ to emphasize the fact that this string of pseudo-random numbers does not have to be the precise consecutive sequence of the numbers x_i discussed previously), the string of numbers $\{e_i\}$, where

$$e_i = -\lambda \ln \nu_i \tag{4.2}$$

will be variates from such a negative exponential (Allen, 1990; Perros, 2009). Of course, unlike in the theoretical exponential distribution, the values of these variates cannot be arbitrarily large since they are constrained by the maximum value of the floating point number on the given machine architecture. Any distribution derived from the exponential, such as Erlang, hyper-exponential, Coxian or phase type in general, can then easily be generated based on the availability of exponentially distributed variates.

For the normal (Gaussian) distribution, a possible generation method might be based on summing K pseudo-random numbers and invoking the Central Limit theorem (Allen, 1990; Trivedi, 2008), which states that such sum tends to a normal distribution as the value of K increases. Once we know how to generate normal variates, distributions derived for the normal can also be generated.

Distributions such as the continuous uniform can be generated using a simple linear transformation of the variates $\{\nu_i\}$ from the basic generator. In particular, variates $\{u_i\}$ for a continuous uniform distribution defined on the interval (a, b) can be obtained as $u_i = a + (b - a) * \nu_i$.

4.3 Example of a simulation system

It is not our objective here to teach any particular simulation system or language. However, in order to make things concrete, we need to use an example of a simulation system. Since it is simple and easy to grasp, we will use the example of ***smpl***, a set of simulation routines developed by a former boss of one of us, Macdougall (1987), specifically as a tool to analyze computer and network performance models. The simulation system in question, written in the programming language C, is really a set of routines, with its own internal data structures, which can be linked as a library invoked by another program.

Keeping in mind the general outline of a discrete-event simulation discussed earlier, let's briefly review the smpl simulation world. The initialization of the simulation data structures is performed by a routine named *smpl()*. A built-in pseudo-random number is returned by a call to the routine *ranf()*. This routine supports 15 different streams of random numbers, and a routine is provided to explicitly set the number of the stream to use *stream(n)*, where $n = 1, \ldots, 15$ is the desired number. Several routines are included for the generation of different distributions, among which *expntl*(x), where x is the mean for the exponential, *erlang(x,s)* where $s < x$ is the standard deviation and x is the mean for the Erlang distribution, *uniform(a,b)*. The smpl simulation system does not employ the notion of customers or servers. Instead, it employs "tokens", active entities, and "facilities", which are essentially the equivalent of servers in queueing models and which can be used by the tokens. It is up to the simulation user to map "tokens" onto whatever the simulation is supposed to handle (e.g., customers).

A facility has a name and a given number of servers. It is created by a call to the routine *facility(name, number of servers)*, which returns a facility identifier for use in facility-related operations. Specifically, three routines handle facilities. As the name indicates, the routine *request(facility #, token, priority)* attempts to request the facility for a given token with a specified priority level. The priority level specification is useful to model the non-preemptive priority (head-of-line) discipline. The routine returns 0 if the facility had a server available and 1 if the request had to be queued. We are confronted here with one of the choices of simulation systems: how much to automate the handling of queueing operations. In smpl, if a token is queued, the event in which this happens will be re-executed next time a server becomes available. This allows a seamless automated handling of the queue(s) associated with the facility. On the flip side, the user has to make sure to structure the simulation code so as to avoid any unexpected and unwanted side effects when the event is re-executed by the simulation system. Analogously, the routine *preempt(facility #, token, priority)* attempts to request the facility for the given token with the specified priority, but, this time, the service of another token of lower priority can be preempted. If the preemption was not successful, i.e., there are no lower priority tokens using the facility, smpl automatically handles the queueing operations at the facility. The same caveat applies regarding the fact that operations in the event from which the preemption was requested will be re-executed "under the covers" when the server becomes available. Finally, a facility can be released on behalf of a token through the routine *release(facility #, tkn)*.

The smpl system also handles "events" which are associated with tokens. The meaning of each event is defined in the logic of the user's code invoking smpl routines. The simulation system provides a routine to schedule an event *schedule(event #, time from now, token)* which puts the event, with its associated token number, on the simulation time schedule, as well as a routine that allows the user to move to the next event. The routine *cause(*event, *token)* serves this purpose. It returns the event number and the associated token number of the next event on the simulation schedule. The user can always query the current simulation time (internal time in the simulation) using the routine *stime()* (we modified slightly the name of this routine to avoid name conflicts with common routines in many programming languages).

We show in Figure 4.1 an example of a very simple simulation of the M/M/1 queuing model in pseudo-code with smpl routine calls.

4.4 Simulation output analysis

Once we understand the idiosyncrasies of the simulation system, the simulation itself seems quite simple. The routine shown in Figure 4.1 tries to obtain the mean number of customers in an M/M/1 queue

```
/* M/M/1 model simulation */
integer variables: nc, n, arrival, server, event, token, completed ;
double variables: ta, ts, sumtn, tn, avn ;
/* simulation parameters */
ta = 10.0;              /* mean time between arrivals */
ts = 8.0;               /* mean service time */
nc = 30000;             /* desired number of customer completions */
/*---- initialization ---*/
        /* instrumentation */
n=0;                    /* current number of customers in system */
sumtn = 0.0;            /* cumulated sum of times for mean number computation */
tn = 0.0;               /* time during which there were n customers */
        /* simulation system
smpl("M/M/1 queue");                /* initial smpl call */
stream(1);                          /* set random numbers stream to use */
server = facility("server", 1);     /* create the server facility */
completed = 0;          /* # of completions */
arrival = 1;            /* starting token number */
schedule(1, 0.0, arrival);          /* schedule initial event */
/*---- simulation loop ---*/
while(completed< nc) {
        cause(&event, &token);              /* move to the next event */
        switch(event) {
                case 1:             /* schedule server request and next arrival */
                        schedule(2.0, 0.0, arrival);  /* server request */
                        arrival++;                    /* update arrival number */
                        schedule(1.0, ta, arrival);   /* next arrival */
                        /* instrumentation */
                        tn = stime()-tn;    sumtn += n*tn;    tn = stime();
                        n++;        /* current number of customers in system */
                        break;
                case 2:             /* request server, schedule completion */
                        if (request(server, token, 0) != 1) {
                                /* got the server */
                                schedule(3.0, expntl(ts), token);/* schedule end of service */
                        }
                        break;
                case 3:             /* release server */
                        release(server, token);
                        completed++;        /* update number of completions */
                        /* instrumentation */
                        tn = stime() - tn; sumtn += tn; tn = stime();
                        n--;        /* current number of customers in the system */
                        break;
        }       /* end event switch */
}       /* end while loop */
/*--- compute simulation results ---*/
avn =sumtn/stime();         /* estimated mean number of customers in system */
/* .... report results ... */
```

Figure 4.1. Pseudo-code for a simple simulation of the M/M/1 model.

with the given parameters, i.e., at 80% server utilization, based on a single simulation run in which 30,000 customer completions occur. But how good or even valid at all is our estimated mean number of customers? Several issues, germane to discrete-event simulations, quickly become apparent here.

First, our simulation starts with an empty system. But this is not a very likely state with the model parameter values used (with traffic intensity $\rho = 0.8$, the mean number of customers is 4). Some simulation texts suggest ignoring a certain number of initial events and starting the collection of simulation instrumentation statistics only after a given "warm-up" period. How many arrivals does it take for the system to "warm up"? Of course, in general, we have no idea how long the warm-up period should be. Another proposed solution is to start the simulation in a "more likely" state. This might work for the M/M/1 queue, but not in a more complex model where we don't really know which states are likely, and setting up a state where the system is not empty can be complicated to implement in complex models.

The second, closely related, issue is whether the selected duration of the simulation, expressed in the number of customer completions in our case, is sufficient. We intuitively feel that the simulation run has to be "long enough" for the results to be meaningful. But what does that really mean?

This brings us to the crux of the matter, which is that the simulation, in fact, does not allow us to compute the value of the mean number of customers (in our case), at least not in the sense in which we can compute the expected number of customers in the steady state from formula (2.11). This is because a discrete-event simulation, when used to solve a model such as the M/M/1 queue, is a statistical experiment, from which we try to estimate a theoretical quantity, the mean number of customers in the steady state, in our example case. If we were to rerun our simulation with a different generator seed, i.e., a different stream of random numbers, we would undoubtedly get a different value for the mean number of customers in the system. The same is true if we decided to run the simulation for a different length, for example, for 50,000 completions.

So, let's assume that we make n simulation runs and take their average to use it as our estimate of the mean number of customers in our case. The question for us is then to assess how close this result obtained from finite-length simulations is to the true value. We follow the down-to-earth spirit of the text by Macdougall (1987) in our discussion of this subject. This problem, referred to as the output analysis problem, is typically cast in the following terms. We view the individual values obtained from our simulation runs as a set of sample values from a distribution with true mean v. Let's denote by $Y_1, Y_2, \ldots, Y_n$ this set of sample values. By taking the average of these values, we compute the sample mean

$$\bar{Y} = \frac{1}{n} \sum_{i=1}^{n} Y_i. \tag{4.3}$$

Since we plan to use the sample mean as an estimate of the true mean v, we would like to know how close $\bar{Y}$ is to v. The answer to this question is cast in terms of how confident we are (or want to be, more exactly) that the true value is in a certain interval around the sample mean. This can be written as a probability statement:

$$\text{Prob}\{\bar{Y} - h \leq v \leq \bar{Y} + h\} = 1 - a. \tag{4.4}$$

The interval $[\bar{Y} - h, \bar{Y} + h]$, whose half-width is h, is referred to as the confidence interval. $1 - a$ is the desired confidence level; typical values might be 90% or 95%, i.e., $a = 0.1$ or $a = 0.05$. The half-width of the confidence interval, h, is determined by the sample values used, the number of samples n and the value of a.

The point to keep in mind in order to properly interpret the meaning of a confidence interval is that a different set of simulation runs will likely yield a different confidence interval. If we were to run a large number of sets of simulation runs and compute the confidence interval for each set, we would expect the fraction of intervals containing the true mean v to approach $1-a$. A confidence interval that contains the true mean is said to "cover" the mean. This is the reason why the confidence level $1-a$ is also called the nominal coverage probability.

If the sample values $Y_1, \ldots, Y_n$ are independent and come from a normal, i.e., Gaussian, distribution, the half-width of the confidence interval can be computed from the Student's t-distribution (Trivedi, 2008), the sample variance and the number of values in the sample (Bhattacharyya & Johnson, 1977; Macdougall, 1987). Specifically, we have

$$h = t_{a/2;n-1} \frac{s}{\sqrt{n}}, \tag{4.5}$$

Table 4.1. Values of the upper $a/2$ quantile of a Student's t-distribution with $n-1$ degrees of freedom.

$n-1$	$t_{0.05;n-1}$	$t_{0.025;n-1}$	$t_{0.005;n-1}$
5	2.02	2.57	4.03
6	1.94	2.45	3.71
7	1.90	2.37	3.50
8	1.86	2.31	3.36
9	1.83	2.26	3.25
10	1.81	2.23	3.17

where $t_{a/2;n-1}$ is the upper $a/2$ quantile of a Student's t-distribution with $n-1$ degrees of freedom, $s^2 = \frac{1}{n-1}\sum_{i=1}^{n}(Y_i - \bar{Y})^2$ is the estimated variance of the distribution of sample values, and the quantity $\frac{s}{\sqrt{n}}$ appearing in formula (4.5) is referred to as the standard error of the mean.

The values of $t_{a/2;n-1}$ are readily available from most statistics texts or software. As an example, we show in Table 4.1 the values of $t_{a/2;n-1}$ at 90%, 95% and 99% confidence levels for several values of n.

It is clearly apparent from the data in Table 4.1 that the values of $t_{a/2;n-1}$ decrease at a slowing pace as the number of degrees of freedom increases. They also become significantly larger as the desired confidence level increases. As an example, with a total of $n = 6$ runs, the confidence interval at 99% confidence level would be almost four times the interval at 90% confidence level. This effect becomes less pronounced for larger values of n. Increasing the number of runs will likely help reduce the standard error of the mean at roughly a pace of $1/\sqrt{n}$. Note that the Student's t-distribution tends to a Gaussian distribution as the number of degrees of freedom n increases.

So, provided the sample values can be viewed as independent and drawn from a normal distribution, we have a (relatively) simple formula to compute the confidence interval for the estimate of the mean value based on a set of simulation runs. The central limit theorem is usually invoked as a justification for the assumed normality of the underlying distribution. This theorem applies to sums of independent identically distributed random variables. Specifically, if $X_1, \ldots, X_i, \ldots$ are such random variables with mean μ and standard deviation $\sigma > 0$, and if we consider a sum of k such variables,

$S_k = X_1 + \cdots + X_k$, this sum approaches a normal distribution with mean $k\mu$ and variance $k\sigma^2$ as k increases (Allen, 1990; Trivedi, 2008). Denoting by $\Phi(.)$ the cumulative distribution function of a standard normal distribution (Gaussian with mean of 0 and variance of 1), this can be written as

$$\lim_{k\to\infty} \text{Prob}\left\{x \leq \frac{S_k - k\mu}{\sigma\sqrt{k}} \leq y\right\} = \Phi(y) - \Phi(x). \tag{4.6}$$

4.5 Independent replications and batch means

The prevailing consensus among simulation experts seems to be that normality typically is not the issue if there is a large enough number of events in each simulation run. This leaves us with the issue of creating independent samples. We note in passing that consecutive samples within a single simulation run tend to be more or less strongly correlated and therefore cannot be used (Macdougall, 1987). There are two main simulation methods to create a set of runs: independent replications and batch means.

In the first method, the simulation of a reasonably large number of events is repeated some number of times (say, between 6 and 15 times), each time with a different stream of random numbers. Each run is referred to as replication. Each replication starts in the same state and has the same length in terms of events such as the number of completions. The mean number of customers, or another mean value of interest to estimate from the simulation, is computed for each replication. These individual replication values are used as the sample values Y_i. Since the random number streams are clearly independent, the replications are mutually independent. As mentioned before, normality is assumed and the approximate confidence intervals are computed using formula (4.5). Obviously, the simulation program needs to be designed appropriately to create independent replications and handle the sample values Y_i.

In the batch means method, we use a single long simulation run. After possibly eliminating some number of "warm-up" events, we divide the run into n sub-runs, called batches, of equal length. The mean values estimated from each batch are used as our sample values Y_i, i.e., the batches are treated as if they were independent.

The advantage of the batch means method is that it incurs a single warm-up period. Its drawback is its weak theoretical underpinning although it seems to produce reasonable results. We mention in passing the regenerative simulation method (Fishman, 1977) which creates independent samples by using "epochs" between regeneration points in a single simulation run. A regeneration point is a point at which the system "forgets" the past so that the results of each epoch are independent random variables. For our simple M/M/1 model, the point at which there are no customers in the system is an obvious regeneration point, however, for a complex system, a regeneration point may be very difficult to find.

The independent replications method has a solid theoretical base (the results of each run are truly independent), is easy to implement, and is our method of choice. Typically, around 6–10 replications may be used, and for a given total simulation length, it may be preferable to opt for a lower number of longer replications than the other way around. This is mostly because each replication incurs its own warm-up period.

We show in Figure 4.2 the pseudo-code for a modified simulation using the replications method.

4.6 More about output analysis and discrete-event simulation in practice

In our discussion so far, we focused on the confidence intervals for a single quantity, the average number of customers in the system. Frequently, in practice, we want to estimate several quantities in a single simulation, e.g., the mean number of customers in the queue and the mean response time. This can make things a bit more complicated. Suppose we estimate confidence intervals at $(1-a)$ confidence level for k quantities. Then, by Bonferroni's inequality (Allen, 1990), the probability that all k confidence intervals contain their respective true means is lower bounded by $1-ka$. Thus, for instance, with four quantities and 95% confidence level, the probability that all four intervals contain the true mean could be as low as $1-4\times 0.05 = 0.8$. This implies, somewhat paradoxically, that if we want an overall confidence level of $(1-a)$, we need to choose confidence levels of a/k for the individual quantities. In reality, the individual confidence levels

```
/* M/M/1 model simulation with independent replications */
integer variables: nc, nrepl, repno, n, arrival, server, event, token, completed, i, ;
double variables: ta, ts, conflevel, avsample[16], sumtn, tn, avn, svar, avhw ;
/* simulation parameters */
ta = 10.0;              /* mean time between arrivals */
ts = 8.0;               /* mean service time */
nc = 30000;             /* desired number of customer completions per replication */
nrepl = 7;              /* number of replications */
conflevel = 0.95;       /* desired confidence level for confidence interval */

for (repno=1; repno <= nrepl; repno++) {
        /*---- replication number repno ---*/
                /* instrumentation */
        n=0;                    /* current number of customers in system */
        sumtn = 0.0;            /* cumulated sum of times for mean number computation */
        tn = 0.0;               /* time during which there were n customers */
                /* simulation system
        smpl("M/M/1 queue");        /* initialize smpl simulation system */
        stream(repno);              /* set random numbers stream to use */
        server = facility("server", 1);     /* create the server facility */
        completed = 0;          /* # of completions */
        arrival = 1;            /* starting token number */
        schedule(1, 0.0, arrival);          /* schedule initial event */
        /*---- simulation loop ---*/
        while(completed< nc) {
                cause(&event, &token);              /* move to the next event */
                switch(event) {
                        case 1:         /* schedule server request and next arrival */
                                schedule(2.0, 0.0, arrival); /* server request */
                                arrival++;                   /* update arrival number */
                                schedule(1.0, ta, arrival);  /* next arrival */
                                /* instrumentation */
                                tn = stime()-tn;     sumtn += n*tn;     tn = stime();
                                n++;            /* current number of customers in system */
                                break;
                        case 2:         /* request server, schedule completion */
                                if (request(server, token, 0) != 1) {
                                        /* got the server */
                                        schedule(3.0, expntl(ts), token);/* schedule end of
                                                service */
                                }
                                break;
                        case 3:         /* release server */
                                release(server, token);
                                completed++;        /* update number of completions */
                                /* instrumentation */
                                tn = stime() - tn; sumtn += tn; tn = stime();
                                n--;            /* current number of customers in the system */
                                break;
                }       /* end event switch */
        }       /* end while loop */
        /*--- compute and store simulation results for this replication ---*/
                avsample[repno] =sumtn/stime(); /* estimated mean number of customers in
                        system on this replication */
        }       /* end replication loop */
```

Figure 4.2. Pseudo-code for a simulation of the M/M/1 model using independent replications.

```
/* compute approximate confidence interval */
avn = 0.0;
for (i=1; i<= nrep; i++) {
        avn += avsample[i];
}
avn /= nrepl;          /* confidence interval midpoint */
svar = 0.0;
for (i=1; i<= nrepl; i++) {
        svar += (avsample[i]-avn)*(avsmaple[i]-avn);
}
svar /= (nrepl-1);            /* sample variance */
/* compute confidence interval half-width */
avhw = studentupperquantile((1.0-conflevel)/2.0, n-1)*sqrt(svar/nrepl) ;
/* .... report results ... */
```

Figure 4.2. *(Continued)*

for the different quantities $(1 - a_j)$ don't have to be equal as long as $\sum_{j=1}^{k} a_j = a$.

A loosely related question is that of comparing two designs. Indeed, the evaluation of two alternative designs is a frequent goal of computer system or network modeling. A paired comparison method is one possible approach to estimating the confidence interval for the difference between two mean values. Let X be the quantity on which we base the comparison of the two designs, and let's denote by $X_{i,A}$ and $X_{i,B}$ the mean values obtained from the ith replication (or batch if using the batch means method) for designs A and B. We can compute the half-width for the difference between the means using the sample values $D_i = X_{i,A} - X_{i,B}$. The overall mean difference is given by

$$\bar{D} = \frac{1}{n} \sum_{i=1}^{n} D_i. \tag{4.7}$$

If we let

$$s_d^2 = \frac{1}{n-1} \sum_{i=1}^{n} (D_i - \bar{D})^2, \tag{4.8}$$

the confidence interval at $(1 - a)$ confidence level $\{\bar{D} - h_d, \bar{D} + h_d\}$ is given by

$$h_d = t_{a/2;n-1} \frac{s_d}{\sqrt{n}}. \tag{4.9}$$

The values $X_{i,A}$ and $X_{i,B}$ don't have to be independent so that a common random number generator can be used for the simulation of each design.

Our discussion of confidence intervals has been specific to estimating a mean value. Sometimes, we are faced with the need to estimate a probability (or a set of probabilities) in a modeling study. The same central limit theorem we invoked above can be used to help us obtain the confidence interval for certain probability estimates. Suppose we wish to estimate the probability p that an incoming message is lost in a system with a finite queueing capacity. We could observe n message arrivals and count the number of times S_n that the arriving messages were lost out of these n submitted messages. As an estimate of the unknown probability p, we would then use the quantity

$$\hat{p} = S_n/n. \tag{4.10}$$

Our confidence interval problem can be formulated as $\text{Prob}\{\hat{p} - \delta \leq p \leq \hat{p} + \delta\} = 1 - a$, where $\delta > 0$ is the half-width of the confidence interval, and $1 - a$ is the desired confidence level. If we view the observation of an arriving message as a Bernoulli trial, where "success" is the loss of a message due to full buffer condition, then S_n becomes a sum of n independent Bernoulli random variables, and as such, by the central limit theorem, tends to a normal random variable. The variance of the ratio S_n/n is given by $p(1-p)/n$. Letting $r = \delta\sqrt{n}/\sqrt{p(1-p)}$, we have

$$\text{Prob}\left\{\left|\frac{S_n}{n} - p\right| \leq \delta\right\} \approx 2\Phi(r) - 1. \tag{4.11}$$

$\Phi(r)$ is the cumulative distribution function of the standard normal distribution at point r. Our problem has been transformed into finding the value of r such that $2\Phi(r) - 1 = 1 - a$, which can be written as $\Phi(r) = 1 - a/2$, i.e., the $100 \times (1 - a/2)$-th percentile of the standard normal distribution. At the 95% confidence level ($a = 0.05$), this is the 97.5th percentile, which is equal to 1.96. It is roughly 2.58 at 99% confidence level. Knowing the value of r, we can approximately compute the confidence interval half-width δ

$$\delta = r\frac{\sqrt{p(1-p)}}{\sqrt{n}} \approx r\frac{\sqrt{\hat{p}(1-\hat{p})}}{\sqrt{n}} \leq r\frac{1}{2\sqrt{n}}. \tag{4.12}$$

The value $\delta = r\frac{1}{2\sqrt{n}}$ provides a conservative estimate for the half-width of the confidence interval for our probability estimate $\hat{p}$. To improve the independence of our Bernoulli trials, we may want to count, say, only every fourth or fifth event. Better yet, we could select the events to take into account at random with a probability of, say, 0.25 or 0.2, using the random number generator built into the simulation system.

This method of estimating the probability of an event does not require that the simulation be divided into independent replications. Since it relies on Bernoulli trials, it seems well suited to estimate the confidence intervals of probabilities seen by an arriving customer. It cannot be applied directly to estimating steady-state probabilities, which are viewed as fractions of the time during which a given state was in effect.

As we just saw, in its principle, discrete-event simulation is simple, and its strong points are its generality and flexibility. For example, changing the assumptions on the service time distribution in our simulations would require little more than replacing one line of code: replacing the call to *expntl()* with a call to a routine generating variates from the desired distribution. This is certainly much simpler than having to change the state description, the nature of the balance equations and the solution approach.

On the flip side, the proper interpretation of discrete-event simulation results requires a level of statistical sophistication. We need to keep in mind that our confidence intervals rely on assumptions, which may not be quite satisfied in a specific situation, such as the normality of the underlying population. A large proportion of studies of the properties of simulation approaches was based on the M/M/1 model for the simple reason that a lot of properties of this model are known analytically. For complex models for which the exact solution is not available, we don't really know how quickly such models reach a steady state, so it is difficult to have a good answer that fits all situations for the elimination of the warm-up period, for example. As is the case in statistics, we can never be 100% sure of our results.

Among the down-to-earth difficulties linked to a simulation is reasoning with (quasi-) asynchronous sets of events. Oftentimes, in a simulation of a complex system, it is not easy to understand what the simulation program is doing, and in particular, whether it is really doing what was intended. Subtle timing errors in simulations do happen and their debugging can be excruciatingly difficult, especially if they happen only for certain sequences of events and manifest themselves after a long, seemingly correct, simulation. Certain model features, easy to handle in an analytical solution, such as state-dependent service rates at a server, are quite difficult to handle in a discrete-event simulation since they require that events that have already been scheduled be canceled and rescheduled for a different time.

We have addressed in our discussion only some topics specific to the fact that a discrete-event simulation is a statistical experiment. We left out difficulties related to simulating rare events as well as dealing with intrinsic high variability in certain models.

4.7 Example of variability issues

As an example of some of the difficulties that may arise when using discrete-event simulation, assume that we wish to study the mean number of customers in the steady state in an M/G/1 queue. We denote the mean service time of a customer by $1/\mu$ and the rate of customer arrivals by λ. We know from the Pollaczek–Khintchine formula (2.6) in Section 2.5 that the mean number of customers in such a queue is given by $E[N] = \frac{\rho^2}{(1-\rho)}(\frac{1+c_s^2}{2})+\rho$, where $\rho = \lambda/\mu$ (with $\rho < 1$) is the server utilization and c_s is the coefficient of variation of the service time. This formula applies regardless of the higher-order moments, i.e., of the shape of the service time distribution (provided that the mean and variance of the service time exist). Consequently, if two M/G/1 queues have the same server utilization and different service time distributions but with equal first two moments, the mean number of customers in these two systems must be the same.

Using a discrete-event simulator, we simulated the steady state of an M/G/1 queue for different types of service time distributions but, as mentioned before, with the same first two moments and the same server utilization. It is clear from the above discussion that all these simulations should produce equivalent values for the mean number of customers, or more precisely, asymptotically tending toward the same value of $E[N]$.

In our experiments (Begin & Brandwajn, 2010), we chose $\rho = 0.75$, which corresponds to a moderate level of server utilization. First, we considered four different service time distributions with a coefficient of variation of the service time $c_s < 1$, viz. Erlang,

Uniform, Log-Normal, and Pareto, which all had $1/\mu = 1$ and $c_s = 1/\sqrt{3}$. Our results shown in Table 4.2 indicate that within less than 100,000 service time completions, the values of $E[N]$ estimated in the simulation are all close to the exact value (relative difference of less than 3%) and gradually converge toward this value if we increase the simulation length.

Next, we set $c_s = 1.2$ while keeping $1/\mu = 1$, and we considered five different service time distributions: hyper-exponential (using the decomposition of Morse, 2004), three different Coxian distributions with two stages, Log-Normal, and Pareto, all with the same values for their first two moments. Note that this set of distribution differs from the set for the case of $c_s < 1$, since Erlang distribution and positive uniform distribution cannot have a coefficient of variation larger than 1 and $1/\sqrt{3}$, respectively. On the other hand, the coefficient of variation for hyper-exponential and Coxian distributions can only be larger than 1 and $1/\sqrt{2}$, respectively. In our results shown in Table 4.3, the convergence toward the exact value of $E[N]$ occurs quickly except for the case of the Pareto distribution, where simulations some 1,000 times longer were needed to attain the same level of accuracy.

Finally, we increased the coefficient of variation of the service time c_s to 2, and we repeated the same experiments. Not surprisingly, we observe in Table 4.4 that more time is needed for the estimated values of $E[N]$ to approach the true value. In fact, in the case of the Pareto distribution, even 1 billion service completions was not enough to come as close as in the other cases (i.e., within less than 3%) to the exact value of $E[N]$. This is related to the difficulty of sampling a Pareto distribution given its heavy tail. In particular, using the

Table 4.2. Relative deviation in percent between the exact value of the mean number of customers in an M/G/1 queue and values obtained from simulation for several simulation lengths and service time distributions with $c_s < 1$.

	Simulated time			
	1,000	10,000	100,000	10^6
Erlang	11.652	2.077	0.301	0.163
Uniform	0.763	2.598	0.968	0.467
Log-normal	9.513	0.529	2.776	0.049
Pareto	19.108	5.526	1.014	0.490

Table 4.3. Relative deviation in percent between the exact value of the mean number of customers in an M/G/1 queue and values obtained from simulation for several simulation lengths and service time distributions with $c_s = 1.2$.

	Simulated time						
	1,000	10,000	100,000	10^6	10^7	10^8	10^9
Hyper-exponential	16.527	5.322	1.547	0.852		<0.5%	
Cox-2 (type 1)	1.424	0.684	0.881	0.629		<0.5%	
Cox-2 (type 2)	7.587	1.792	4.159	0.418		<0.5%	
Cox-2 (type 3)	32.729	38.740	4.846	4.846		<0.5%	
Log-Normal	35.062	1.568	5.773	0.087		<0.5%	
Pareto	10.357	0.463	6.890	2.273	3.020	2.650	<0.5%

Table 4.4. Relative deviation in percent between the exact value of the mean number of customers in an M/G/1 queue and values obtained from simulation for several simulation lengths and service time distributions with $c_s = 2$.

	Simulated time						
	1,000	10,000	100,000	10^6	10^7	10^8	10^9
Hyper-Exponential	57.346	9.472	0.708	0.334		<0.5%	
Cox-2 (type 1)	24.449	22.804	2.923	0.326		<0.5%	
Cox-2 (type 2)	51.045	15.451	6.366	0.262	<0.5%		
Cox-2 (type 3)	95.841	17.241	71.504	22.990	3.116	3.751	<0.5%
Log-Normal	57.957	0.533	12.242	1.815		<0.5%	
Pareto	1.214	9.119	5.269	5.794	9.308	7.435	3.480

pseudo-random generators included in MATLAB (2016), SMPL, or the simulation platform ns-2 (Issariyakul & Hossain, 2009), 2 billion samples were found to not be enough to correctly estimate the true coefficient of variation of the distribution.

Overall, gathering a representative set of samples for certain distributions (e.g., Pareto distribution) may require a very large number of samples. This in turn implies that we may easily obtain strongly biased results in simulated models if these models involve highly variable distributions (e.g., IP flow size). Although discrete-event simulation can be a great solution method, we need to keep in mind that its use requires caution and, ideally, a fine post-mortem analysis of the actual distributions generated in the simulation. In some cases, despite its obvious help, even the use of confidence intervals may not be enough to avoid the pitfalls of discrete-event simulations. For this last point, we refer the interested user to a discussion by Law and Kelton (1984).

What we need to retain from the examples in this section is that simulating systems with high variability requires additional effort and a healthy amount of caution when interpreting simulation results.

4.8 Ready-to-use simulation environments: Pros and cons

When real experimentations are out of reach and analytical models do not seem suited, the discrete-event simulation may appear as a good avenue to evaluate the performance of a new distributed algorithm, of a protocol, or more generally, of a solution. Computer networks are a case in point where simulations are frequently used. Due to the increasing complexity of networks, researchers are often conducting their performance studies by means of computer simulation. To put things in perspective, in 2003, out of more than 2000 papers published in renowned journals, a majority of them reported the results obtained by means of simulation, with the rest of the papers relying on two other approaches: analytical models and experimentation (Pawlikowski, 2003). It is therefore no wonder that pre-packaged simulation tools have been introduced for specific problem areas, such as computer networks. Common network simulators include ns-2 (Issariyakul & Hossain, 2009), ns-3 (Riley & Henderson, 2010), OMNeT++ (Varga & Hornig, 2010), and Packet Tracer (Jesin, 2014). Similarly, in the field of cloud computing, performance studies are often conducted using simulators like CloudSim (Calheiros *et al.*, 2011) and SimGrid (Casanova *et al.*, 2008).

Discrete-event simulation packages typically come with a number of assets. Most simulators involve a great level of detail as they often aim at reproducing as closely as possible the behavior of a real system. In the case of ns-3, the five layers of networks are implemented with most of their respective protocols (e.g., 802.11, ARP, DHCP, IPv6, TCP, HTTP). Attaining this goal represents a formidable effort of development with the ultimate ambition of making the outcome of the simulations trustworthy. Among the other important assets of simulators is their native reproducibility, which facilitates peer validation and reuse of existing work, and comes at a relatively low cost since running simulations does not require purchasing any specialized hardware.

However, for all these important advantages, leaving aside the potential issue of their execution time, simulations are prone to pitfalls that one should be aware of. First, as discussed in the previous section, some probability distributions may be difficult to simulate causing potential bias (errors) in the results. Second, a simulator may be very accurate regarding aspects that are only of marginal (or none at all) relevance for the actual matter of a given study and conversely represent other aspects that really matter at a too coarse-grained level. This risk is particularly high when using an existing simulator framework, which offers the end-users the convenience of a quick start. Indeed, this convenience of a quick start evades and obfuscates the often necessary (and always enlightening) questions regarding the underlying modeling assumptions that were made to allow such a quick start. Said differently, an end-user can easily produce simulation results without knowing what assumptions were made on the different components (e.g., protocols, algorithms) that compose the simulator. It is not difficult to see that the outcome can be misleading conclusions when analyzing the results. Of course, the embedded level of detail in an area-specific ready-to-use simulation package may require exceedingly long simulation runs to approach anything resembling a steady state (if our goal is to study the steady-state behavior of a system).

References

Allen, A. O. (1990). *Probability, Statistics, and Queueing Theory.* Gulf Professional Publishing.

Begin, T. & Brandwajn, A. (2010, May). Note sur la simulation d'une file M/G/1 selon la distribution du temps de service. In *12èmes Rencontres Francophones sur les Aspects Algorithmiques de Télécommunications (AlgoTel).*

Bhattacharyya, G. K. & Johnson, R. A. (1977). *Statistical Concepts and Methods.* John Wiley & Sons.

Calheiros, R. N., Ranjan, R., Beloglazov, A., De Rose, C. A., & Buyya, R. (2011). CloudSim: A toolkit for modeling and simulation of cloud computing environments and evaluation of resource provisioning algorithms. *Software: Practice and experience*, 41(1), 23–50.

Casanova, H., Legrand, A., & Quinson, M. (2008, April). Simgrid: A generic framework for large-scale distributed experiments. In *Tenth International Conference on Computer Modeling and Simulation (uksim 2008)* (pp. 126–131). IEEE.

Fishman, G. S. (1977). Achieving specific accuracy in simulation output analysis. *Communications of the ACM*, 20(5), 310–315.

Heidelberger, P. & Lavenberg, S. (1984). Computer performance evaluation methodology. *IEEE Transactions on Computers*, 100(12), 1195–1220.

Issariyakul, T. & Hossain, E. (2009). *Introduction to Network Simulator 2 (NS2)* (pp. 1–18). Springer, US.

Jesin, A. (2014). *Packet Tracer Network Simulator.* Packt Publishing Ltd.

Knuth, D. (1981). Seminumerical algorithms. *The Art of Computer Programming, 2.*

Law, A. M. & Kelton, W. D. (1984). Confidence intervals for steady-state simulations: I. A survey of fixed sample size procedures. *Operations Research*, 32(6), 1221–1239.

L'Ecuyer, P. (2017, December). History of uniform random number generation. In *2017 Winter Simulation Conference (WSC)* (pp. 202–230). IEEE.

MacDougall, M. H. (1987). *Simulating Computer Systems: Techniques and Tools.* MIT press.

MATLAB. (2016). Natick, Massachusetts: The MathWorks Inc. https://www.mathworks.com/.

Morse, P. M. (2004). *Queues, Inventories and Maintenance: The Analysis of Operational Systems with Variable Demand and Supply.* Courier Corporation.

Pawlikowski, K. (2003). Do not trust all simulation studies of telecommunication networks. In *Information Networking: International Conference, ICOIN 2003, Cheju Island, Korea, February 12–14, 2003. Revised Selected Papers* (pp. 899–908). Springer Berlin Heidelberg.

Perros, H. (2009). *Computer Simulation Techniques: The Definitive Introduction!* Raleigh, NC: Harry Perros.

Riley, G. F. & Henderson, T. R. (2010). *The ns-3 Network Simulator. In Modeling and Tools for Network Simulation* (pp. 15–34). Berlin, Heidelberg: Springer.

Steele Jr, G. L. & Vigna, S. (2022). Computationally easy, spectrally good multipliers for congruential pseudorandom number generators. *Software: Practice and Experience*, 52(2), 443–458.

Trivedi, K. S. (2008). *Probability & Statistics with Reliability, Queuing and Computer Science Applications.* John Wiley & Sons.

Varga, A. & Hornig, R. (2010, May). An overview of the OMNeT++ simulation environment. In *1st International ICST Conference on Simulation Tools and Techniques for Communications, Networks and Systems.*

Chapter 5

Numerical Methods

5.1 Characteristics of balance equations

We saw earlier that the steady-state behavior of a large number of models is described by a system of balance equations. Together with the normalizing condition that all probabilities must sum to one, these equations form a system of linear equations for the unknown steady-state probabilities. There exist well-established methods to solve systems of linear equations numerically (Henrici, 1964) so one might think any method will do. However, certain characteristics of the balance equations may make their numerical solution a bit challenging.

First, as we saw when discussing Jacksonian and BCMP networks, there is a potentially very large number of states, i.e., of unknowns. Thus, the size of the system is of concern in terms of computer storage needed, time required to solve the system of equations, and numerical stability. We need to keep in mind that floating-point computation is, generally speaking, only approximate and has a limited dynamic range, i.e., the range between the smallest and largest numbers that can be represented. Of course, for open systems with infinite capacity, the size of the system space is infinite, so something has to be done if one adopts such an open model. One simple and obvious approach is to truncate the system dimensions to some manageable value.

Second, in terms of the matrix of coefficients of the unknowns, balance equations tend to be quite sparse in the sense that an equation for a given state tends to only involve a small subset of all the unknown probabilities. This is due to the very nature of common distributional assumptions; for instance, in very many models, only transitions between neighboring states are allowed. This sparsity means that standard matrix methods storing a full matrix of coefficients for the unknowns would be quite wasteful since so many of the coefficients would be equal to zero.

An additional characteristic particular to balance equations for finite-capacity models is the fact that they are actually redundant. Indeed, consider our simple queue with capacity limited to K customers studied in Chapter 2. We have a total of $K+1$ unknowns and seemingly $K+1$ balance equations given by formulas (2.17)–(2.19). But, as we saw during the derivation of the analytical solution for this model, one of the balance equations is redundant, meaning carries no new information. To provide the missing equation, one possibility is to use an equation like, for example, $p_K = 1 - \sum_{k=0}^{K-1} p(k)$. The issue with this approach is the possibility of a significant loss of accuracy in floating-point computation, especially for larger state spaces (the reasons for this loss of accuracy are discussed in Section 5.4). Another possibility is to try to use all the balance equations and add a normalization step to get the values obtained from the balance equations to sum to one.

Among the many numerical methods, the so-called matrix-geometric method (Neuts, 1994; Latouche & Ramaswami, 1993, 1999) leverages the repetitive structure of the balance equations to avoid arbitrary truncation in open networks. A number of other methods developed specifically for or applicable to the numerical solution of balance equations are discussed by Mitrani and Chakka (1995), Stewart (1995), and Wallace (1974). We are going to discuss in more detail two solution methods. Both methods are iterative in nature.

5.2 The γ-method

The first method, which we refer to as the γ-method, uses all the balance equations for a model and has the interesting feature of preserving the normalization of the probabilities across iterations (Brandwajn, 1979). We introduce it using the example of the simple queue with limited capacity of Chapter 2 (M/M/1/K with state-dependent arrival and service rates) whose balance equations are reproduced in the following. Recall that in this example $p(n)$ denotes the steady-state probability that there are n customers in the system, $\lambda(n)$ is the instantaneous rate of arrivals, and $\mu(n)$ is the rate of service when there are n customers. The number of customers in the systems (queued and in service) is limited to K.

$$0 = -\lambda(0)p(0) + \mu(1)p(1), \tag{5.1}$$

$$0 = -[\lambda(n) + \mu(n)]p(n) + \lambda(n-1)p(n-1) + \mu(n+1)p(n+1),$$
$$n = 1, \ldots, K-1, \tag{5.2}$$

$$0 = -\mu(K)p(K) + \lambda(K-1)p(K-1). \tag{5.3}$$

We have re-written these equations with the left-hand side equal to zero. Our goal is to solve these balance equations numerically, as if we didn't know the system's analytical solution. We use an iterative approach and denote by $p^i(n)$ the value for $p(n)$ computed at iteration number i, $i = 1, 2, \ldots$ We start with a feasible set of initial values, denoted by $p^0(n)$, for instance, equally likely, i.e., $p^0(n) = \frac{1}{K+1}$ for all $n = 0, 1, \ldots, K$. We introduce a constant $\gamma > 0$ (hence our name for the method), and, at iteration i we compute a new set of approximate values for the state probabilities, $p^i(n)$, using the following equations:

$$p^i(0) = p^{i-1}(0) + \gamma\{-\lambda(0)p^i(0) + \mu(1)p^{i-1}(1)\}, \tag{5.4}$$

$$p^i(n) = p^{i-1}(n) + \gamma\{-\lambda(n)p^i(n) - \mu(n)p^{i-1}(n) + \lambda(n-1)p^i(n-1)$$
$$+ \mu(n+1)p^{i-1}(n+1)\}, \quad n = 1, \ldots, K-1, \tag{5.5}$$

$$p^i(K) = p^{i-1}(K) + \gamma\{-\mu(K)p^{i-1}(K) + \lambda(K-1)p^i(K-1)\}. \tag{5.6}$$

You will note that the terms in curly brackets represent in fact the right hand side of the balance equations (5.1)–(5.3). With $\gamma > 0$, this means that, if we find a set of values such that $p^i(n) = p^{i-1}(n)$ for all $n = 0, \ldots, K$, such a set will satisfy our balance equations. We discuss later in this section suggestions on how to select the value of the constant γ.

Let's first deal with the question of convergence of our iterative scheme. Denote by $\hat{p}^i$ the vector of the values $p^i(n)$ obtained at iteration number i. For such an iterative scheme to converge, we must have what is known as a contracting mapping (Khamsi and Kirk, 2011), i.e.,

$$\|\hat{p}^{i+1} - \hat{p}^{i}\| < \|\hat{p}^{i} - \hat{p}^{i-1}\|, \quad i > i^*. \tag{5.7}$$

In other words, starting with some iteration i^*, the distance between consecutive iterates must be decreasing. If this is the case, as $i \to \infty$ we will have $p^i(n) \to p^{i-1}(n)$ for all values of n. This in turn implies that the values computed will converge to the solution of the balance equations. One can prove that the iterative scheme described is indeed a contracting mapping (Brandwajn, 1979).

Let us examine the iteration equations more closely. We enumerate system states in the order of increasing values of n. You may have noticed that in the computation of "new" values $p^i(n)$ the term corresponding to $-[\lambda(n) + \mu(n)]p(n)$ in equation (5.2) has been split into a part computed at the current iteration: $-\lambda(n)p^i(n)$ and a part kept from the preceding iteration: $-\mu(n)p^{i-1}(n)$. There is a reason for this madness. Indeed, let us evaluate the sum of all "new" values. We get, using equations (5.4)–(5.6),

$$\sum_{n=0}^{K} p^i(n) = \sum_{n=0}^{K} p^{i-1}(n) + \gamma \left\{ - \sum_{n=0}^{K-1} \lambda(n)p^i(n) - \sum_{n=1}^{K} \mu(n)p^{i-1}(n) \right.$$
$$\left. + \sum_{n=0}^{K-1} \lambda(n)p^i(n) + \sum_{n=1}^{K} \mu(n)p^{i-1}(n) \right\}. \tag{5.8}$$

We see that our iterative scheme preserves the sum of the iterates regardless of the value of γ since the terms in the curly braces cancel out. This means that, if the initial values $p^0(n)$ are normalized, so are the values computed at later iterations. This eliminates the need for normalization at each iteration. Since we enumerate states in the order of increasing values of n, the terms corresponding to departures from the system are the ones taken from the preceding iteration.

So far, we have not said much about how to select a value for the constant γ. Clearly, with a value of zero, the iteration would never converge, so on intuitive grounds one could expect that increasing the value of γ might result in faster convergence, i.e., a smaller number of iterations needed to attain convergence. Rearranging equation (5.4)–(5.6), we have,

$$p^i(0)[1+\lambda(0)\gamma] = p^{i-1}(0) + \gamma\{\mu(1)p^{i-1}(1)\}, \tag{5.9}$$

$$p^i(n)[1+\lambda(n)\gamma] = p^{i-1}(n)[1-\gamma\mu(n)] + \gamma\{\lambda(n-1)p^i(n-1) + \mu(n+1)p^{i-1}(n+1)\}, \quad n = 1, \ldots, K-1, \tag{5.10}$$

$$p^i(K) = p^{i-1}(K)[1-\gamma\mu(K)] + \gamma\{\lambda(K-1)p^i(K-1)\}. \tag{5.11}$$

We note that if we make sure that $[1-\gamma\mu(n)] \geq 0$ for $n = 1, \ldots, K$, all values for $p^i(n)$ are guaranteed to be non-negative. This is certainly a desirable feature for the values computed at each iteration to be probabilities (no negative results regardless at the number of iterations we run the scheme for). Hence, a reasonable choice for the factor γ is

$$\gamma \leq 1/\max_n \mu(n). \tag{5.12}$$

With these considerations, here is how we would proceed in practice. We start by selecting a value for the constant γ. A good choice is suggested by formula (5.12). We need a single array of $K+1$ values to hold our iterates $p^i(n)$. We initialize this array to our initial values. Then we use formulas (5.9)–(5.11) to obtain the values $p^i(n)$ for all $n = 0, \ldots, K$ at iteration $i = 1, 2, \ldots$. To know when to stop iterating, a simple approach is as follows. For each value of n, we compute the absolute value of the difference between the value obtained at the preceding iteration $p^{i-1}(n)$ and the newly computed value $p^i(n)$, and we stop when

$$\max_n |p^i(n) - p^{i-1}(n)| < \varepsilon. \tag{5.13}$$

ε is the desired convergence stringency. What is a reasonable value for ε depends on the purpose of the model. A value of ε between 10^{-5} and 10^{-7} is likely to work fine in many applications. One may want a greater convergence stringency, i.e., a smaller value for ε, if the numerical solution is used to assess the accuracy of an approximation method. A characteristic of the stopping criterion (5.13) is that it treats smaller and larger probability values equally. If such an equal treatment is not desirable, we can replace (5.13) with a relative criterion

$$\max_n \left| 1 - \frac{p^i(n)}{p^{i-1}(n)} \right| < \varepsilon. \tag{5.14}$$

In a practical implementation, of course, care must be taken to avoid divisions by zero or overflows if one uses formula (5.14) and $p^{i-1}(n)$ happens to be too small. As a final point, let us note that we could decide to enumerate the states in the order of decreasing values of $n = K, K-1, \ldots, 0$. The iterative scheme would have to be modified so that the terms corresponding to arrivals are now the ones taken from the previous iteration. Also, the value for the constant γ might have to be modified so that $\gamma \leq 1/\max_n \lambda(n)$.

Obviously, we don't need a numerical method to obtain the steady-state solution for the simple system used in our example. This is not the case for the model of two classes of customers sharing a resource depicted in Figure 5.1.

In this model, there are two classes of customers, labeled 1 and 2. We denote by n_r the current number of class r customers in the system. M_r is the maximum number of customers of class r that can be present in our model, i.e., $n_r = 0, \ldots, M_r$, $r = 1, 2$. Customers of each class arrive from their own Poisson source with rate λ_r. We assume that the service times for customers have memoryless distributions. Since customers of both classes share a common resource, and they don't use the resource equally, their instantaneous

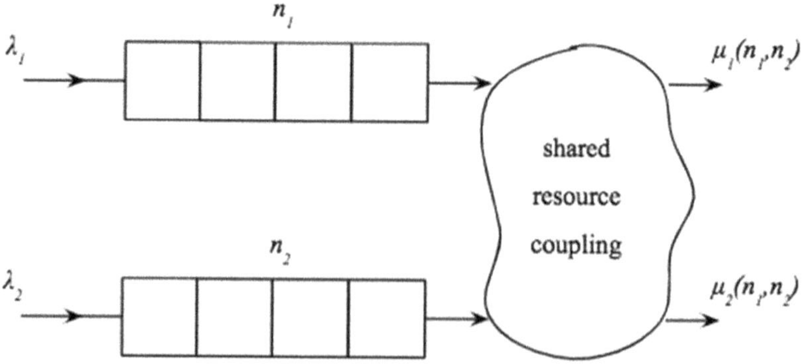

Figure 5.1. Two customer classes sharing a resource.

service rates depend on the current numbers of users of both classes. We denote by $\mu_1(n_1, n_2)$ and by $\mu_2(n_1, n_2)$ the instantaneous service rates for classes 1 and 2, respectively.

A simple example of such a coupling between the two classes is a single server where class 1 gets to use a fraction α and class 2 a fraction $\beta = 1 - \alpha$ of the server, except that each class gets to use the server fully when there are no users of the other class. The respective service rates for each class when using the server alone are μ_1 and μ_2. A quick review of the BCMP theorem convinces us that, because of the dependency of the service rates of a class on the number of customers of the other class, this model is not a BCMP network. Denote by $p(n_1, n_2)$ the steady-state probabilities of the number of customers of each class in the system. It is not difficult to see that no obvious local balance holds. For the simple example of a server shared by two classes mentioned above, the balance equations include the following:

$$p(0,0)(\lambda_1 + \lambda_2) = \mu_1 p(1,0) + \mu_2 p(0,1), \tag{5.15}$$

$$\begin{aligned} p(n_1,0)(\lambda_1 + \lambda_2 + \mu_1) &= \lambda_1 p(n_1 - 1, 0) + \mu_1 p(n_1 + 1, 0) \\ &\quad + \beta\mu_2 p(n_1, 1), \quad n_1 = 1, \ldots, M_1 - 1, \end{aligned} \tag{5.16}$$

$$\begin{aligned} p(0,n_2)(\lambda_1 + \lambda_2 + \mu_2) &= \alpha\mu_1 p(1, n_2) + \lambda_2 p(0, n_2 - 1) \\ &\quad + \mu_2 p(0, n_2 + 1), \quad n_2 = 1, \ldots, M_2 - 1. \end{aligned} \tag{5.17}$$

Local balance would imply a solution that is a product of two factors, one for each class. If one tried a factor of the form $\frac{\lambda_1}{\mu_1}$ as suggested by equations (5.15) and (5.16) for class 1, this factor wouldn't work in equation (5.17). Similarly for class 2, a factor of the form $\frac{\lambda_2}{\mu_2}$, suggested by equations (5.15) and (5.17), is contradicted by equation (5.16).

So, to obtain $p(n_1, n_2)$ or performance metrics derived from it, such as, for example, the expected time each class spends in the system, we could use simulation or solve the model numerically. Since this chapter is devoted to numerical methods, it is the latter approach that we explore here. The balance equations for our model can be written as

$$
\begin{aligned}
0 = &-[\lambda_1\chi(n_1) + \lambda_2\chi(n_2) + \mu_1(n_1, n_2) + \mu_2(n_1, n_2)]p(n_1, n_2) \\
&+ \lambda_1 p(n_1 - 1, n_2) + \lambda_2 p(n_1, n_2 - 1) + \mu_1(n_1 + 1, n_2)p(n_1 + 1, n_2) \\
&+ \mu_2(n_1, n_2 + 1)p(n_1, n_2 + 1). \qquad (5.18)
\end{aligned}
$$

The function $\chi(n_r)$ is defined as follows:

$$
\chi(n_r) = \begin{cases} 1, & \text{if } n_r < M_r \\ 0, & \text{if } n_r = M_r \end{cases}. \qquad (5.19)
$$

The service rates $\mu_r(n_1, n_2)$ are taken to be equal to zero if $n_r = 0$, $r = 1, 2$. To obtain the appropriate balance equations for the boundary cases from (5.18), we assume that for $n_r = 0$, the term involving $n_r - 1$ vanishes and the term involving $n_r + 1$ vanishes for $n_r = M_r$. For example, for $n_1 = 0$ and $n_2 = M_2$, we have

$$
0 = -[\lambda_1 + \mu_2(0, M_2)]p(0, M_2) + \lambda_2 p(0, M_2 - 1) + \mu_1(1, M_2)p(1, M_2). \qquad (5.20)
$$

We apply the γ-method to the balance equation for our system with two classes of customers, i.e., we introduce a constant $\gamma > 0$. We enumerate system states in the order of increasing values of n_1 and, for each n_1, in the order of increasing values of n_2. In other words, we enumerate states in the order $n_1 = 0$ and $n_2 = 0, \ldots, M_2$, then $n_1 = 1$ and $n_2 = 0, \ldots, M_2$, and so on. We select a feasible set of initial values $p^0(n_1, n_2)$, for example, all probabilities equally likely, i.e., equal to $1/[(M_1+1)(M_2+1)]$. At iteration number i, $i = 1, 2, \ldots$, we use the following general computation scheme:

$$
\begin{aligned}
p^i(n_1, n_2) = p^{i-1}(n_1, n_2) + \gamma\{&-[\lambda_1\chi(n_1) + \lambda_2\chi(n_2)]p^i(n_1, n_2) \\
&- [\mu_1(n_1, n_2) + \mu_2(n_1, n_2)]p^{i-1}(n_1, n_2) + \lambda_1 p^i(n_1 - 1, n_2) \\
&+ \lambda_2 p^i(n_1, n_2 - 1) + \mu_1(n_1 + 1, n_2)p^{i-1}(n_1 + 1, n_2) \\
&+ \mu_2(n_1, n_2 + 1)p^{i-1}(n_1, n_2 + 1)\}. \qquad (5.21)
\end{aligned}
$$

This scheme applies also to the appropriate boundary equations in which $n_r = 0$ or $n_r = M_r$ for one or both classes of customers.

Using our example case $n_1 = 0$ and $n_2 = M_2$, we get

$$p^i(0, M_2) = p^{i-1}(0, M_2) + \gamma\{-\lambda_1 p^i(0, M_2) - \mu_2(0, M_2)p^{i-1}(0, M_2) + \lambda_2 p^i(0, M_2 - 1) + \mu_1(1, M_2)p^{i-1}(1, M_2)\}. \quad (5.22)$$

As usual, values with the superscript $i-1$ are those computed from the preceding iteration (or the initial values for $i = 1$).

It is not difficult to show that our computation scheme preserves the sum of the iterates

$$\sum_{n_1=0}^{M_1} \sum_{n_2=0}^{M_2} p^i(n_1, n_2) = \sum_{n_1=0}^{M_1} \sum_{n_2=0}^{M_2} p^{i-1}(n_1, n_2). \quad (5.23)$$

In other words, if the initial values are normalized (i.e., sum to one), so are the values computed at subsequent iterations. One can also prove that the iterative scheme described above converges (Brandwajn, 1979). It is clear that it must therefore converge to the solution of the balance equations for our model. Indeed, the terms in the curly braces in equations (5.21) and (5.22) correspond to the right-hand side of the balance equations (5.18) and (5.20), respectively.

Rearranging the terms in (5.21), we obtain

$$\begin{aligned} p^i(n_1, n_2)[1 + \gamma[\lambda_1\chi(n_1) + \lambda_2\chi(n_2)]] &= p^{i-1}(n_1, n_2)\{1 - \gamma[\mu_1(n_1, n_2) + \mu_2(n_1, n_2)]\} \\ &+ \gamma[\lambda_1 p^i(n_1 - 1, n_2) + \lambda_2 p^i(n_1, n_2 - 1) \\ &+ \mu_1(n_1 + 1, n_2)p^{i-1}(n_1 + 1, n_2) \\ &+ \mu_2(n_1, n_2 + 1)p^{i-1}(n_1, n_2 + 1)]. \end{aligned} \quad (5.24)$$

To ensure that no values $p^i(n_1, n_2)$ are negative during the iteration, we can select the constant γ such that $1-\gamma[\mu_1(n_1, n_2)+\mu_2(n_1, n_2)] \geq 0$ for all n_1 and n_2. This gives

$$\gamma \leq 1/\max_{n_1,n_2}[\mu_1(n_1, n_2) + \mu_2(n_1, n_2)]. \quad (5.25)$$

From equation (5.24), we then compute the "new" values $p^i(n_1, n_2)$ as follows:

$$\begin{aligned} p^i(n_1, n_2) = &\frac{1}{1 + \gamma[\lambda_1 \chi(n_1) + \lambda_2 \chi(n_2)]} \{p^{i-1}(n_1, n_2)\{1 - \gamma[\mu_1(n_1, n_2) \\ &+ \mu_2(n_1, n_2)]\} + \gamma[\lambda_1 p^i(n_1 - 1, n_2) + \lambda_2 p^i(n_1, n_2 - 1) \\ &+ \mu_1(n_1 + 1, n_2) p^{i-1}(n_1 + 1, n_2) \\ &+ \mu_2(n_1, n_2 + 1) p^{i-1}(n_1, n_2 + 1)]\}. \end{aligned} \tag{5.26}$$

The iterative solution stops when convergence has been attained as defined, for example, by the criterion

$$\max_{n_1, n_2} |p^i(n_1, n_2) - p^{i-1}(n_1, n_2)| < \varepsilon. \tag{5.27}$$

As before, ε is the desired convergence stringency.

The γ-method is theoretically guaranteed to converge. The number of iterations required to reach convergence depends on the particular values of model parameters, including the size of the state space, the initial distribution $p^0(n_1, n_2)$, the value of the constant γ, and the convergence stringency required. It may vary from a few tens to several thousand iterations. Fortunately, the complexity of the computation at each iteration is relatively low.

5.3 The method of conditionals

We now use the same model of a system with two customer classes shown in Figure 5.1 to introduce another numerical approach, which we refer to as the method of conditionals. This is a semi-numerical (or semi-analytical, depending on your point of view) approach because it exploits the known form of the marginal probability distribution for n_r. Denote by $p(n_1)$ the steady-state probability that there are n_1 customers of class 1. Denote also by $p(n_2|n_1)$ the conditional probability that there are n_2 customers of class 2 given that the number of customers of class 1 is n_1. From the very definition of conditional probabilities, we have the following identity:

$$p(n_1, n_2) = p(n_1) p(n_2|n_1). \tag{5.28}$$

This identity means that we must also have

$$\sum_{n_2=0}^{M_2} p(n_2|n_1) = 1, \quad \forall n_1 = 0, 1, \dots, M_1. \tag{5.29}$$

In other words, the conditional probabilities $p(n_2|n_1)$ are normalized for each value of n_1.

Let's now concentrate on $p(n_1)$. If there are n_1 customers of class 1 in the system, the instantaneous rate with which n_1 increases is obviously $\lambda_1\chi(n_1)$. The factor $\chi(n_1)$, defined in formula (5.19), is there to indicate that n_1 cannot increase beyond its maximum value M_1. What is the instantaneous rate of decrease for n_1? It is $\mu_1(n_1, n_2)$ if the number of customers of class 2 is n_2. The probability that this is the case when there are n_1 customers of class 1 is given by the conditional probability $p(n_2|n_1)$. So, the instantaneous rate of decrease for n_1 is then $u_1(n_1)$ given by

$$u_1(n_1) = \sum_{n_2=0}^{M_2} \mu_1(n_1, n_2)p(n_2|n_1), \quad n_1 = 1, \dots, M_1. \tag{5.30}$$

Another, more formal, way to derive formula (5.30) is to substitute the identity (5.28) in the balance equations for $p(n_1, n_2)$ and, for each n_1, to sum the resulting equations over all values of n_2.

It follows from our reasoning that, with respect to $p(n_1)$, our system behaves like a simple state-dependent M/M/1 queue with service rate given by formula (5.3) and capacity limited to M_1. This equivalent queue is shown in Figure 5.2.

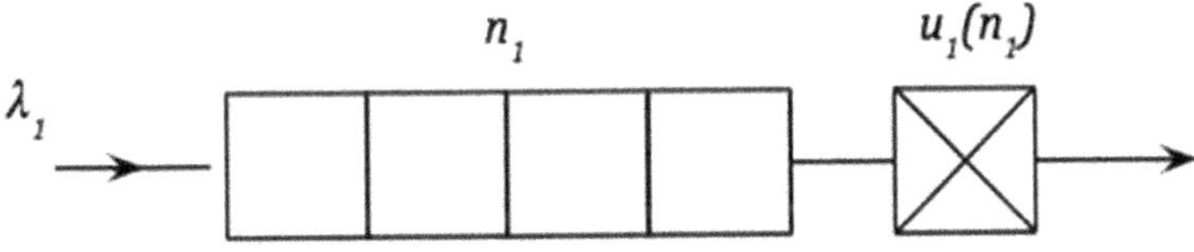

Figure 5.2. Equivalent queue for $p(n_1)$ derived from the model of Figure 5.1.

The steady-state probability $p(n_1)$ is therefore given by the familiar formula

$$p(n_1) = \frac{1}{G}\prod_{i=1}^{n_1}\frac{\lambda_1}{u_1(i)}, \qquad n_1 = 0, 1, \ldots, M_1. \tag{5.31}$$

G is a normalizing constant, and, by convention, empty products are equal to one. From formula (5.31), we have

$$\frac{p(n_1-1)}{p(n_1)} = \frac{u_1(n_1)}{\lambda_1}; \qquad \text{for } n_1 = 1, \ldots, M_1, \tag{5.32}$$

and

$$\frac{p(n_1+1)}{p(n_1)} = \frac{\lambda_1}{u_1(n_1+1)}, \qquad \text{for } n_1 = 0, \ldots, M_1 - 1. \tag{5.33}$$

The balance equations for our model we are seeking to solve can be written as

$$\begin{aligned}
&p(n_1, n_2)[\lambda_1\chi(n_1) + \lambda_2\chi(n_2) + \mu_1(n_1, n_2) + \mu_2(n_1, n_2)] \\
&\quad = p(n_1, n_2 - 1)\lambda_2 + p(n_1, n_2 + 1)\mu_2(n_1, n_2 + 1) \\
&\qquad + p(n_1 - 1, n_2)\lambda_1 + p(n_1 + 1, n_2)\mu_1(n_1 + 1, n_2).
\end{aligned} \tag{5.34}$$

As before, the service rates $\mu_r(n_1, n_2)$ are taken to be equal to zero if $n_r = 0$, $r = 1, 2$. To obtain the appropriate balance equations for the boundary cases from (5.34) we assume that for $n_r = 0$, the term involving $n_r - 1$ vanishes and the term involving $n_r + 1$ vanishes for $n_r = M_r$. Here, in re-writing the balance equations we rearranged the terms so as to group them by the value of n_1. We now use the identity (5.28) in equation (5.34), and we divide both sides of the equation by $p(n_1)$. This transformation yields

$$\begin{aligned}
&p(n_2|n_1)[\lambda_1\chi(n_1) + \lambda_2\chi(n_2) + \mu_1(n_1, n_2) + \mu_2(n_1, n_2)] \\
&\quad = p(n_2 - 1|n_1)\lambda_2 + p(n_2 + 1|n_1)\mu_2(n_1, n_2 + 1) \\
&\qquad + p(n_2|n_1 - 1)\lambda_1\frac{p(n_1 - 1)}{p(n_1)} \\
&\qquad + p(n_2|n_1 + 1)\mu_1(n_1 + 1, n_2)\frac{p(n_1 + 1)}{p(n_1)}.
\end{aligned} \tag{5.35}$$

We then use formulas (5.32) and (5.33) to transform these equations into equations involving only the conditional probabilities $p(n_2|n_1)$:

$$\begin{aligned} p(n_2|n_1)&[\lambda_1\chi(n_1) + \lambda_2\chi(n_2) + \mu_1(n_1, n_2) + \mu_2(n_1, n_2)] \\ &= p(n_2 - 1|n_1)\lambda_2 + p(n_2 + 1|n_1)\mu_2(n_1, n_2 + 1) \\ &\quad + p(n_2|n_1 - 1)u_1(n_1) + p(n_2|n_1 + 1)\mu_1(n_1 + 1, n_2)\frac{\lambda_1}{u_1(n_1 + 1)}. \end{aligned} \tag{5.36}$$

It is this set of equations for $p(n_2|n_1)$, subject to the normalizing condition $\sum_{n_2=0}^{M_2} p(n_2|n_1) = 1\ \forall n_1$, that we solve numerically in the method of conditionals.

We use a superscript to denote the iteration number. We use a single array to hold the iterates $p^i(n_2|n_2)$ and another array to hold the corresponding instantaneous rates of departures for customers of class 1 $u_1^i(n_1)$. We start with a set of feasible conditional probabilities $p^0(n_2|n_1)$ for all n_1. A simple possibility is to select all probabilities equally likely, i.e., $p^0(n_2|n_1) = \frac{1}{M_2+1}, \forall n_1$. We compute the corresponding values for $u_1^0(n_1)$ using formula (5.30). We choose to enumerate system states in the order of increasing $n_1 = 0, 1, \ldots, M_1$ and, for each n_1, in the order of increasing $n_2 = 0, 1, \ldots, M_2$. At iteration $i = 1, 2, \ldots,$ we consider consecutive values of n_1, and for each n_1, we compute a set of new values $\tilde{p}^i(n_2|n_1)$ (we use $\tilde{p}^i(.)$ to denote non-normalized values at iterations i) from

$$\begin{aligned} \tilde{p}^i(n_2|n_1) &= \frac{1}{\lambda_1\chi(n_1) + \lambda_2\chi(n_2) + \mu_1(n_1, n_2) + \mu_2(n_1, n_2)} \\ &\quad \times \Big\{\tilde{p}^i(n_2 - 1|n_1)\lambda_2 + p^{i-1}(n_2 + 1|n_1)\mu_2(n_1, n_2 + 1) \\ &\quad + p^i(n_2|n_1 - 1)u_1^{i-1}(n_1) + p^{i-1}(n_2|n_1 + 1)\mu_1(n_1 + 1, n_2) \\ &\quad \times \frac{\lambda_1}{u_1^{i-1}(n_1 + 1)}\Big\}, \quad n_2 = 0, 1, \ldots, M_2. \end{aligned} \tag{5.37}$$

The values $\tilde{p}^i(n_2|n_1)$ are then normalized to sum to one per (5.29), and we get

$$p^i(n_2|n_1) = \frac{\tilde{p}^i(n_2|n_1)}{\sum_{n_2=0}^{M_2} \tilde{p}^i(n_2|n_1)}. \tag{5.38}$$

Before moving on to the next value of n_1, or to the next iteration if $n_1 = M_1$ and convergence has not been attained, we compute a new value for $u_1^i(n_1)$:

$$u_1^i(n_1) = \sum_{n_2=0}^{M_2} \mu_1(n_1, n_2)p^i(n_2|n_1). \tag{5.39}$$

The above iterative solution stops when convergence has been attained, as defined, for example, by the criterion

$$\max_{n_1,n_2} |p^i(n_2|n_1) - p^{i-1}(n_2|n_1)| < \varepsilon. \tag{5.40}$$

Or, if we prefer a relative convergence criterion,

$$\max_{n_1,n_2} \left| 1 - \frac{p^i(n_2|n_1)}{p^{i-1}(n_2|n_1)} \right| < \varepsilon. \tag{5.41}$$

As usual, ε is the desired convergence stringency and $\varepsilon \in [10^{-5}, 10^{-7}]$ seems a reasonable choice for many applications. The reader interested only in the principle of the method of conditionals can skip to the paragraph following formula (5.53).

The computation of the next iterate using formulas (5.37) and (5.38) is "easy-going" in the sense that it involves little work beyond using values as they become available. We can be more "ambitious" and try to compute more quantities at the same iteration i in the equation for each n_1. The more ambitious goal is

$$\begin{aligned} 0 = &-p^i(n_2|n_1)[\lambda_1\chi(n_1) + \lambda_2\chi(n_2) + \mu_1(n_1, n_2) + \mu_2(n_1, n_2)] \\ &+ p^i(n_2 - 1|n_1)\lambda_2 + p^i(n_2 + 1|n_1)\mu_2(n_1, n_2 + 1) \\ &+ p^i(n_2|n_1 - 1)u_1^i(n_1) + p^{i-1}(n_2|n_1 + 1)\mu_1(n_1 + 1, n_2) \\ &\times \frac{\lambda_1}{u_1^{i-1}(n_1 + 1)}. \end{aligned} \tag{5.42}$$

As we consider consecutive values of $n_1 = 0, 1, \ldots,$ we would like to take from the preceding iteration only the values that pertain to $n_1 + 1$. Intuitively, the more of the equation to be solved is included in each iteration, the fewer iterations should be needed. This seems particularly true if the terms taken from the preceding iteration are small relative to the remainder of the equation being solved. Let's rearrange slightly the terms in equation (5.42) so as to have the term $p^i(n_2+1|n_1)$ on the left-hand side:

$$p^i(n_2+1|n_1) = \frac{1}{\mu_2(n_1,n_2+1)}\Bigg\{p^i(n_2|n_1)[\lambda_1\chi(n_1)+\lambda_2\chi(n_2) + \mu_1(n_1,n_2)+\mu_2(n_1,n_2)] - \Bigg[p^i(n_2-1|n_1)\lambda_2 + p^i(n_2|n_1-1)u_1^i(n_1) + p^{i-1}(n_2|n_1+1)\mu_1 \times(n_1+1,n_2)\frac{\lambda_1}{u_1^{i-1}(n_1+1)}\Bigg]\Bigg\}, \quad n_2 = 0,\dots,M_2-1. \tag{5.43}$$

We note that, for each n_1, (5.43) can be viewed as a relatively simple recurrence for the $p^i(n_2|n_1)$ of the form

$$p^i(n_2|n_1) = a(n_2)p^i(0|n_1) + b(n_2)u_1^i(n_1) + c(n_2). \tag{5.44}$$

The coefficients $a(n_2), b(n_2)$, and $c(n_2)$ are in general different for each value of n_1 and have to be recomputed for $n_1 = 0, 1, \dots$. We have $a(0) = 1, b(0) = c(0) = 0$. Then, coefficients $a(n_2), b(n_2)$, and $c(n_2)$ for $n_2 = 1,\dots,M_2$ are determined from formula (5.43). We have, for example, from formula (5.43) for $n_2 = 0$

$$a(1) = \frac{1}{\mu_2(n_1,1)}a(0)[\lambda_1\chi(n_1)+\lambda_2+\mu_1(n_1,0)], \tag{5.45}$$

$$b(1) = -\frac{1}{\mu_2(n_1,1)}p^i(0|n_1-1), \tag{5.46}$$

$$c(1) = -\frac{1}{\mu_2(n_1,1)}p^{i-1}(0|n_1+1)\mu_1(n_1+1,0)\frac{\lambda_1}{u_1^{i-1}(n_1+1)}. \tag{5.47}$$

Next, from formula (5.43) we get for $n_2 = 1,\dots,M_2-1$

$$a(n_2+1) = \frac{1}{\mu_2(n_1,n_2+1)}\{a(n_2)[\lambda_1\chi(n_1)+\lambda_2\chi(n_2)+\mu_1(n_1,n_2) + \mu_2(n_1,n_2)] - a(n_2-1)\lambda_2\}, \tag{5.48}$$

$$b(n_2+1) = \frac{1}{\mu_2(n_1,n_2+1)}\{b(n_2)[\lambda_1\chi(n_1)+\lambda_2\chi(n_2)+\mu_1(n_1,n_2)$$
$$+\mu_2(n_1,n_2)] - b(n_2-1)\lambda_2 - p^i(n_2|n_1-1)\}, \quad (5.49)$$
$$c(n_2+1) = \frac{1}{\mu_2(n_1,n_2+1)}\Big\{c(n_2)[\lambda_1\chi(n_1)+\lambda_2\chi(n_2)+\mu_1(n_1,n_2)$$
$$+\mu_2(n_1,n_2)] - c(n_2-1)\lambda_2 - p^{i-1}(n_2|n_1+1)\mu_1$$
$$\times(n_1+1,n_2)\frac{\lambda_1}{u_1^{i-1}(n_1+1)}\Big\}. \quad (5.50)$$

Formula (5.44) has two unknowns $p^i(0|n_1)$ and $u_1^i(n_1)$, which we denote by x and y, respectively. These unknowns are determined from the definition of $u_1(n_1)$ (5.30) and from the normalizing condition (5.29). Specifically, we have

$$y = x\sum_{n_2=0}^{M_2}\mu_1(n_1,n_2)a(n_2) + y\sum_{n_2=0}^{M_2}\mu_1(n_1,n_2)b(n_2)$$
$$+\sum_{n_2=0}^{M_2}\mu_1(n_1,n_2)c(n_2) \quad (5.51)$$

and

$$1 = x\sum_{n_2=0}^{M_2}a(n_2) + y\sum_{n_2=0}^{M_2}b(n_2) + \sum_{n_2=0}^{M_2}c(n_2). \quad (5.52)$$

Having computed x and y from equations (5.51) and (5.52), we get our new iterates $p^i(n_2|n_1)$ for the given value of n_1 as

$$p^i(n_2|n_1) = xa(n_2) + yb(n_2) + c(n_2), \quad \text{for } n_2 = 0,\ldots,M_2. \quad (5.53)$$

We note that this "more ambitious" version of the method of conditionals uses more storage than our first version since we need storage for the arrays $a(n_2)$, $b(n_2)$ and $c(n_2)$, in addition to $p^i(n_2|n_1)$ and $u_1^i(n_1)$. As always, the iteration continues until we reach convergence or we exhaust a pre-specified maximum number of iterations.

Unlike the γ-method, the method of conditionals is not theoretically guaranteed to converge. In practice, when it converges, it tends to converge in considerably fewer iterations than the γ-method. Its convergence tends to be particularly fast when the rates of transitions that change the condition are smaller than the rates of transitions that do not change the condition. In our case, this would be λ_1 and μ_1 much smaller than λ_2 and μ_2.

Upon convergence of the method of conditionals, in addition to the conditional probabilities $p(n_2|n_1)$, we get the rate of customer departures $u_1(n_1)$. This allows us to obtain the probability $p(n_1)$ from formula (5.31) and any performance indices derived from it. Performance measures for customers of class 2 can be obtained using the conditional probability $p(n_2|n_1)$ together with $p(n_1)$. For example, the mean number of customers of class 2 can be expressed as $\sum_{n_1=0}^{M_1} p(n_1) \sum_{n_2=1}^{M_2} n_2 p(n_2|n_1)$. Of course, the model parameters may be such that it would be preferable to use $p(n_1, n_2) = p(n_2)p(n_1|n_2)$ as the basis for the transformation of the balance equations into equations for conditional probabilities. In this case, one can easily derive the appropriate equations or simply renumber the classes.

As mentioned above, there is no theoretical guarantee that the method of conditionals will converge in a particular case, and there is no easy general test to decide a priori the convergence issue. Nonetheless, the method of conditionals is well worth the effort of trying. Through the use of conditional probabilities, it effectively partitions the state space into independently normalized subspaces. This may be the reason why it tends to be numerically stable and it handles particularly well cases where the rates of transitions that change the condition are much smaller than other transition rates in the model.

So, how can we deal in practice with a situation where the method may not converge? We simply implement the method using the appropriate conditionals and set a reasonable maximum number of iterations the method is allowed to run, e.g., one thousand. In most cases, if the method converges, it will terminate before the maximum number of iterations is exhausted. If, after the maximum number of iterations has been attained, the convergence criterion is close to being met, this most likely means that we need to increase the maximum number of iterations. On the other hand, if the convergence criterion is very far from having reached an acceptable value, the method terminates unsuccessfully. One also can build in a more sophisticated convergence monitoring into the method. As an example, after some number of iterations (say, fifty), we check if differences between consecutive iterates do decrease. If this is not the case, it is likely the method will not converge, and we can terminate the iteration. Fortunately, if the appropriate conditionals are used, lack of convergence is not a frequent occurrence.

Additionally, if we consider a system with infinite queue capacities, the method of conditionals can leverage the convergence of conditional probabilities to their limiting values as the number of customers in the condition grows. This is a convenient way to avoid arbitrary truncation and we show examples of its application in Chapters 6 and 7.

5.4 Numerical considerations in practice

We now look briefly at a few practical considerations related to numerical methods and even to the numerical evaluation of simple analytical formulas such as the steady-state probabilities in the state-dependent M/M/1 queue (2.21) and (2.22). As mentioned earlier, we need to keep in mind that floating-point computation on classical computers is not exact and that floating-point numbers have a range of maximum and minimum values that can be represented on a given computer architecture IEEE Computer Society (2019). As an example, consider precisely the solution given in formula (2.21) for the single-server model with memoryless state-dependent arrival and service rates:

$$p(n) = p(0) \prod_{i=1}^{n} \frac{\lambda(i-1)}{\mu(i)}, \quad \text{for } n = 0, 1, \ldots, K. \tag{5.54}$$

For the sake of our discussion, assume that we want also to compute the expected number of customers in the system given by

$$E[N] = \sum_{n=1}^{K} np(n). \tag{5.55}$$

Let's just consider how we could evaluate these quantities in practice. As we know, the probabilities $p(n)$ must be normalized, i.e., we must have

$$\sum_{n=0}^{K} p(n) = 1. \tag{5.56}$$

Using the symbol "$\sim$" to denote non-normalized values, a simple approach is to start with non-normalized value $\tilde{p}(0) = v_0$, together with the normalizing constant G initialized to $G = v_0$, where $v_0 = 1$ seems like an obvious choice. We would then compute non-normalized values $\tilde{p}(n)$ and $\tilde{E}[N]$ in a single loop for $n = 1, \ldots, K$ as follows:

$$\tilde{p}(n) = \tilde{p}(n-1)\frac{\lambda(n-1)}{\mu(n)}, \tag{5.57}$$

$$\tilde{E}[N] = \sum_{n=1}^{K} n\tilde{p}(n), \tag{5.58}$$

$$G = \tilde{p}(0) + \sum_{n=1}^{K} \tilde{p}(n). \tag{5.59}$$

The normalized quantities $p(n)$ and $E[N]$ can then be computed as

$$p(n) = \frac{1}{G}\tilde{p}(n), \quad \text{for } n = 0, 1, \ldots, K, \tag{5.60}$$

and

$$E[N] = \frac{1}{G}\tilde{E}[N]. \tag{5.61}$$

In practice, this approach will work in many cases, but whether it will work or not will depend on the particular values of the arrival and service rates $\lambda(n)$ $(n = 0, \ldots, K-1)$ and $\mu(n)$ $(n = 1, \ldots, K)$ and the value of K. Indeed, if the ratio $\frac{\lambda(n-1)}{\mu(n)}$ is very large, we can quite rapidly encounter a floating-point overflow, in increasing order of likelihood, in the computation of $\tilde{p}(n)$, G, or $\tilde{E}[N]$. In many cases, selecting the starting point v_o very small might solve the issue but not always. In some cases, it may be necessary to switch the order of the computation and start from $\tilde{p}(K) = v_0$ and enumerate the states in the order $n = K-1, K-2, \ldots, 0$:

$$\tilde{p}(n) = \tilde{p}(n+1)\frac{\mu(n+1)}{\lambda(n)}, \tag{5.62}$$

$$\tilde{E}[N] = K\tilde{p}(K) + \sum_{n=K-1}^{1} n\tilde{p}(n), \tag{5.63}$$

$$G = \tilde{p}(K) + \sum_{n=K-1}^{0} \tilde{p}(n). \tag{5.64}$$

Here, we would try to select v_0 large enough so that we don't experience a floating-point underflow, which, in many computing environments, might go undetected but would cause some loss of accuracy since the values of $\tilde{p}(n)$ would simply be set to 0 starting with some value of n.

Conversely, if the ratio $\frac{\lambda(n-1)}{\mu(n)}$ is very small, we would need to try starting with a large value for v_o to enumerate the states in the order $n = 1, \ldots, K$ if we want to avoid premature underflows and consequent loss of accuracy. We may need to reverse the order of computation, start with a small value for $\tilde{p}(K)$, and enumerate the states in decreasing order. Of course, the danger then is that we might encounter an overflow for some value of $\tilde{p}(n)$, G or $\tilde{E}(N)$. In extreme cases, we may need to start the evaluation somewhere for n_0 in the interval $(0, K)$ and proceed in both directions from there. As you can see, the evaluation of the simple formula (5.54) may require a significant amount of code to detect and treat numerical problems.

To the extent that it requires the computation of the probabilities $p(n_1)$ using the equivalent rate of service $u_1(n_1)$ obtained from the iteration, the method of conditionals may be subject to the issues discussed above. The γ-method operates directly on the balance equations of the model and thus avoids this particular difficulty. On the flip side, it might exhibit its own issues if there are many orders of magnitude differences in the values of some transition rates.

Let's mention that, besides over and underflows, loss of accuracy can be expected anytime we add or subtract two floating-point numbers which differ by several orders of magnitude in scale, or we subtract two numbers close to each other. We need to be mindful of these issues, in particular, when dealing with recurrent computations as rounding errors may cumulate rapidly from one step of the recurrence to another.

As a final point, we note that both the γ method and the method of conditionals involve fixed-point iterations. Several approximate solutions discussed in Chapters 6 and 7 of this book also resort to fixed-point procedures. Speaking in general, not specifically about any of the fixed-point iterations discussed in this text, occasionally, in some iterative schemes, for certain values of model parameters, oscillations may be observed during iteration, i.e., the values obtained oscillate from one iteration to the other (or every few iterations, the same values are repeated). When this happens, usually, the situation can be remedied through "dampening" the iteration by taking a linear combination of the old and the newly computed value as the new value in the iteration. Thus, if such oscillations were to happen in a fixed-point iteration used to compute some probabilities $p(n)$, we could use $dp^j(n) + (1-d)p^{j-1}(n)$, where $0 < d \leq 1$, to create the new value to be retained at iteration j. Of course, we must have $d \gg \varepsilon$, and if d is very small, the convergence will be very slow (we use a subscript to denote the iteration number and ε is the desired convergence stringency of the iteration).

References

Brandwajn, A. (1979). An iterative solution of two-dimensional birth and death processes. *Operations Research*, 27(3), 595–605.

Henrici, P. (1964). *Elements of Numerical Analysis*. John Wiley & Sons.

IEEE Computer Society (2019-07-22). IEEE Standard for Floating-Point Arithmetic. IEEE STD 754-2019. IEEE. pp. 1–84. doi:10.1109/IEEESTD.2019.8766229. ISBN 978-1-5044-5924-2. IEEE Std 754-2019.

Khamsi, M. A., & Kirk, W. A. (2011). *An Introduction to Metric Spaces and Fixed Point Theory*. John Wiley & Sons.

Latouche, G., & Ramaswami, V. (1993). A logarithmic reduction algorithm for quasi-birth-death processes. *Journal of Applied Probability*, 30(3), 650–674.

Latouche, G., & Ramaswami, V. (1999). *Introduction to Matrix Analytic Methods in Stochastic Modeling*. Society for Industrial and Applied Mathematics.

Mitrani, I., & Chakka, R. (1995). Spectral expansion solution for a class of Markov models: Application and comparison with the matrix-geometric method. *Performance Evaluation*, 23(3), 241–260.

Neuts, M. F. (1994). *Matrix-geometric Solutions in Stochastic Models: An Algorithmic Approach.* Courier Corporation.

Stewart, W. J. (1995). *Introduction to the Numerical Solution of Markov Chains.* Princeton University Press.

Wallace, V. L. (1974). Algebraic techniques for numerical solution of queueing networks. In *Mathematical Methods in Queueing Theory: Proceedings of a Conference at Western Michigan University, May 10–12, 1973* (pp. 295–305). Berlin, Heidelberg: Springer.

Chapter 6

Selected Approximation Methods

6.1 Equivalence and decomposition for global dependencies

With few exceptions, queueing network models for which their analytical solution is known are separable, i.e., their steady-state solution has a product form. This is directly tied to the existence of local balance in the corresponding balance equations. As mentioned before, many model features may destroy local balance. One such feature involves dependencies of service rates and routing probabilities on the state of other servers or subsets of servers in the system. As an example, consider the model of a virtual memory system in an interactive environment represented in Figure 6.1.

Transactions are generated by users at a set of N_t terminals. The idle time between transaction requests at a terminal ("think time") is exponentially distributed with mean $1/\alpha$. The system itself consists of a CPU and two peripherals. The mean service times at the peripherals, numbered 1 and 2, are $1/u_1$ and $1/u_2$, respectively. Service times at the peripheral devices and at the CPU are assumed to have memoryless distributions. In our model, a service period at the CPU ends for one of three reasons: the transaction completes its execution, the transaction generates a page fault and thus needs service from peripheral 1 or the transaction needs to access peripheral 2 for non-paging I/O. We denote by n ($n = 0, 1, \ldots, N_t$) the current number of transactions in the system. Completions occur at instantaneous rates v_o. Page faults happen with instantaneous rates

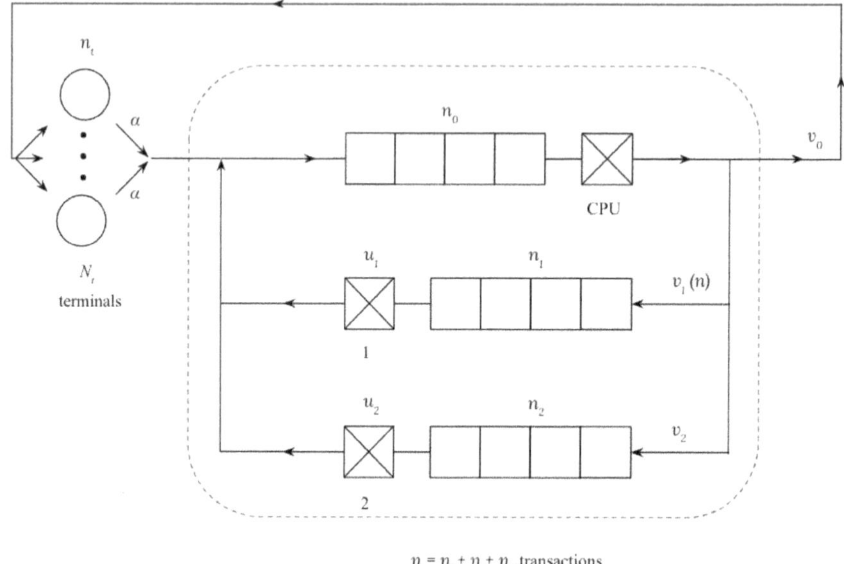

Figure 6.1. A model of an interactive virtual memory system.

$v_1(n)$, a function of the current number of transactions (jobs) in the system, i.e., sharing real memory (typically, as n increases, so does the page fault rate). Finally, non-paging I/O operations, i.e., requests for access to peripheral 2, occur with instantaneous rate v_2. This formulation of the CPU service is equivalent to saying that the instantaneous rate of service at the CPU is $u_0(n) = v_0 + v_1(n) + v_2$ and that the probability that a transaction is complete after a CPU service period is $p_0(n) = v_0/u_0(n)$, the probability that the transaction needs access to peripheral 1 or 2 is given by $p_1(n) = v_1(n)/u_0(n)$ and $p_2(n) = v_2/u_0(n)$, respectively.

Our goal is to obtain the expected response time of a transaction, denoted by $E[W]$, as a function of the number of terminals and of the mean service times of the peripherals. The state of the system can be described by the vector (n_0, n_1, n_2, n_t) of the current numbers of transactions at the CPU, at peripheral 1, peripheral 2, and at the terminals, respectively. We must have $n_0 + n_1 + n_2 + n_t = N_t$ so that one of the variables, say n_2, can be omitted.

Because of the dependence of the routing probabilities at the CPU on the current number of customers in the system, our seemingly simple model is not a Jacksonian or BCMP queueing network and it does not possess a known analytical solution. As we already know, we could use simulation or a numerical method to try to obtain the desired performance metric. A potential difficulty for the discrete-event simulation approach is that transitions internal to the system, i.e., corresponding to a page fault or an I/O, typically take place at time scales orders of magnitude smaller than user think time or transaction response time. This means that for a single system interaction, we may have to simulate many internal events, resulting in a very large number of events, in order to obtain a good estimate for the mean transaction response time. This difference in time scales of events might also create difficulties for some numerical methods (albeit not necessarily for the method of conditionals) if we decide to try to solve the problem numerically. But, as it turns out, this very time scale difference becomes an asset in an approximate solution using the equivalence and decomposition method.

Instead of the "classical" state description (n_0, n_1, n_2, n_t), we use the state description (n_0, n_1, n_2, n), which carries the same information since $n = n_0 + n_1 + n_2 = N_t - n_t$ and we denote by $p(n_0, n_1, n_2, n)$ the corresponding steady-state probability. The method proceeds in two distinct steps. First, we consider the current number of transactions in the system. We denote by $p(n)$ the steady-state probability that there are n transactions in the system. Clearly, the instantaneous rate of increase in n is

$$\lambda(n) = (N_t - n)\alpha, \quad \text{for } n = 0, \ldots, N_t - 1. \tag{6.1}$$

This corresponds to $n_t\alpha$. The instantaneous rate of decrease is v_0 provided that the CPU is active. We denote by $A(n)$ the conditional probability that the CPU is active given that there are n transactions in the system. We refer to $A(n)$ as the conditional CPU utilization. Thus, the instantaneous rate of decrease in n can be written as

$$u(n) = A(n)v_0, \quad \text{for } n = 1, \ldots, N_t. \tag{6.2}$$

It is immediately apparent that, with respect to the steady-state probability $p(n)$, our model is equivalent to the simple queueing network shown in Figure 6.2, which can be viewed as a single state-dependent queue.

In this equivalent queueing network, the system part of our initial model is replaced by a state-dependent server whose rate is $u(n)$.

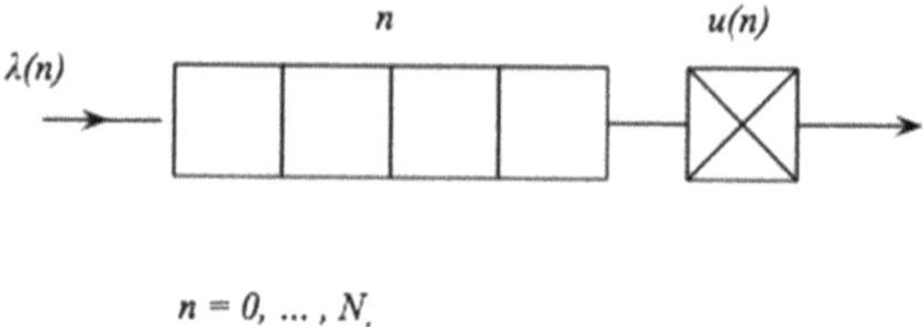

Figure 6.2. Equivalent queueing network (state-dependent queue).

Although the steady-state solution for our original model is not known, we can easily obtain the formal analytical solution for $p(n)$. Indeed, we have

$$p(n) = \frac{1}{\mathrm{G}} \prod_{i=1}^{n} \frac{\lambda(i-1)}{u(i)}, \quad n = 0, 1, \ldots, N_t. \tag{6.3}$$

The normalizing constant G is given by

$$\mathrm{G} = \sum_{n=0}^{N_t} \prod_{i=1}^{n} \frac{\lambda(i-1)}{u(i)}. \tag{6.4}$$

As usual, in the above formulas, empty products are taken to be equal to 1.

It is important to note that these formulas in themselves are exact in the sense that, if we knew the instantaneous rate $u(n)$, we would have the exact solution for $p(n)$. Since, in general, we don't know the conditional probability that the CPU is active given that there are n jobs in the system, $A(n)$, we use a decomposition method to find a good approximation for it. If the rates of transitions internal to the system are significantly higher than the rates of transitions that change the number of transactions in the system, it makes intuitively sense to assume that $A(n)$ must be close to the CPU utilization in the subnetwork obtained by isolating the system with n jobs in it as shown in Figure 6.3. We denote this utilization by A_n and by $p_n(n_0, n_1, n_2)$ the corresponding steady-state probability of the state (n_0, n_1, n_2) in the isolated subnetwork with n jobs.

It is clear that the subnetwork in Figure 6.3 is a Jacksonian network (since the routing probabilities in the subnetwork are constant for each value of n), and, therefore, its solution can be readily

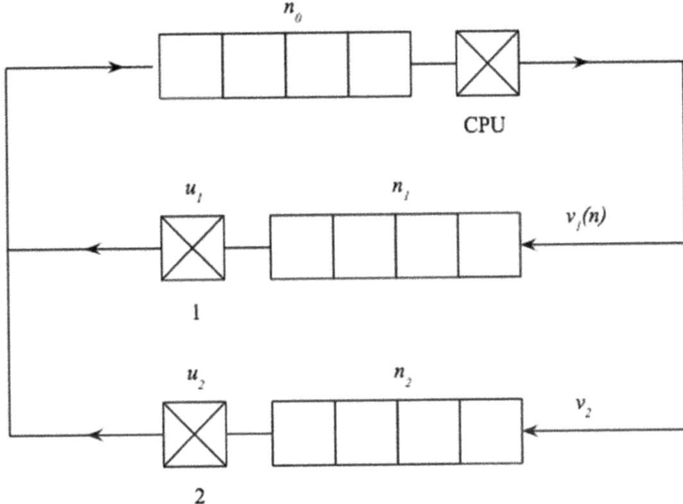

Figure 6.3. Decomposed subnetwork with a population of n jobs (transactions).

obtained. In fact, our decomposed subnetwork is identical to the virtual memory model considered in Figure 3.3 in Section 3 with ν_0 set to 0.

Using decomposition, we compute approximate values for $u(n)$ by solving the subnetwork in isolation to obtain A_n for all values of $n = 1, \ldots, N_t$, and we set

$$u(n) \approx A_n v_0. \tag{6.5}$$

This allows us to compute approximate values for the steady-state probabilities $p(n)$ using formula (6.3), and, from there, the mean number of transactions in the system $E[N]$

$$E[N] = \sum_{n=1}^{N_t} np(n). \tag{6.6}$$

The mean rate of transaction arrivals to the system, $E[\Lambda]$, is not difficult to obtain

$$E[\Lambda] = \sum_{n=0}^{N_t} \lambda(n)p(n) = \sum_{n=0}^{N_t} (N_t - n)\alpha p(n) = (N_t - E[N])\alpha. \tag{6.7}$$

The mean response time is then given by Little's formula as

$$E[W] = \frac{E[N]}{(N_t - E[N])\alpha}. \tag{6.8}$$

In practice, we start by solving the subnetwork shown in Figure 6.3 to obtain the values of A_n for $n = 1, \ldots, N_t$. Then, we use these values to obtain approximate values for the equivalent rate of transaction completions $u(n)$ from formula (6.5). Formula (6.1) gives us the equivalent rate of transaction arrivals. It suffices then to use formula (6.3) to compute the steady-state distribution of the number of transactions (jobs) in the system and hence any derived performance indices as discussed above.

To understand the underpinnings of this intuitively appealing approach, let us consider the balance equations for the selected state description for our model of Figure 6.1. To denote a term that vanishes when the number of transactions is not greater than zero, we use the function $\zeta(.)$ defined as follows:

$$\zeta(n_i) = \begin{cases} 0, & \text{if } n_i \le 0 \\ 1, & \text{if } n_i > 0 \end{cases}. \tag{6.9}$$

We have for $n = 0, 1, \ldots, N_t$, $n_0 = 0, \ldots, n$, $n_1 = 0, \ldots, n - n_0$, and $n_2 = n - n_1 - n_2$:

$$\begin{aligned} 0 = &-p(n_0, n_1, n_2, n)\{\zeta(n_0)[v_1(n) + v_2] + \zeta(n_1)u_1 + \zeta(n_2)u_2 + \lambda(n) \\ &+ \zeta(n_0)v_0\} + \zeta(n_0 - 1)p(n_0 - 1, n_1 + 1, n_2, n)u_1 \\ &+ \zeta(n_0 - 1)p(n_0 - 1, n_1, n_2 + 1, n)u_2 \\ &+ \zeta(n_1 - 1)p(n_0 + 1, n_1 - 1, n_2, n)v_1(n) \\ &+ \zeta(n_2 - 1)p(n_0 + 1, n_1, n_2 - 1, n)v_2 \\ &+ \zeta(n_0 - 1)p(n_0 - 1, n_1, n_2, n - 1)\lambda(n - 1) \\ &+ \zeta(N_t - n - 1)p(n_0 + 1, n_1, n_2, n + 1)v_0. \end{aligned} \tag{6.10}$$

Obviously, in the above equation, any terms for non-feasible states with negative arguments or arguments exceeding N_t are equal to zero. We denote by $p(n_0, n_1, n_2|n)$ the conditional probability that there are n_0, n_1 and n_2 transactions at the CPU, at peripheral 1 and at peripheral 2, respectively, given that the total number of transactions in the system is n. The conditional CPU utilization $A(n)$ appearing in the formula for $u(n)$ can be expressed in terms of the conditional probability $p(n_0, n_1, n_2|n)$ as

$$A(n) = \sum_{n_0=1}^{n} \sum_{n_1=0}^{n-n_0} p(n_0, n_1, n_2 = n - n_0 - n_1|n). \tag{6.11}$$

Similarly, our approximate value A_n can be expressed in terms of $p_n(n_0, n_1, n_2)$

$$A_n = \sum_{n_0=1}^{n} \sum_{n_1=0}^{n-n_0} p_n(n_0, n_1, n_2 = n - n_0 - n_1). \tag{6.12}$$

In the decomposition step, we assumed that $A_n \approx A(n)$. This would certainly be true if for all feasible states we had

$$p_n(n_0, n_1, n_2) \approx p(n_0, n_1, n_2|n). \tag{6.13}$$

Therefore, it makes sense to examine equations for the conditional probabilities $p(n_0, n_1, n_2|n)$ and see how the above approximation fits into them. In order to transform the balance equations (6.10) into the desired equations for the conditional probabilities, we note that we have from the definition of conditional probabilities

$$p(n_0, n_1, n_2, n) = p(n)p(n_0, n_1, n_2|n), \quad \text{for all feasible states } (n_0, n_1, n_2, n). \tag{6.14}$$

The form of the solution for $p(n)$ in formula (6.3) implies that we must have

$$\frac{p(n-1)}{p(n)} = \frac{u(n)}{\lambda(n-1)} \quad \text{for } n = 1, \ldots, N_t, \quad \text{and}$$
$$\frac{p(n+1)}{p(n)} = \frac{\lambda(n)}{u(n+1)} \quad \text{for } n = 0, \ldots, N_t - 1. \tag{6.15}$$

To obtain equations for the conditional probabilities $p(n_0, n_1, n_2|n)$, we substitute formula (6.14) in equation (6.10), divide by $p(n)$ and use the relationships in (6.15). This yields the following equation for $n = 0, 1, \ldots, N_t$, $n_0 = 0, \ldots, n$, $n_1 = 0, \ldots, n - n_0$, and $n_2 = n - n_1 - n_2$:

$$\begin{aligned}0 = &-p(n_0, n_1, n_2|n)\{\zeta(n_0)[v_1(n) + v_2] + \zeta(n_1)u_1 + \zeta(n_2)u_2 + \lambda(n)\\&+\zeta(n_0)v_0\} + \zeta(n_0 - 1)p(n_0 - 1, n_1 + 1, n_2|n)u_1\\&+\zeta(n_0 - 1)p(n_0 - 1, n_1, n_2 + 1|n)u_2\\&+\zeta(n_1 - 1)p(n_0 + 1, n_1 - 1, n_2|n)v_1(n)\\&+\zeta(n_2 - 1)p(n_0 + 1, n_1, n_2 - 1|n)v_2\\&+\zeta(n_0 - 1)p(n_0 - 1, n_1, n_2|n - 1)u(n)\\&+\zeta(N_t - n - 1)p(n_0 + 1, n_1, n_2|n + 1)v_0\lambda(n)/u(n + 1). \qquad (6.16)\end{aligned}$$

By now, it should be quite straightforward for the reader to derive the balance equations for the probabilities $p_n(n_0, n_1, n_2)$, i.e., for the Jacksonian network of Figure 6.3. We use these balance equations together with the postulated approximation (6.13) in equation (6.16), which yields

$$\begin{aligned}0 \approx &-p_n(n_0, n_1, n_2)\{\lambda(n) + \zeta(n_0)v_0\}\\&+\zeta(n_0 - 1)p_{n-1}(n_0 - 1, n_1, n_2)A_n v_0\\&+\zeta(N_t - n - 1)p_{n+1}(n_0 + 1, n_1, n_2)\lambda(n)/A_{n+1}. \qquad (6.17)\end{aligned}$$

We notice that all terms related to internal system transitions have vanished and the remaining terms are upper-bounded by $\lambda(n) + \zeta(n_0)v_0$. So, in agreement with our intuitive argument, if these rates are much smaller than the rates of internal system transitions, the terms on the right-hand side of equation (6.17) are indeed very small (compared to the majority of terms in equation (6.16)). Similar arguments have been used for nearly decomposable systems in other areas of science (Simon & Ando, 1961; Courtois, 2014).

The terms in (6.17) can be rearranged as

$$\begin{aligned}&p_n(n_0, n_1, n_2)\{\lambda(n) + \zeta(n_0)v_0\}\\&\quad\approx \zeta(n_0 - 1)p_{n-1}(n_0 - 1, n_1, n_2)A_n v_0\\&\qquad+\zeta(N_t - n - 1)p_{n+1}(n_0 + 1, n_1, n_2)\lambda(n)/A_{n+1}. \qquad (6.18)\end{aligned}$$

Regardless of the values of the rates of transaction (job) arrivals and completions ($\lambda(n)$ and v_0), we have a sum of two terms on the left-hand side of the equation and two terms on the right-hand side of (6.18).

Overall, how well the approximate equality in (6.18) is satisfied depends on the characteristics of the original model of Figure 6.1. In particular, if the rate $v_1(n)$ is a constant, i.e., does not depend on the number of jobs in the system; these terms balance exactly whether $\lambda(n)$ and v_0 are smaller than the rates of other transitions or not. With constant $v_1(n)$, the model of Figure 6.1 becomes a separable network and our equivalence and decomposition approach actually yields the exact solution regardless of other considerations, which is an appealing feature. Formula (6.18) implies that, in general, the accuracy of the equivalence and decomposition approximation will depend both on differences in time scales in transition rates and the degree of departure of the model from a separable network. The above examination of how well our approximate solution satisfies the balance equations of our model gives us an insight into the factors affecting its accuracy.

Since we already have the exact balance equations for our original model of Figure 6.1, it seems appropriate to mention at this point that we can easily obtain an exact numerical solution of these equations via the method of conditionals. Clearly, our approximate decomposition solution would provide an excellent initial solution in this case. We use a superscript to denote the iteration number so that $p^i(n_0, n_1, n_2|n)$ denotes the value computed at iteration $= 1, \ldots$. The initial distribution for the iteration becomes $p^0(n_0, n_1, n_2|n) = p_n(n_0, n_1, n_2)$, and we let

$$A^i(n) = \sum_{n_0=1}^{n} \sum_{n_1=0}^{n-n_0} p^i(n_0, n_1, n_2 = n - n_0 - n_1|n), \quad \text{for } i = 0, 1, \ldots. \tag{6.19}$$

A simple iterative scheme could be as follows. Enumerate system states in the order $n = 1, \ldots, N_t$, and for each value of n, consider consecutive values $n_0 = 0, \ldots, n$, for each value of n_0, consecutive values $n_1 = 0, \ldots, n - n_0$ (and hence $n_2 = n - n_1 - n_2$):

$$\begin{aligned}
p^i(n_0, n_1, n_2|n)\{&\zeta(n_0)[v_1(n) + v_2] + \zeta(n_1)u_1 + \zeta(n_2)u_2 + \lambda(n) \\
&+ \zeta(n_0)v_0\} = \zeta(n_0 - 1)p^i(n_0 - 1, n_1 + 1, n_2|n)u_1 \\
&+ \zeta(n_0 - 1)p^i(n_0 - 1, n_1, n_2 + 1|n)u_2 \\
&+ \zeta(n_1 - 1)p^{i-1}(n_0 + 1, n_1 - 1, n_2|n)v_1(n) \\
&+ \zeta(n_2 - 1)p^{i-1}(n_0 + 1, n_1, n_2 - 1|n)v_2 \\
&+ \zeta(n_0 - 1)p^i(n_0 - 1, n_1, n_2|n - 1)v_0 A^{i-1}(n) \\
&+ \zeta(N_t - n - 1)p^{i-1}(n_0 + 1, n_1, n_2|n + 1)\lambda(n)/A^{i-1}(n + 1).
\end{aligned} \tag{6.20}$$

For each value of $n > 1$, once all new non-normalized values have been computed for all $n_0 = 0, \ldots, n$, $n_1 = 0, \ldots, n - n_0$, $n_2 = n - n_1 - n_2$, normalize these values so that

$$\sum_{n_0=0}^{n} \sum_{n_1=0}^{n-n_0} p^i(n_0, n_1, n_2 = n - n_0 - n_1|n) = 1. \tag{6.21}$$

When the new values for $p^i(n_0, n_1, n_2|n)$ have been normalized, compute a new value for $A^i(n)$ using formula (6.19). This iteration exits when a reasonable convergence criterion has been satisfied or a reasonable maximum number of iterations has been exhausted. Such an iterative scheme is likely to converge within just a few iterations if the rates $\lambda(n)$ and v_0 are much smaller than the rates of other transitions.

The equivalence and decomposition approximation, in its different "flavors", has been used successfully for a large number of models (Brandwajn, 1975, 1980, 1985). The strength of this method lies in its ability to decompose a difficult model into smaller models that are more easily handled than the original model, as we saw for the system of Figure 6.1. In this case, we were able to obtain an (approximate) analytical solution for the decomposed model of the system, which could then be used in the equivalent model of Figure 6.2. As a matter of terminology, sometimes, the decomposed model of the system is referred to as the "inner model" and the equivalent model is referred to as the "outer model" (Chandy *et al.*, 1975; Chandy & Sauer, 1978; Balsamo & Iazeolla, 1982).

6.2 Fixed-point iteration and equivalence for approximate tandem network analysis

Such a neat hierarchical decomposition is not always possible. As an example, consider the tandem network represented in Figure 6.4.

The network consists of K nodes, labeled 1 through K. The nodes have a limited capacity, and we denote by M_i the capacity of node i, $i = 1, \ldots, K$. Arrivals to the network are assumed to come from a Poisson source with rate λ, and service times at nodes are assumed to be exponentially distributed. Arrivals to the network that happen when node 1 is at capacity are simply lost (equivalently, we can assume that the outside source of customers shuts down when the first node is full). The service rate for a customer at node i, $i = 1, \ldots, K-1$ is μ_i provided that the number of customers at node $i+1$ is less than M_{i+1}, i.e., provided that the immediately downstream node is not at capacity. Otherwise, node i is blocked until a customer at node $i+1$ leaves the node. The type of inter-node blocking considered in our network is often referred to as communications blocking, the idea being that the service at a node may represent message or packet transmission, and available storage is needed at the downstream node to store the incoming message.

We study the model in steady state. Our goal is to obtain the attained throughput of customers in this network, denoted by θ, as well as the expected time customers spend at each node, denoted by $E[W_i]$ for node $i, i = 1, \ldots, K$. We let n_i be the current number of customers at node i and we denote by $p(n_1, n_2, \ldots, n_K)$ the joint steady-state probability for the number of customers at the nodes of our tandem network. Clearly, if we knew $p(n_1, n_2, \ldots, n_k)$, we would have no trouble computing the performance measures of interest. Unfortunately, despite its simple topology and the exponential assumptions on times between arrivals and on service times, our model is not a Jacksonian (or BCMP) network. This

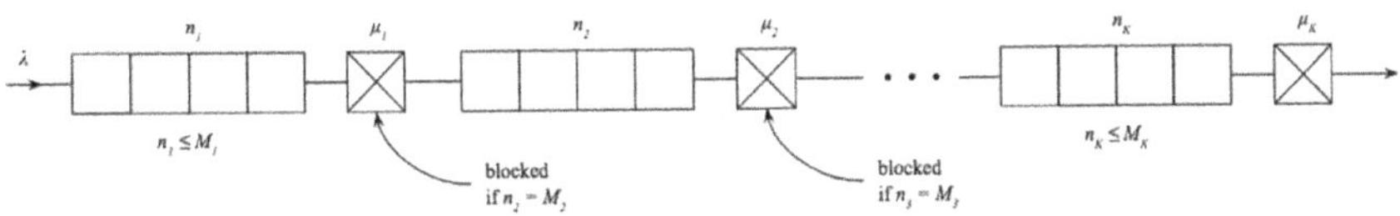

Figure 6.4. K-node tandem network with blocking.

means that there is no solution of the form $p(n_1, n_2, \ldots, n_K) = f_1(n_1)f_2(n_2)\cdots f_K(n_K)$ to the balance equations for our tandem network. The reason for this is the type of blocking present in the model.

We will now develop a couple of simple approximations for this deceptively simple model. The simplest idea would seem to be to analyze the network one node at a time. Let's try this approach. We denote by $p_i(n_i)$ the steady-state probability that there are n_i customers at node i, and we use $\lambda_i(n_i)$ and $u_i(n_i)$ to denote the instantaneous rates of increase and decrease in n_i, respectively. We denote by $p(n_{i+1}|n_i)$ the conditional probability that there are n_{i+1} customers at node $i+1$ given that the number of customers at node i, $i = 0, \ldots, K-1$, is n_i Similarly, we denote by $p(n_{i-1}|n_i)$ the conditional probability that there are n_{i-1} customers at node $i-1$ given that there are n_i customers at node i, $i = 1, \ldots, K$.

Arrivals at the first node come from the outside Poisson source, while arrivals at the remaining nodes are the result of a departure from the immediately preceding node. This yields for the instantaneous rate of arrivals at a node

$$\lambda_i(n_i) = \begin{cases} \lambda, & \text{for } i = 1 \\ \mu_{i-1}p(n_{i-1} > 0|n_i), & \text{for } i = 2, \ldots, K \end{cases},$$
$$n_i = 0, 1, \ldots, M_i - 1. \tag{6.22}$$

Customer departures at nodes other than the last one can only happen if the following node is not full. For the last node in the tandem, there is no such restriction. Thus, the instantaneous rate of customer departure at node i is given by

$$u_i(n_i) = \begin{cases} \mu_i p(n_{i+1} < M_{i+1}|n_i), & \text{for } i = 1, \ldots, K-1 \\ \mu_K, & \text{for } i = K \end{cases},$$
$$n_i = 1, \ldots, M_i. \tag{6.23}$$

Formally, the steady-state probability for the number of customers at node i can be expressed as

$$p_i(n_i) = \frac{1}{\mathrm{G}} \prod_{j=1}^{n_i} \frac{\lambda_i(j-1)}{u_i(j)}, \quad n_i = 0, 1, \ldots, M_i. \tag{6.24}$$

As usual, G is normalizing constant (possibly different for each node) such that $\sum_{n_i=0}^{M_i} p(n_i) = 1$, and empty products are assumed to be equal to one. The attained customer throughput can be determined as

$$\theta = \lambda[1 - p_1(M_1)]. \tag{6.25}$$

θ is also the throughput of customers at each node so that the expected time a customer spends at node i can be computed from Little's formula as

$$E[W_i] = \frac{\sum_{n_i=1}^{M_i} n_i p_i(n_i)}{\theta}. \tag{6.26}$$

Formula (6.24) is exact provided we know the instantaneous rates $\lambda_i(n_i)$ and $u_i(n_i)$. Unfortunately, these rates involve conditional probabilities of the state of neighboring nodes, which cannot be obtained by analyzing one node at a time. As a simple, rough approximation, we can replace the conditionals with the corresponding nonconditional, i.e., marginal, probabilities as follows:

$$p(n_{i+1}|n_i) \approx p_{i+1}(n_{i+1}), \quad i = 1, \ldots, K-1, \tag{6.27}$$

$$p(n_{i-1}|n_i) \approx p_{i-1}(n_{i-1}), \quad i = 2, \ldots, K. \tag{6.28}$$

We still have a wrinkle to iron out, viz., if we assume that we analyze the nodes in their order in the network, for nodes $i = 1, \ldots, K-1$, we need the marginal probabilities of the state of the next node $p_{i+1}(n_{i+1})$, which have not been computed yet. To solve this problem, we use a fixed-point iteration. Specifically, using a superscript to denote the iteration number, we start with a reasonable set of values $p_{i+1}^0(n_{i+1})$, $i = 1, \ldots, K-1$. Such "reasonable" initial values could be obtained, for example, by assuming $p(n_{i-1} > 0|n_i) = 1$ for $i = 2, \ldots, K$, and $p(n_{i+1} < M_{i+1}|n_i) = 1$ for $i = 2, \ldots, K-1$, and then using formulas (6.22) and (6.23) in formula (6.24) to compute the desired initial values for the probabilities $p_i(n_i)$, $i = 2, \ldots, K$. Then, at iteration $j = 1, \ldots,$ we use

$$\lambda_i(n_i) \approx \begin{cases} \lambda, & \text{for } i = 1 \\ \mu_{i-1} p_{i-1}^j(n_{i-1} > 0), & \text{for } i = 2, \ldots, K \end{cases},$$

$$n_i = 0, 1, \ldots, M_i - 1, \tag{6.29}$$

$$u_i(n_i) \approx \begin{cases} \mu_i p_{i+1}^{j-1}(n_{i+1} < M_{i+1}), & \text{for } i = 1, \dots, K-1 \\ \mu_K, & \text{for } i = K \end{cases},$$

$$n_i = 1, \dots, M_i, \tag{6.30}$$

as the instantaneous arrival and departure rates to compute consecutive approximate values $p_i^j(n_i)$ for $i = 1, \dots, K$ using formula (6.24). We iterate like that until the approximate values stabilize, i.e., no longer change to some reasonable degree. As was the case for numerical iterative methods, a possible stopping criterion might be $\max_{n_i,i} |p_i^j(n_i) - p_i^{j-1}(n_i)| < \varepsilon$ or $\max_{n_i,i} |1 - \frac{p_i^j(n_i)}{p_i^{j-1}(n_i)}| < \varepsilon$ with some reasonable value for ε. The iteration is summarized in Figure 6.5.

Such fixed-point iterative schemes are not necessarily guaranteed to converge to a unique solution, so, theoretically, we would need a proof that our fixed-point iteration converges and that there is a single fixed point. In practice, proofs of these convergence properties tend to be exceedingly hard to obtain, and for the very many fixed-point iterations that have been successfully used in computer performance modeling over the years, there are few, if any, such proofs

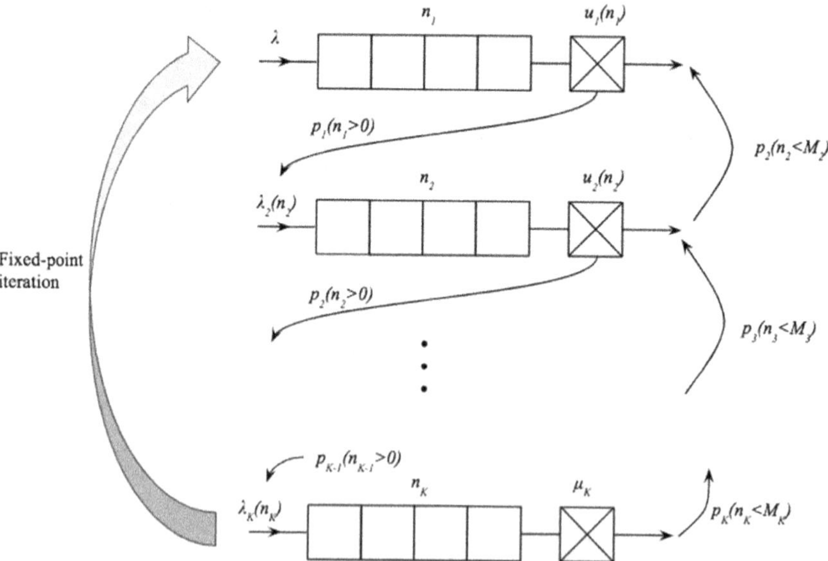

Figure 6.5. Fixed-point iteration single node at a time.

available. In the majority of cases, the fixed-point iterations proposed in the literature tend to converge "as advertised".

Our simple fixed-point approximation, in which we consider a single node at a time, does not capture directly the blocking between neighboring nodes, which is a defining feature of our tandem model. Therefore, intuitively, we may expect it to not be very good. A potentially better approach might be to consider network nodes in adjacent pairs $i, i+1$ $(i = 1, \ldots, K-1)$. This way, blocking between neighboring nodes can be taken into account more directly.

Let's try this approach. We denote by $p(n_i, n_{i+1})$ the joint probability that there are n_i and n_{i+1} customers at nodes i and $i+1$, respectively. The instantaneous rate with which n_i increases is given by

$$a_i(n_i, n_{i+1}) = \begin{cases} \lambda, & \text{for } i = 1 \\ \mu_{i-1} p(n_{i-1} > 0 | n_i, n_{i+1}), & \text{for } i > 1 \end{cases},$$
$$n_i = 0, \ldots, M_i - 1. \tag{6.31}$$

In the above expression, $p(n_{i-1}|n_i, n_{i+1})$ is the conditional probability that there are n_{i-1} customers at the preceding node, given that there are (n_i, n_{i+1}) customers at the selected pair. Inside the selected node pair, the instantaneous rate of departures from node i to node $i+1$ is μ_i, provided that $n_{i+1} < M_{i+1}$.

The rate with which n_{i+1} decreases is given by

$$u_{i+1}(n_i, n_{i+1}) = \begin{cases} \mu_{i+1} p(n_{i+2} < M_{i+2} | n_i, n_{i+1}), & \text{for } i = 1, \ldots, K-2 \\ \mu_K, & \text{for } i = K-1 \end{cases},$$
$$n_{i+1} = 1, \ldots, M_{i+1}. \tag{6.32}$$

Here, $p(n_{i+2}|n_i, n_{i+1})$ is the conditional probability that there are n_{i+2} customers at the following node, given that there are (n_i, n_{i+1}) customers at the selected pair. These expressions can also be derived from the balance equations for the full state description $p(n_1, n_2, \ldots, n_k)$ for our tandem network (Brandwajn & Jow, 1988). This means that, with respect to $p(n_i, n_{i+1})$, the selected pair of neighboring nodes behaves like the two-node tandem network shown in Figure 6.6 with a state-dependent arrival rate $a_i(n_i, n_{i+1})$ and the rate of departures from the second node $u_{i+1}(n_i, n_{i+1})$.

In this equivalent system, the two nodes have the same limited capacity as in our original network. The times between arrivals to

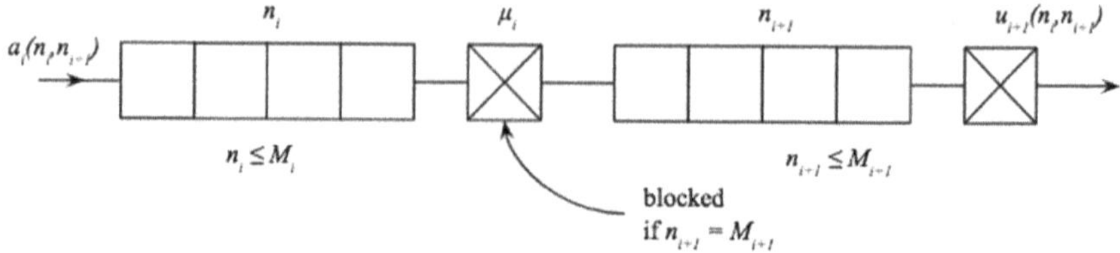

Figure 6.6. Equivalent system for a pair of neighboring nodes.

this two-node system, as well as the service at the second node, have memoryless distributions, and the service at the first node can start only when the second node is not full. This equivalent two-node network is a representation of the marginal joint probability $p(n_i, n_{i+1})$, and, as such, involves no approximation.

Such a two-node system can be solved using a numerical method, for example, the γ-method discussed in Chapter 5. This would allow us to obtain the joint probability distribution $p(n_i, n_{i+1})$ if we knew the rates of arrivals and service completions, $a_i(n_i, n_{i+1})$ and $u_{i+1}(n_i, n_{i+1})$. These rates involve the probability of the number of users at nodes $i-1$ and $i+2$ given the state of the selected pair of nodes (n_i, n_{i+1}). In order to approximate these probabilities, we assume that only the state of the immediate neighbor matters in the condition so that

$$p(n_{i-1}|n_i, n_{i+1}) \approx p(n_{i-1}|n_i), \quad i = 2, \ldots, K-1,$$

and

$$p(n_{i+2}|n_i, n_{i+1}) \approx p(n_{i+2}|n_{i+1}), \quad i = 1, \ldots, K-2. \tag{6.33}$$

Hence, we get the following approximation for the arrival and service rates in the two-node system:

$$a_i(n_i, n_{i+1}) \approx \hat{a}_i(n_i) = \begin{cases} \lambda, & \text{for } i = 1 \\ \mu_{i-1} p(n_{i-1} > 0|n_i), & \text{for } i > 1 \end{cases},$$
$$n_i = 0, \ldots, M_i - 1, \tag{6.34}$$

and

$$u_{i+1}(n_i, n_{i+1}) \approx \hat{u}_{i+1}(n_{i+1})$$
$$= \begin{cases} \mu_{i+1} p(n_{i+2} < M_{i+2}|n_{i+1}), & \text{for } i = 1, \ldots, K-2 \\ \mu_K, & \text{for } i = K-1 \end{cases},$$
$$n_{i+1} = 1, \ldots, M_{i+1}. \tag{6.35}$$

Intuitively, assumption (6.33) states that nearly all the influence of node $i+1$ on node $i-1$ is contained in the state of node i or, in other words, that the state of the immediate neighbor is much more important than the state of a more remote neighbor. This assumption allows us to develop an iterative approach in which we consider the nodes in the original tandem network of Figure 6.4 two-by-two. We use the probability $p(n_i, n_{i+1})$ obtained from the solution of the pair of nodes i and $i+1$, reduced to the simpler system shown in Figure 6.7, to compute $\hat{u}_i(n_i)$ for the pair $i-1$, i and $\hat{a}_{i+1}(n_{i+1})$ for the pair $i+1$, $i+2$.

Since this approach requires values that have not yet been computed as we explore the nodes, we use a fixed-point iteration with the following steps (we use a superscript to denote the iteration number):

1. Select an initial approximation for the conditional probabilities $p^0(n_{i+1}|n_i)$, $i = 2, \ldots, K-1$. A simple possibility is $p^0(n_{i+1}|n_i) = \frac{1}{(M_{i+1}+1)}$, $\forall n_{i+1}$, n_i. We must have $\sum_{n_{i+1}=0}^{M_{i+1}} p^j(n_{i+1}|n_i) = 1$, for $n_i = 0, \ldots, M_i$ for $j = 0, 1, \ldots$.
2. At iteration j ($j = 1, 2, \ldots$), solve $K-1$ pairs of neighboring nodes $i, i+1$, for $i = 1, \ldots, K-1$. In the solution, use $p^j(n_{i-1}|n_i)$ to obtain $\hat{a}_i(n_i)$, and $p^{j-1}(n_{i+2}|n_{i+1})$ for $\hat{u}_{i+1}(n_{i+1})$. The solution of the pair yields $p^j(n_i, n_{i+1})$, which we use to compute $p^j(n_i|n_{i+1})$ and $p^j(n_{i+1}|n_i)$ needed for other pairs, as well as any performance measures for nodes i and $i+1$.
3. If a convergence criterion, for example $\max_{n_i, i} |p^j(n_i, n_{i+1}) - p^{j-1}(n_i, n_{i+1})| < \varepsilon$, for some reasonable $\varepsilon > 0$, has been met, stop. Otherwise, return to Step 2 for iteration $j+1$.

Step 2 of the iteration calls for the computation of the conditional probabilities $p(n_i|n_{i+1})$ and $p(n_{i+1}|n_i)$ from the joint probability

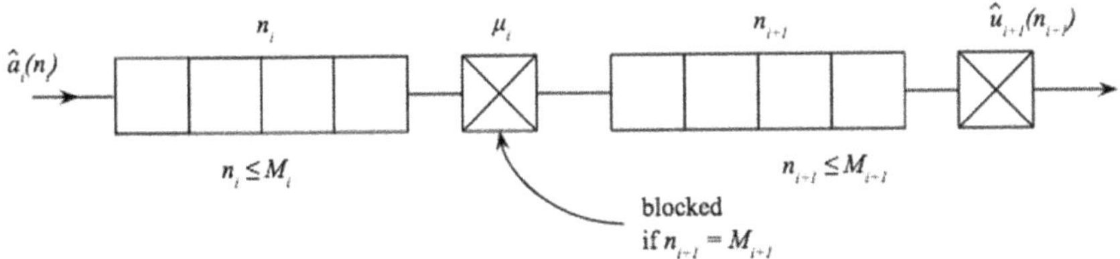

Figure 6.7. Simplified two-node cell.

$p(n_i, n_{i+1})$. This is easily accomplished since we have

$$p(n_i|n_{i+1}) = \frac{p(n_i, n_{i+1})}{\sum_{n_i=0}^{M_i} p(n_i, n_{i+1})},$$
$$n_i = 0, \ldots, M_i, \quad i = 1, \ldots, K-1$$

and

$$p(n_{i+1}|n_i) = \frac{p(n_i, n_{i+1})}{\sum_{n_{i+1}=0}^{M_{i+1}} p(n_i, n_{i+1})},$$
$$n_{i+1} = 0, \ldots, M_{i+1}, \quad i = 1, \ldots, K-1. \tag{6.36}$$

To illustrate the approximation procedure, let's apply it to a system with $K = 4$ nodes:

- First, we select $p^0(n_2|n_1)$, $p^0(n_3|n_2)$, and $p^0(n_4|n_3)$. We set the iteration number j to 1.
- At iteration j, we solve three pairs of the two-node system of Figure 6.7 as follows:
  1. For the pair of nodes 1 and 2, $\hat{a}_1(n_1) = \lambda$ is of course known, and we use $p^{j-1}(n_3|n_2)$ to compute $\hat{u}_2(n_2)$. Then, we solve for $p^j(n_1, n_2)$.
  2. For the pair of nodes 2 and 3, we use $p^j(n_1|n_2)$ obtained from the preceding pair to compute $\hat{a}_2(n_2)$ and $p^{j-1}(n_4|n_3)$ to compute $\hat{u}_3(n_3)$. Then, we solve for $p^j(n_2, n_3)$.
  3. For the pair of nodes 3 and 4, we use $p^j(n_2|n_3)$ obtained from the preceding pair to compute $\hat{a}_3(n_3)$; $\hat{u}_4(n_4) = \mu_4$ is known. Then, we solve for $p^j(n_3, n_4)$.
- If the convergence criterion is satisfied, we stop. Otherwise, we set j to $j + 1$ and repeat the iteration.

Each iteration of our fixed-point procedure requires $K - 1$ solutions of the two-node cell of Figure 6.7. This iterative procedure has been found to converge in a reasonable number of iterations (typically, a few tens of iterations) (Brandwajn & Jow, 1988). As could be expected, the number of iterations required for convergence increases somewhat with the number of nodes in the network, so that the complexity of this approximation grows relatively moderately with the number of nodes. This is to be compared with a combinatorial growth of the complexity if one considers a direct numerical solution for the full state description $p(n_1, n_2, \ldots, n_k)$.

An obvious question that arises when considering an approximate solution is: how accurate is it and what is its domain of applicability, i.e., when should it and when should it not be applied? Even though the underlying idea of the approximation may offer good clues to help answer these questions, it is always a good idea to study the approximation's behavior as other factors may influence its accuracy. As an example, for the equivalence and decomposition approximation, time-scale differences turn out to be only one of the factors affecting the accuracy of the results.

To illustrate the behavior of our fixed-point iterations for the tandem network with blocking, we use several examples borrowed from a published paper (Brandwajn & Jow, 1988). As a basis for comparison, the examples use either the "exact" numerical solution of the balance equations for the tandem network or, for networks with more nodes, the results of discrete-event simulations. We look at the expected total number of customers in the system, attained throughput, and expected time in the system, as well as the expected numbers of customers at each node. In some examples, we also compare for each node the probability that the node is idle, $p_i(0)$, and the probability that the node is at capacity, $p_i(M_i)$.

We start with a simple three-node tandem network with the following parameters: $M_1 = M_2 = M_3 = 2$, $\mu_1 = \mu_2 = \mu_3 = 1$, and $\lambda = 1$. This network, studied as an example in Boxma and Konheim (1980), has the interesting property that the expected total number of customers in the system is equal to the number of nodes. For this example, both fixed-point approximations we have considered for our tandem network, i.e., single node at a time and two nodes at a time, converge within just a few iterations. We show in Table 6.1 the results obtained. In addition to the actual values, we have included the relative errors computed as $\frac{\text{approximate}-\text{exact}}{\text{exact}} * 100\%$. The labels "Appr. 1" and "Appr. 2" refer to our single-node and two-node fixed-point approximations, respectively.

Quite remarkably, for this first example, both approximation methods produce the correct expected total number of customers in the system (Table 6.1(a)) and reasonably good values for the mean numbers of customers at each node (Table 6.1(b)). The two-node iteration does come across as somewhat more accurate, but since the observed relative errors for the simpler method are below 3%, there does not seem to be much to make fuss about.

Table 6.1. Accuracy of approximate solutions: results for first example.

	Exact	Appr. 1	Appr. 2	Relative error (%) Appr. 1	Relative error (%) Appr. 2
(a) *System-level quantities*					
Number in system	3.0000	3.0000	3.0000	0.00	0.00
Throughput	0.5362	0.5263	0.5342	−1.84	−0.37
Sojourn time	5.5362	5.7000	5.6159	2.96	1.44
(b) *Expected number of customers at each node*					
Node 1	1.2538	1.2632	1.2562	0.75	0.19
Node 2	1.0000	1.0000	1.0000	0.00	0.00
Node 3	0.7462	0.7368	0.7438	−1.25	−0.32

In our second example, we consider the performance of our two approximation methods for another three-node tandem network with the following parameters: $M_1 = 11$, $M_2 = 2$, $M_3 = 3$, $\mu_1 = 2.5$, $\mu_2 = 10$, $\mu_3 = 1$ and $\lambda = 1$. The parameters of this model have been selected so as to create a significant blocking stemming from the last node and propagating back to the head of the network. Our fixed-point approximations converge without issues, and we show in Table 6.2 the numerical results produced by both methods.

Here, if we look at the system-level results (Table 6.2(a)), it is immediately apparent that, in general, the single-node approximation is significantly less accurate than the two-node fixed-point iteration, with relative errors exceeding 30% for the expected total number of customers in the system and for the time a customer spends in the system. By contrast, for the two-node approximation the relative errors for these quantities stay below 6%. If we look at the expected number of customers at each node (Table 6.2(b)), we note that the largest relative errors happen for the first node, where our first approximation fails miserably with an error of almost 63% versus 13% for our second approximation. When examining the probabilities of the node being empty, $p_i(0)$ (Table 6.2(c)), and at capacity, $p_i(M_i)$ (Table 6.2(d)), we note that for small probability values, both approximations have large relative errors, but large relative errors

Table 6.2. Accuracy of approximate solutions: results for second example.

	Exact	Appr. 1	Appr. 2	Relative error (%) Appr. 1	Relative error (%) Appr. 2
(a) *System-level quantities*					
Number in system	8.1581	5.5898	7.9353	−31.48	−2.73
Throughput	0.9327	0.9975	0.9597	6.95	2.89
Sojourn time	8.7467	5.6036	8.2685	−35.94	−5.47
(b) *Expected number at each node*					
Node 1	4.5204	1.6835	3.9357	−62.76	−12.93
Node 2	1.1510	1.0709	1.1620	−6.96	0.96
Node 3	2.4867	2.8354	2.5823	14.02	3.84
(c) *Marginal probability*, $p_i(0)$					
Node 1	0.1965	0.3672	0.2030	86.89	3.31
Node 2	0.3058	0.2985	0.2914	−2.38	−4.71
Node 3	0.0673	0.0025	0.0403	−96.31	−40.12
(d) *Marginal probability*, $p_i(M_i)m$					
Node 1	0.0673	0.0025	0.0403	−96.35	−40.12
Node 2	0.4568	0.3694	0.4534	−19.13	−0.74
Node 3	0.7019	0.8578	0.7321	22.21	4.30

on small quantities happen easily in approximations. How important they are depends on how important it is for us to know whether the probability of a node being full is 0.0673 or 0.0403, as predicted by the two-node approximation. The values produced by our single-node approach are in this case an order of magnitude off. Overall, with one exception for $p_2(0)$, our second approximation performs significantly better. The results for node 2 in Tables 2.6(b) and 2.6(c) teach us also an important lesson. Although the relative error in the mean number of customers at this node is less than 1% for our two-node approximation, the relative error for $p_2(0)$ is not far from 5%. Thus, we can have excellent approximate results for the expected number of customers and, nonetheless, much less accurate results for individual probabilities.

The fact that the accuracy of the approximate solutions in our second example is worse for the first network node whose capacity is larger is surprising because, as the node capacity increases, the approximations become asymptotically exact since the tandem network then has a product-form solution. Therefore, as our third example, we consider a three-node network with the same node capacities but service rates chosen to remove the large jump in blocking probabilities present in our second example: $M_1 = 11$, $M_2 = 2$, $M_3 = 3$, $\mu_1 = 1.111\ldots$, $\mu_2 = 3.333$, $\mu_3 = 1.666\ldots$, and $\lambda = 1$. We show in Table 2.3 the results for this example.

The results in Table 6.3 clearly demonstrate the superiority of the two-node fixed-point approximation over the single-node approximation. We note that the latter tends to be an order or two orders of magnitude less accurate than the two-node approach.

The tandem networks we studied so far had only three nodes and it is important to find out how the approximations are faring for a larger number of nodes. This is what we are exploring in our last example with the following parameters: $K = 10$, $M_1 = M_3 = M_6 = M_7 = M_9 = 3$, $M_2 = M_5 = M_8 = 4$, $M_4 = M_{10} = 2$, $\mu_1 = 1.125$, $\mu_2 = \mu_4 = \mu_6 = 1$, $\mu_3 = \mu_7 = 0.833\ldots$, $\mu_5 = \mu_8 = \mu_9 = 0.666\ldots$, $\mu_{10} = 1.111\ldots$, and $\lambda = 1$. In this example, we use the results of discrete-event simulations as the basis to

Table 6.3. Accuracy of approximations: results for third example.

				Relative error (%)	
	Exact	Appr. 1	Appr. 2	Appr. 1	Appr. 2
(a) *System-level quantities*					
Number in system	6.9703	6.5553	6.9530	−5.95	−0.25
Throughput	0.9217	0.9270	0.9206	0.57	−0.12
Sojourn time	7.5624	7.0719	7.5526	−6.49	−0.13
(b) *Expected number of customers at each node*					
Node 1	5.5876	5.2219	5.5727	−6.54	−0.27
Node 2	0.4375	0.3903	0.4365	−10.80	−0.23
Node 3	0.9452	0.9432	0.9438	−0.22	−0.15

Table 6.4. Accuracy of approximations: fourth example.

	Simulation				Relative error (%)	
	Lower bound	Upper bound	Appr. 1	Appr. 2	Appr. 1	Appr. 2
(a) *System-level quantities*						
Number in system	19.7491	19.8399	19.9700	19.8392	0.89	0.23
Throughput	0.4472	0.4490	0.4686	0.4477	4.57	−0.09
Sojourn time	43.9987	44.3451	42.6181	44.3136	−3.52	0.32
(b) *Expected number of customers at each node*						
Node 1	2.3050	2.3130	2.2624	2.3009	−2.02	−0.35
Node 2	3.3169	3.3251	3.2584	3.3933	−1.88	2.18
Node 3	2.3055	2.3145	2.2365	2.3180	−3.18	0.35
Node 4	1.1758	1.1822	1.1180	1.1848	−5.17	0.49
Node 5	2.6884	2.7116	2.6484	2.7004	−1.91	0.01
Node 6	1.5283	1.5457	1.4226	1.5347	−7.44	−0.15
Node 7	1.8041	1.8259	1.8211	1.8028	0.34	−0.67
Node 8	2.5553	2.5807	2.5612	2.5553	−0.27	−0.49
Node 9	1.5250	1.5330	2.0811	1.5224	36.11	−0.43
Node 10	0.5253	0.5287	0.5603	0.5272	6.32	0.04

assess the accuracy of the approximations. For the mean numbers, the simulation results were obtained using the independent replications method with 10 replications of 50,000 completions each. Simulation results in Table 6.4 are reported as confidence intervals at 95% confidence level.

Comparisons of approximation results with those of a discrete-event simulation can be tricky. In Table 6.4, the relative errors are computed versus mid-points of the confidence intervals estimated in the simulation. This may lead to assigning different accuracies to two values that both fall within a confidence interval, while there is little reason to assume that the interval mid-point is more "true" than any other point within the confidence interval. The results for the expected number of customers at node 8 provide a case in point. Of course, this becomes a non-issue if the confidence intervals are very narrow. We note that, with few exceptions, the results of the two-node approximation tend to be much closer to the simulation results than those of the single-node approach, with considerably fewer large deviations, i.e., large relative errors.

The original paper presenting the two-node approximation (Brandwajn & Jow, 1988) provides a different set of confidence intervals for the expected numbers of customers at each node. These confidence intervals, obtained using the batch means approach, as opposed to our independent replications method, generally overlap our intervals, with the exception of the results for node 2. Indeed, the confidence interval reported in the original paper (Brandwajn & Jow, 1988) for the expected number of customers at this node is (3.4005, 3.4206) versus our (3.3169, 3.3251). Roughly speaking, there is a difference of about 2% between these two estimates, which is not a lot but may be vexing when evaluating the accuracy of an approximation method with relative errors of the same order of magnitude. If we increase the number of completions per replication to 500,000, we get only a slightly different confidence interval (3.3206, 3.3234), so this is not much help in our case. We are going to chalk it up to either an error in the paper or one of the mysteries of the discrete-event simulation.

6.3 Approximate class aggregation to model a system with non-preemptive priorities

A somewhat related, but ultimately different, approximation method has been successfully used in systems with multiple classes of requests (customers). The basic idea of the approach is to reduce the complexity of the initial problem by aggregating, i.e., lumping together, some customer classes. As an example, we will use a simple model of SCSI (Small Computer System Interface) bus acquisition by devices on the bus. SCSI connections were widely used for interconnecting disks in workstations and larger storage controllers before the advent of optical fiber connections. In the SCSI standard, the unique IDs of devices on the bus define a fixed priority in situations where several devices compete for the use of the bus (we neglect certain optional features of the standard). This priority is non-preemptive in nature, i.e., a bus transfer already in progress is not interrupted by an arriving higher-priority request. Such a non-preemptive priority is often referred to as Head-of-Line or HOL priority. The concern in such

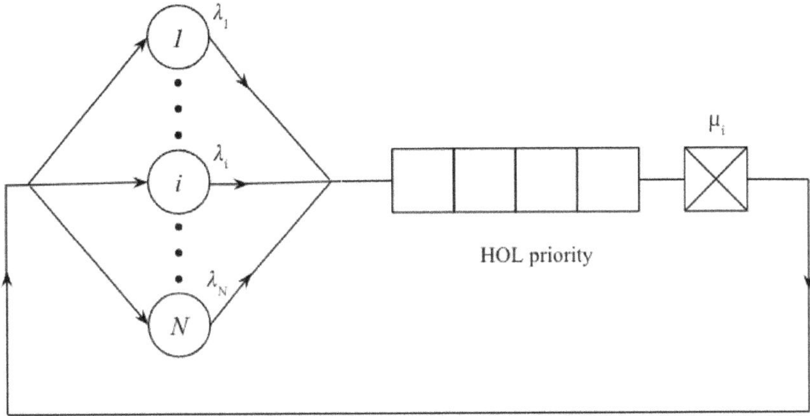

Figure 6.8. Finite-source model of bus priorities.

a fixed-priority bus arbitration scheme is the potential unfairness for lower-priority devices, especially if higher-priority devices happen to have higher I/O rates.

Let us consider a simple model of SCSI bus acquisition in which the devices on the bus are represented as sources of requests and the bus as a single server. Our model is shown in Figure 6.8. There are N devices, numbered 1 through N, competing for the use of the server (the bus). Each device issues requests at its priority level, which corresponds to its number. There can be only one request per device at a time, i.e., there can be at most $N-1$ requests queued waiting for the use of the bus. We assume that device 1 has the highest priority. As mentioned above, the priority is non-preemptive, i.e., it determines only which request is selected to use the server when the bus finishes serving a request and there are several requests waiting. We let $1/\mu_i$ denote the mean service time of a request issued by device i, and we assume that the service times (which include bus setup and cleanup overheads) are exponentially distributed. We denote by $1/\lambda_i$ the mean time during which device i is idle, i.e., has no request for bus use, between two consecutive requests, and we assume that device idle times are exponentially distributed. In other words, λ_i is the rate with which an idle device issues a new request for bus use.

We consider the system in steady state, and our goal is to study the expected time a device has to wait to acquire the bus under a given attained rate of I/Os for each device. We let Θ_i denote this attained I/O rate for device i, and we denote by $E[W_i]$ the corresponding expected bus wait time. Such formulation of the problem, in which one studies the performance of a system for a specified attained rate of I/Os (or transactions, as the case may be), is frequent in the computer industry. This implies that the server (bus) utilization is also known for each request class, i.e., each device on the bus. We let $u_i = \Theta_i/\mu_i$ denote this known server utilization. We also let $E[N_{qi}]$ denote the expected number of class i requests waiting for the bus. Since each device can have only a single bus request outstanding, $E[N_{qi}]$ is also the fraction of time device i spends waiting for the bus. By Little's law, we have

$$E[W_i] = E[N_{qi}]/\Theta_i. \tag{6.37}$$

But how can the known attained rate of I/Os for class i be related to the model parameter λ_i? A device can issue requests only when it is idle, i.e., not using the bus or waiting to use it. Hence, we must have

$$(1 - E[N_{qi}] - u_i)\lambda_i = \Theta_i. \tag{6.38}$$

When solving the model, we can use a fixed-point iteration to determine the proper values for λ_i, $i = 1, \ldots, N$, such that we have the desired attained I/O rates Θ_i for all devices. For a given set of values of the rates $\Lambda = \{\lambda_1, \ldots \lambda_N\}$, we denote by $\rho_i(\Lambda)$ the fraction of time during which the server is busy serving device i requests. Our goal is then to determine a set of λ_i, $i = 1, \ldots, N$, such that $\rho_i(\Lambda) = u_i$ for all devices. Let's use a superscript to denote the iteration number. A reasonable initial value for the λ_i can be $\lambda_i^0 = \frac{\Theta_i}{1-u_i}$ since we must have $\lambda_i \geq \frac{\Theta_i}{1-u_i}$ from formula (6.38). Then, a very simple fixed-point iteration to determine the λ_i might be

$$\lambda_i^k = \lambda_i^{k-1} \frac{u_i}{\rho_i(\Lambda^{k-1})}, \quad i = 1, \ldots, N. \tag{6.39}$$

In the above formula, we simply increase the rate λ_i if the utilization of the device i is below the target value u_i, and decrease this rate if the utilization exceeds the target value. One can also determine the values for the λ_i using different methods; the point is it is not very difficult to do.

Closing this parenthesis regarding the determination of the model parameters from the quantities assumed known, we turn our attention to the actual solution of the model with non-preemptive priorities shown in Figure 6.8. This type of priority is not included in the BCMP networks so that the exact solution is not known for our model. The system has N priority levels corresponding to the N devices, i.e., classes of requests. With three or less priority levels, one certainly can solve the balance equations numerically, but for a larger number of devices sharing the bus, a good approximation may be in order.

If we focus on priority level i, $i = 2, \ldots, N-1$, we must account for requests at priority levels both higher and lower than i since both affect the selected focus level. A simple idea is then to aggregate devices with priority higher than i into a single priority level, and similarly for devices with priority lower than i. This leads to a system with three priority levels. We use the subscripts h and l, respectively, to refer to priority levels higher and lower than the selected level. We denote by n_i ($n_i = 0, 1$) the current number of requests of class i in the system (waiting or using the bus) by n_h ($n_h = 0, \ldots, i-1$) and n_l ($n_l = 0, \ldots, N-i$), respectively, the numbers of higher and lower-priority requests in the system. Let $s = i, h, l$ be the priority of the device currently using the bus, except when the bus is idle ($n_i = n_h = n_l = 0$). We use the vector $(n_i,\ n_h, n_l, s)$ to describe the state of the system in the steady state, and we denote by $p(n_i, n_h, n_l, s)$ the corresponding steady-state probability that there are n_i requests at level i, n_h requests of higher priority, n_l requests of lower priority, and the device currently using the bus is of priority s. The rates with which n_i increases and decreases are known: λ_i and μ_i. To be able to write the balance equations for the selected state description, we need also the conditional rates of increase and decrease for n_h, as well as for n_l.

For n_h, the conditional rate of increase, which we denote by $\alpha_h(n_i, n_h, n_l, s)$ is given by

$$\alpha_h(n_i, n_h, n_l, s) = \sum_{j=1}^{i-1} [1 - \overline{m}_j(n_i, n_h, n_l, s)]\lambda_j,$$

where $\overline{m}_j(n_i, n_h, n_l, s)$ is the conditional expected number of class j requests in the system given that the state is (n_i, n_h, n_l, s).

The conditional rate of decrease can be expressed as

$$\beta_h(n_i, n_h, n_l, s) = \sum_{j=1}^{i-1} q_j(n_i, n_h, n_l, s)\mu_j,$$

where $q_j(n_i, n_h, n_l, s)$ denotes the conditional probability that a higher (than i) priority device j is using the bus given n_i, n_h, n_l, and s.

Similarly, for the number of requests on lower priority levels, n_l, the rate of increase is given by

$$\alpha_l(n_i, n_h, n_l, s) = \sum_{j=i+1}^{N} [1 - \overline{m}_j(n_i, n_h, n_l, s)]\lambda_j,$$

and the rate of decrease can be expressed as

$$\beta_l(n_i, n_h, n_l, s) = \sum_{j=i+1}^{N} r_j(n_i, n_h, n_l, s)\mu_j,$$

where $r_j(n_i, n_h, n_l, s)$ is the conditional probability that a lower-priority device j is using the bus.

We note that the rate of decrease in n_h only matters when $s = h$, and analogously for n_l, the decrease is only possible when $s = l$. As an approximation, we assume that quantities pertaining to higher-priority classes depend only on n_h, and those pertaining to lower-priority classes depend only on n_l. With our focus on class i, we denote by $N_h = i - 1$ the total number of higher-priority devices, and by $N_l = N - i$ the total number of lower-priority devices. Since we know the server utilization by each device, a simple approximation is

$$q_j(n_h, s = h) \approx u_j \Big/ \sum_{k=1}^{i-1} u_k, \quad n_h = 1, \ldots, N_h \quad \text{and}$$

$$r_j(n_l, s = l) \approx u_j \Big/ \sum_{k=i+1}^{N} u_k, \quad n_l = 1, \ldots, N_l. \qquad (6.40)$$

To approximate $\alpha_h(n_h)$, we postulate an arrival rate of the form

$$\alpha_h(n_h) = (N_h - n_h)\gamma_h. \tag{6.41}$$

We can set the value of the fictitious unitary rate of arrivals γ_h so as to maintain the correct expected rate of arrivals for higher-priority requests using the following relationship

$$\left\{N_h - \sum_{j=1}^{i-1}(E[N_{qj}] + u_j)\right\}\gamma_h = \sum_{j=1}^{i-1}\Theta_j. \tag{6.42}$$

Similarly, we approximate $\alpha_l(n_l)$ as

$$\alpha_l(n_l) = (N_l - n_l)\gamma_l. \tag{6.43}$$

Again, γ_l, the fictitious unitary rate of arrivals can be set so as to maintain the correct expected rate of arrivals for lower-priority requests using the relationship

$$\left\{N_l - \sum_{j=i+1}^{N}(E[N_{qj}] + u_j)\right\}\gamma_l = \sum_{j=i+1}^{N}\Theta_j. \tag{6.44}$$

The expected number of requests waiting to use the bus, $E[N_{qj}]$, are not known until we solve our models for priority level $i = 2, \ldots, N-1$, while at the same time, we need them to solve these models. The way around this problem is a fixed-point iterative scheme where we solve the set of $N-2$ $(i = 2, \ldots, N-1)$ three-level models until the values of $E[N_{qj}]$ stabilize. As it happens, in practice, this tends to occur in a small number of iterations.

The balance equations for $p(n_i, n_h, n_l, s)$ can be derived using our standard "rate of flow out = rate of flow in" principle. Together with the normalizing condition $\sum_{n_i=0}^{1}\sum_{n_h=0}^{N_h}\sum_{n_l=0}^{N_l}\sum_s p(n_i, n_h, n_l, s) = 1$, these balance equations form a system of linear equations of moderate dimensions, which can be solved numerically.

In the case of our model, the class aggregation approximation described above appears to exhibit good accuracy. We borrow an example from the original paper (Brandwajn, 2002). In this example, there are four devices on the bus. The mean transfer lengths are

Table 6.5. Example of the accuracy of class aggregation approximation.

	Simulation		Class aggregation appr.	Relative error (%)
	Lower bound	Upper bound		
(a) *Mean waiting time*				
Class 1	0.681	0.767	0.708	−2.21
Class 2	1.267	1.365	1.304	−0.91
Class 3	1.736	1.834	1.758	−1.51
Class 4	2.135	2.239	2.148	−1.78
(b) *Mean number in system*				
Class 1	0.247	0.253	0.249	−0.40
Class 2	0.195	0.201	0.197	−0.51
Class 3	0.177	0.185	0.179	−1.10
Class 4	0.177	0.183	0.178	−1.11

64 Kbytes, 32 Kbytes, 16 Kbytes and 8 Kbytes for devices 1, 2, 3 and 4, respectively. The bus data rate is taken to be 20 MB/s, and the bus setup and cleanup overhead, added to the bus use time, are assumed to be 0.5 ms each per I/O operation. The attained I/O rates Θ_i are kept the same for all four devices and are such that the total bus utilization is approximately 50%. Table 6.5 compares the results obtained using the class aggregation approximation with those of a discrete-event simulation. Simulation results are reported as confidence intervals at 95% confidence level, estimated using the independent replications method.

The percentage values show the relative difference between midpoints of the simulation results and the values produced by the class aggregation approximation. As we mentioned before, the downside of this approach is that it reports approximation "errors" even when the approximation results fall inside the simulation confidence intervals. Be it as it may, the accuracy of the approximation seems very good despite the simplicity of the approximation in formulas (6.40), (6.41), and (6.43). It seems likely that it might be in part due to the attained throughput approach, which forces the correct server utilizations and I/O rates, i.e., class throughputs.

6.4 Partial state description to reduce complexity

An obvious objective of approximation methods is to reduce the difficulty or complexity of the problem at hand. One class of approximations accomplishes this objective through the use of partial state description. Let us consider the single-server queue shown in Figure 6.9.

The system features a phase-type distribution of times between arrivals and a phase-type distribution of the service times at the single server. The system capacity is limited to a total of K customers, and arrivals that find the system at capacity are simply lost. Let $n = 0, \ldots, K$ denote the current number of customers in system. We assume that the times between arrivals are represented as an acyclic phase-type distribution with a phases. We denote by τ_j the probability that the arrival process starts in phase j, $j = 1, \ldots, a$, by λ_j the completion rate of phase j, by r_{jl} $(l > j)$ probability that phase j is followed by phase l, and by $\hat{r}_j$ the probability that the arrival process terminates after phase j. The service time distribution is represented as an acyclic phase-time distribution with b exponential phases. We denote by σ_i the probability that the service process starts in phase i, $i = 1, \ldots, b$, by μ_i the intensity (completion rate) of phase i, by q_{ik} $(k > i)$ probability that phase i is followed by phase k, and by $\hat{q}_i$ the probability that the customer service completes after phase i.

The Ph/Ph/1/K queue described above is a quasi-general model of a single-server device with uncorrelated arrivals and service times. As such, it has a number of potential applications, for example to model I/O devices. Clearly, it would be interesting to have a fast and accurate solution for this model. The system state is fully described

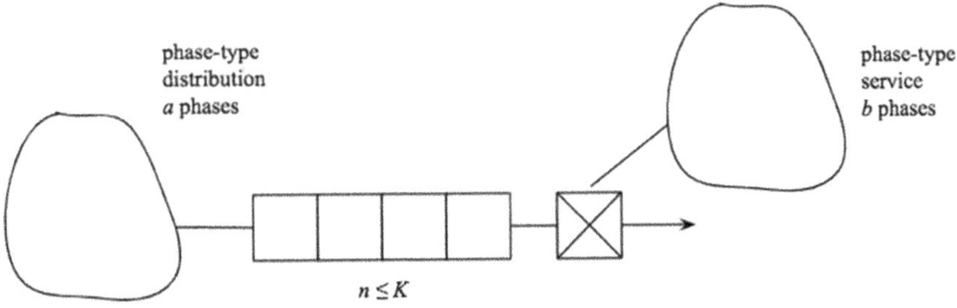

Figure 6.9. Ph/Ph/1/K queue.

by the vector (n, i, j) where j is the current phase of the arrival process, i is the current phase of the service process for the customer in service, and n is the current number of customers in the system (of course, i is meaningless if there are no customers in the system). We consider the system in equilibrium and we denote by $p(n, i, j)$ the corresponding steady-state probability.

It would be nice if there was a simple recurrent solution to our model similar to the recurrence for the Ph/M/1/K-like queues. Unfortunately, this doesn't seem to be the case for $p(n, i, j)$ directly. However, a simple approximation can be developed as follows. Let us consider only the probability that the current number of customers is n and that the arrival process is in its phase j, which we denote by $p_1(n, j)$, $j = 1, \ldots, a; n = 0, \ldots, K$. We also denote by $p(i|n, j)$ the conditional probability that the service is in phase i given (n, j), $i = 1, \ldots, a$; $n > 0$. It is clear that the rate with which n decreases can be expressed as

$$u_1(n, j) = \sum_{i=1}^{b} p(i|n, j)\mu_i \hat{q}_i, \quad \text{for } n > 0. \tag{6.45}$$

As an approximation, assume that the current phase of service depends primarily on n, the number of customers in the system, and not on the phase of the arrival process, j. Consequently, we get

$$u_1(n, j) \approx u(n) = \sum_{i=1}^{b} p(i|n)\mu_i \hat{q}_i, \quad \text{for } n > 0. \tag{6.46}$$

In formula (6.46) $p(i|n)$ is the conditional probability that the service phase is i given that there are n customers in the system. With this approximation, if we knew $u(n)$, we could solve for the probability $p_1(n, j)$ using the recurrence formulas (2.63) and (2.64). If you recall, this recurrence also produces the conditional rate of customer arrivals, denoted by $\alpha(n)$ and given by formula (2.65).

Now, let us consider only the probability that there are n customers in the system and that the current service phase is i, which we denote by $p_2(n, i)$, $i = 1, \ldots, b$; $n = 0, 1, \ldots, K$ (i is meaningless for $n = 0$). We also denote by $p(j|n, i)$ the conditional probability that the phase of the arrival process is j given (n, i). The rate with which arrivals happen (and n increases if $n < K$) can be expressed as

$$\alpha_2(n,i) = \sum_{i=1}^{a} p(j|n,i)\lambda_j \hat{r}_j. \tag{6.47}$$

As an approximation, we assume that the current phase of the arrival process depends less on the phase of service of the user being served than on the number of customers in the system so that

$$\alpha_2(n,i) \approx \alpha(n) = \sum_{i=1}^{a} p(j|n)\lambda_j \hat{r}_j, \quad \text{for } n = 0, \ldots, K. \tag{6.48}$$

With the help of this approximation, if we knew $\alpha(n)$, we could solve for the probability $p_2(n,i)$ using a simple recurrence akin to the one used for the Ph/M/1/K queue. The outline of such a simple recurrent solution for the M/Ph/1/K queue is given in Section 2.8.

Returning now to our initial Ph/Ph/1/K model, with the approximation (6.46), the model can be viewed as a Ph/M/1/K-like queue, provided we know the conditional rate of customer departures $u(n)$. The recurrent solution of the Ph/M/1/K queue produces the conditional rate of customer arrivals $\alpha(n)$. On the other hand, with the approximation (6.48), our Ph/Ph/1/K model can be viewed as an M/Ph/1/K-like queue, provided we know the conditional rate of arrivals $\alpha(n)$. The recurrent solution of the M/Ph/1/K queue (given in Chapter 2 in formulas (2.66–2.77)) produces the conditional rate of completions $u(n)$. We wind up with two models, each needing the results of the other one. A fixed-point iteration is a preferred "go-to" solution in such a case. Let us use a superscript to denote the iteration number. We can start with an initial value for $u(n)$, $u^0(n)$, $n = 1, \ldots, K$. A reasonable choice is to set $u^0(n)$ to the inverse of the mean service time for all $n > 1$. Then, at iteration $k = 1, \ldots,$ we solve the Ph/M/1/K model with $u^{k-1}(n)$. This solution yields the set of $\alpha^k(n)$, which we use to solve the M/Ph/1/K model, which, in turn, yields $u^k(n)$. We continue iterating like that until the values for the conditional rate of customer completions stabilize, i.e., for example, $\left|1 - \frac{u^{k-1}(n)}{u^k(n)}\right| < \varepsilon \; \forall n > 0$ for some reasonable value of ε (for instance, $\varepsilon = 10^{-5}$). Figure 6.10 summarizes such a fixed-point iteration. Again, in principle, one would need a theoretical proof of

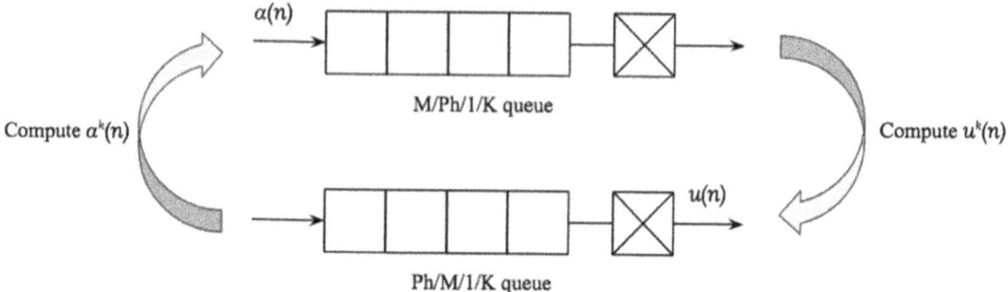

Figure 6.10. Ph/M/1/K-M/Ph/1/K approximation for the Ph/Ph/1/K queue.

convergence to a unique fixed point. In practice, the iteration, summarized in Figure 6.10, is easy to implement and tends to converge within a small number of iterations.

We refer to the type of approximation described above as a reduced-state approximation since it relies on the use of only a selected part of the full state description. As we will see later in this book, the reduced-state approach has been successfully used to derive fast approximations for several difficult models (Brandwajn & Begin, 2016, 2017, 2019, 2023). At the same time, the Ph/Ph/1/K model considered here uses a fixed-point iteration between two simpler models to obtain an approximate solution of a more involved model.

The approximation described above produces exact results if the service times are exponential and arrivals come from a Poisson source. That's a nice feature, but how accurate is the approximation when the times between arrivals and the service times are not exponentially distributed? Table 6.6 shows an example of the accuracy for a model with $K = 20$, the mean service time of 1, the distribution of the time between arrivals consisting of a two-phase hyper-exponential with a coefficient of variation of 3, and the service time distribution consisting of a two-phase hyper-exponential with a coefficient of variation of 2. We have included in our table the mean number of customers in the system, as well as the mean time a customer spends in the system (mean sojourn time), as the latter performance measure takes into account the attained customer throughput.

The results are shown at different load levels, defined by the mean rate of customers arriving at the system, including customers lost due to the system being at capacity. In addition to the actual exact and approximate values, we have included the relative errors and the

Table 6.6. Example of accuracy of the reduced-state approximation for the Ph/Ph/1/K model.

Arrival rate	Exact	Reduced-state appr.	Relative error (%)	Iterations
(a) *Mean number in system*				
0.2	0.5493	0.5623	2.31	10
0.4	2.2052	2.2802	3.29	17
0.6	4.6400	4.7480	2.27	22
0.8	7.0704	7.1257	0.78	24
1	9.1792	9.1409	−0.42	24
1.2	10.9122	10.7820	−1.21	22
1.4	12.3052	12.1026	−1.67	25
(b) *Mean sojourn time*				
0.2	2.7467	2.8118	2.32	10
0.4	5.5719	5.7674	3.39	17
0.6	8.2177	8.4544	2.80	22
0.8	10.2067	10.3889	1.75	24
1	11.7057	11.8019	0.82	24
1.2	15.4458	15.4597	0.09	22
1.4	19.3207	19.2368	−0.44	25

number of iterations needed for the fixed-point iteration to converge with $\varepsilon = 10^{-8}$. Clearly, in the example shown in Table 6.6, the relative errors are quite small, and the number of iterations is modest. The gains in execution times depend, of course, on the size of the full state space (n, i, j). The more phases in the arrival and the service distributions, and the larger the system capacity K, the bigger the gain, potentially reaching several orders of magnitude.

6.5 Approximate mean value analysis

In the models discussed so far in this chapter, the need for approximation arises from the fact that the exact solution for the model is not known. There are cases, however, when the exact analytical solution is known and people still resort to approximations. This is the case, for example, with closed Jacksonian or BMCP networks and the Mean Value Analysis (which, by the way, was introduced as a simpler and faster alternative to computing the normalizing constant in such networks). Let's consider again the Central Server model of Figure 3.4. Recall that the model comprises a Central Server and L peripheral servers, and there are n jobs distributed among these $L+1$ servers. We use the same notation as in Chapter 3. Letting $\varphi_0 = 1/(1-p_0)$ for the central server and $\varphi_i = p_i//(1-p_0)$ for the peripheral servers $i = 1, \ldots, L$, for our system with population level n, we have the following relationships:

$$E[W_i(n)] = E[N_i(n-1)]t_i + t_i, \quad i = 0, 1, \ldots, L, \tag{6.49}$$

$$E[W(n)] = E[W_0(n)]\varphi_0 + \sum_{i=1}^{L} \varphi_i E[W_i(n)], \tag{6.50}$$

$$\Theta(n) = \frac{n}{E[W(n)]}, \tag{6.51}$$

$$\theta_0(n) = \Theta(n)/(1-p_0) \quad \text{and}$$

$$\theta_i(n) = \varphi_i \Theta(n), \quad i = 1, \ldots, L, \tag{6.52}$$

$$E[N_i(n)] = E[W_i(n)]\theta_i(n), \quad i = 0, 1, \ldots, L. \tag{6.53}$$

As discussed before, formula (6.49) comes from the state found upon arrival in Jacksonian networks. It is the basis for the recurrent computation in the Mean Value Analysis, starting from population level $j = 1$. The problem is that, for a very high number of jobs in the system (say, thousands) and a large number of devices (peripheral servers), this represents a large number of computations. This leads to the idea to approximate $E[N_i(n-1)]$ so as to avoid this large number of computations. A simple approximation is (Bard, 1980; Schweitzer, 1981)

$$E[N_i(n-1)] \approx \frac{n-1}{n} E[N_i(n)], \quad i = 0, 1, \ldots, L. \tag{6.54}$$

With the help of this approximation, the set of equations (6.49–6.53) involves only quantities pertaining to the desired population level n. Obviously, the mean number of jobs $E[N_i(n)]$ is not known so that we use a fixed-point iteration (yet again) to solve this set of equations. In practice, we start with a feasible set of values for $E[N_i(n)]$, $i = 0, 1, \ldots, L$, such that $\sum_{i=0}^{L} E[N_i(n)] = n$, and iterate on the set of equations (6.49) (with approximation (6.54) to (6.53)) until the values of the mean numbers of jobs at each server stabilize:

$$E[W_i(n)] = \frac{n-1}{n} E[N_i(n)]t_i + t_i, \quad i = 0, 1, \ldots, L, \tag{6.55}$$

$$E[W(n)] = E[W_0(n)]\varphi_0 + \sum_{i=1}^{L} \varphi_i E[W_i(n)], \tag{6.56}$$

$$\Theta(n) = \frac{n}{E[W(n)]}, \tag{6.57}$$

$$\theta_0(n) = \Theta(n)/(1 - p_0) \quad \text{and}$$

$$\theta_i(n) = \varphi_i \Theta(n), \quad i = 0, \ldots, L, \tag{6.58}$$

$$E[N_i(n)] = E[W_i(n)]\theta_i(n), \quad i = 0, 1, \ldots, L. \tag{6.59}$$

This approximate mean value analysis replaces n solutions of the set of equations (6.49–6.53) by x solutions of the modified set of equations, where x is the number of iterations needed to converge to a fixed point with some reasonable degree of stringency. Of course, this approximation makes sense only if $x \ll n$ and the accuracy of the results is satisfactory.

As a very simple example, let's consider again Scherr's model in Figure 3.9. Recall that in this model, transactions are generated by a set of N_t exponential sources. The mean idle time of a source is $t_t = 1/\alpha$. There is a single server with a mean service time $t_s = 1/\mu$ to process the transactions in FCFS order and the service time is exponentially distributed. We can easily apply Mean Value Analysis to this model. Referring to the single server in this network we have, at population level $n = 0, \ldots, N_t$ $E[W_s(n)] = E[N_s(n-1]t_s + t_s$, $\Theta(n) = \frac{n}{E[W_s(n)]+t_t}$, and $E[N_s(n)] = E[W_s(n)]\Theta(n)$. The mean time spent at the source on each transaction is always t_t so we don't need to expend much energy on this part of the model.

In the case of approximate MVA, we set $E[N_s(n-1)] \approx \frac{n-1}{n}E[N_s(n)]$ with $n = N_t$. If we use $N_t = 1000$, $t_t = 10,000$, and for the mean service time $t_s = 5$, the approximate MVA yields 0.9971 for the expected number of customers at the server versus the exact value of 0.9960. For a mean service time $t_s = 7$, we get 2.3097 versus the exact value of 2.2982, and for $t_s = 9$, the approximate value is

8.2349 versus 7.8244 from the exact MVA. We used a starting value of 1 for the mean number of customers at the server. The respective numbers of iterations in the approximate MVA were 4, 21, and 57 for a straightforward, simple-minded implementation. Clearly, the approximation in this example is quite good and comes at a cost much lower than 1,000 exact MVA steps.

The term approximate mean value analysis has been used in other contexts to denote approximations derived by assuming that the theorem of the state upon arrival in closed queueing networks holds approximately in specific systems that don't possess a product-form solution (Bryant *et al.*, 1984; Eager & Lipscomb, 1988; Zahorjan *et al.*, 1988). We take a closer look at one specific case in the following chapter.

6.6 The X-model approach

As mentioned before, oftentimes, the reason for resorting to approximations is the desire to "divide and conquer", i.e., to decompose a complex model into simpler, more manageable sub-models. Suppose, for example, that we are dealing with an involved model of a larger system where, on the CPU side, we have many classes of processes with a number of priority levels, and on the I/O subsystem side, there is a large number of storage devices with their cache controllers. Figure 6.11 shows a high-level view of such a system.

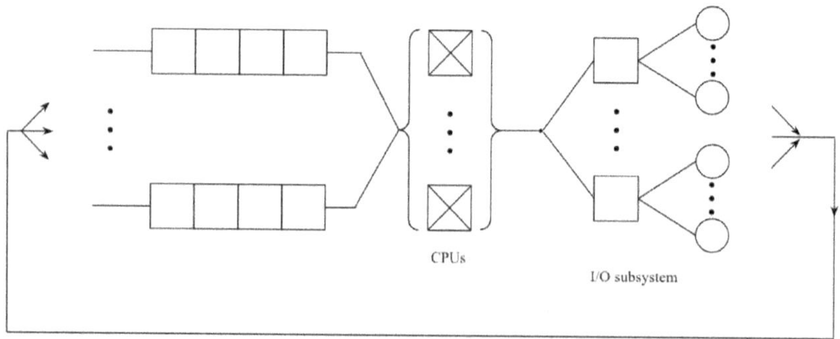

Figure 6.11. A complex model of a large system.

Clearly, especially with a large number of jobs (transactions) in the system, solving such a model is a daunting task. Things could be simplified significantly if we could, somehow, separate the CPU part and the I/O part. In particular, I/O subsystem analysis for a given attained I/O rate is a well-established practice, yielding the expected total I/O times for each device. The CPU side becomes considerably easier to analyze if the I/Os are represented only as pure delays corresponding to the mean time a job spends in the I/O subsystem. This, in a nutshell, is the "X-model" approximation (Herzog, 1974), in which one iterates between the solution of the CPU part with pure-delay I/Os so as to obtain the attained I/O throughput and the solution of the I/O subsystem for that throughput, so as to obtain the mean I/O times. As in many of the approximations discussed in this chapter, the iteration continues until a fixed point is reached. Figure 6.12 summarizes this approach.

The link between this approximation and the equivalence and decomposition approach has been explored in the literature (Brandwajn, 1985), where it is shown that, instead of the state-dependent rates of I/O completion in the equivalence and decomposition, the X-model uses in essence the "average" of such completion rate.

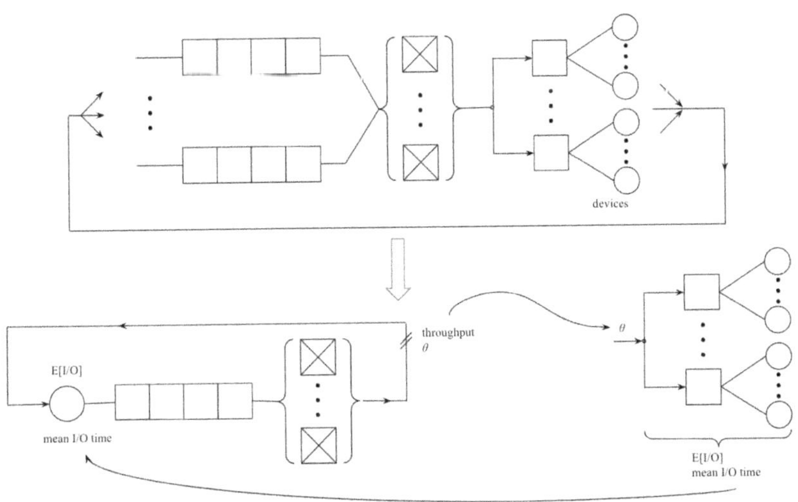

Figure 6.12. Example of X-model approach.

6.7 Function of the average instead of the average of a function

As is apparent by now, an important motivation behind a number of approximate solutions is the desire to simplify the development of models of complex systems. With that respect, an interesting approach was proposed some time ago in the context of modeling large systems (Brandwajn, 1982; Lazowska & Zahorjan, 1982). As an example, consider the system shown in Figure 6.13.

The system has L classes of customers generated by L distinct sets of sources of requests (transactions, jobs). We denote by N_i the number of sources for customers of class ($i = 1, \ldots, L$). The maximum number of customers of class i admitted into the processing part of the system is limited to M_i. We denote by n_i the current total number of customers of class i in the system (including in the admission queue), and by m_i the current number of customers of this class in the processing part of system. Since the maximum number of class i customers admitted into the processing part of the system is M_i, we must have at all times $m_i = \min(n_i, M_i)$, where $n_i = 0, \ldots, N_i$. Clearly, the number of customers of class i awaiting admission is given by $k_i = n_i - m_i$. Let's denote by $1/\alpha_i$ the mean time it takes for an idle source of class i to generate a new transaction. An idle source is one without a transaction currently in the

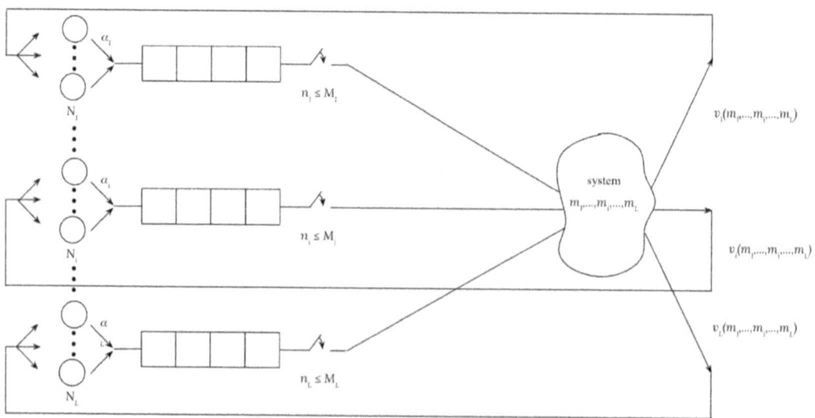

Figure 6.13. A large multiprogramming system with L classes of customers.

system. We assume that the times to generate a new transaction are exponentially distributed.

Let's describe the state of the system by the set of vectors $(n_1, \ldots, n_L)$, which is the same as describing the system by the set of couples $((k_1, m_1), \ldots, (k_L, m_L))$. Without being too specific about the internals of the processing part of the system, we assume that the conditional rate with which customers of class i $(i = 1, \ldots, L)$ complete the processing of a transaction is given by $v_i(m_1, m_2, \ldots, m_L)$. If we apply standard decomposition to evaluate — possibly approximately — the conditional rates $v_i(m_1, m_2, \ldots, m_L)$, we will need to solve the model of the processing part of the system (the inner model) for all possible vectors $(m_1, m_2, \ldots, m_L)$. Recall that in this type of decomposition, conditional rates are evaluated by maintaining the condition constant, as discussed in Section 6.1. For a large system with hundreds or thousands of transactions in the processing part of the system, this can be a significant computational effort.

Knowing these conditional rates, we could generate the appropriate balance equations and solve the equivalent (outer) model for the joint steady-state probability $p(n_1, \ldots, n_L)$. The number of states to consider here might be even larger than for all possible vectors $(m_1, m_2, \ldots, m_L)$ if $N_i > M_i$. Luckily, in reality, we are rarely, if ever, interested in the joint probability of configurations of transactions in the system. This means that we would be virtually always satisfied if we could obtain the steady-state probability $p_i(n_i)$ that there are n_i transactions of class i in the system, $i = 1, \ldots, L$, individually for each customer class. To obtain this probability, we will need a conditional rate of completion of transactions for class i customers, which we denote by $\hat{v}_i(m_i)$. In a large number of systems, it seems reasonable to assume that, for a given class, the number of customers of this class in the processing part of the system matters most and that the influence of customers of other classes can be represented through their mean numbers (Brandwajn, 1982) (or the most likely numbers (Lazowska & Zahorjan, 1982)). This leads to the following approximation:

$$\hat{v}_i(m_i) \approx v_i(\hat{m}_1, \ldots, m_i, \ldots, \hat{m}_L). \tag{6.60}$$

In formula (6.60), $\hat{m}_j$ denotes the integer closest to the mean number of customers of class j in the processing part of the system, i.e., the mean value of m_j. Alternatively, $\hat{m}_j$ might refer to the most probable value of m_j.

If we denote by $p(m_1, \ldots, m_{i-1}, m_{i+1}, \ldots, m_L|m_i)$ the conditional probability of the number of transactions of classes other than i in the processing part of the system given m_i transactions of class i, the exact value of $\hat{v}_i(m_i)$ can be expressed as

$$\hat{v}_i(m_i) = \sum_{m_1} \cdots \sum_{m_{i-1}} \sum_{m_{i+1}} \cdots \sum_{m_L} p(m_1, \ldots, m_{i-1}, m_{i+1}, \ldots, m_L|m_i) v_i \times (m_1, \ldots, m_L). \tag{6.61}$$

So, in a sense, formula (6.60) approximates a conditional average of a function by a function of average argument values.

With the help of the approximation in the formula (6.60), we can now devise a simple fixed-point iteration to solve our system. We start with some arbitrary but feasible values for the quantities $\hat{m}_j$, $j = 2, \ldots, L$. A possible choice might be, for instance, $\hat{m}_j = M_j/2$. At each iteration, we consider customer classes in the order $i = 1, \ldots, L$. For each class, we solve the simple equivalent model shown in Figure 6.14.

The solution of this model for class i produces the steady-state probability $p_i(n_i)$:

$$p_i(n_i) = \frac{1}{\mathrm{G}} \prod_{j=1}^{n_i} \frac{(N_i - j + 1)\alpha}{\hat{v}_i(m_i(j))}, \quad n_i = 0, \ldots, N_i; \quad m_i(j) = \min(j, M_i). \tag{6.62}$$

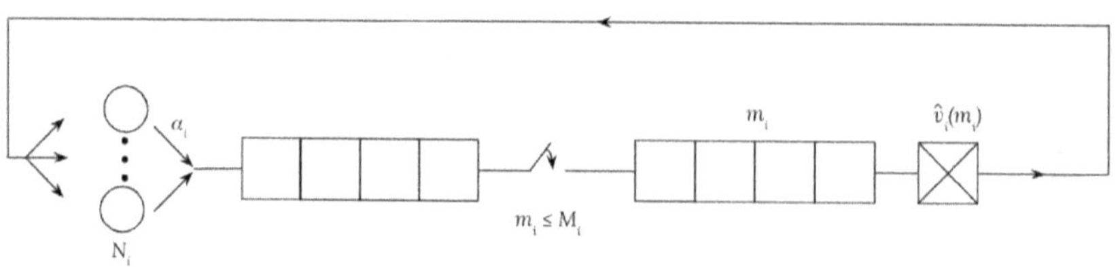

Figure 6.14. Equivalent model for customers of class i.

Knowing $p_i(n_i)$ we compute the probability that there are m_i customers of class i in the processing part of the system

$$p(m_i) = \begin{cases} p_i(n_i), & \text{if } m_i < M_i \\ \sum_{n_i=M_i}^{N_i} p_i(n_i), & \text{if } m_i = M_i \end{cases}. \tag{6.63}$$

The mean number of class i customers in the processing part of the system is given by $\sum_{m_i=1}^{M_i} m_i p(m_i)$. This gives us an updated value for $\hat{m}_i$ to use in the fixed-point iteration. Similarly, if we use the most likely value for $\hat{m}_i$, we would look for the value of m_i that maximizes $p(m_i)$. Our iterative procedure continues until the values for $\hat{m}_i$ stabilize to a pre-specified accuracy (or we exhaust some pre-set maximum number of iterations). Although there is no theoretical proof of convergence to a unique solution of this fixed-point iteration, it works well in practice, in particular for models of large systems. The number of iterations needed to attain a reasonable level of convergence tends to be quite moderate. To summarize, this approximation method replaces a solution of a problem with a potentially very large state space (which grows combinatorially with the number of transaction classes and the number of sources) by a set of solutions of much simpler models. This simplification comes at the price of the loss of the knowledge of joint behavior of customer classes, as well as some loss of accuracy.

6.8 Guided state sampling approximation to reduce complexity

If the maximum number of customers of each class in the system is large, we can oftentimes achieve substantial additional savings in computational effort using a "sampling" approach in decomposition. To explain the method, let's consider the model of a transaction processing system with admission control, similar to the one considered above, but with a single class of transactions depicted in Figure 6.15.

The number of transaction sources is denoted by N_t, the time for an idle source (not currently associated with a transaction) is

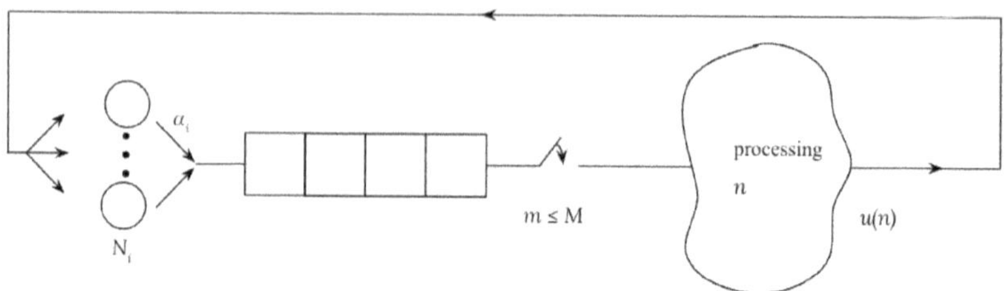

Figure 6.15. A transaction processing model with admission control and a single class of customers.

exponentially distributed with mean $1/\alpha$, and the number of customers, i.e., transactions, in the processing system is limited to M. As we know, in the equivalence and decomposition approach, we would evaluate the model of the processing part of the system (inner model) for all population levels $m = 1, \ldots, M$ in order to obtain the equivalent conditional rate of transaction completions given m, which we denote by $u(m)$. Since we make no special assumptions about the processing part of the system, in general, the decomposition would produce only approximate results. We would then use $u(m)$ to solve the equivalent model (similar to the model we used for each class in Figure 6.14) for the probability that there are n customers in the system $p(n)$.

In very many cases, the probability $p(n)$ is likely to concentrate within a subset of possible values of the number of customers n. This subset corresponds to the most probable "operating region(s)" of the system. This means that it is important to evaluate precisely the conditional rate $u(.)$ in the vicinity of the operating points of the system. However, farther from these points, it should be sufficient to provide a reasonable approximation for $u(.)$. Of course, in general, we don't know *a priori* where these most likely operating regions are. We can, however, design a simple iterative procedure that quickly finds the most important points. Thus, we solve the inner model for these points, and outside these points, we can use a simple linear approximation for the conditional rate $u(.)$. Specifically, we assume that the values of $u(m)$ have been evaluated for the points $m = 1$ and $m = M$, as well as at least two intermediate points, which define k intervals, $k \geq 3$. We denote by x_i and x_{i+1} the bounds of interval i, and we denote by c_i the chord through these two points. Outside of the known evaluation points, we approximate the function $u(m)$ by the cord c_i through the ending points of the (smallest) interval containing the point m, i.e., $u(m) = c_i(m)$ for $m \in (x_i, x_{i+1})$.

At each iteration of this procedure, we use the current values, obtained by evaluating the subnetwork or through our linear approximation, for the conditional rate $u(.)$ to compute the steady-state probabilities $p(n)$. Additional evaluation points are determined by the most likely values of n (and, hence, m) as indicated by the current values of $p(n)$. This procedure continues until no new points are added or until the distributions $p(n)$ from two consecutive iterations become sufficiently close. In essence, this procedure amounts to a sampling of the values of the conditional rate $u(.)$ guided by the values of the state probabilities. In practice, with this "guided sampling" (Brandwajn, 1998), the inner model tends to be solved for a small fraction of the total number of evaluations that would be required with classical decomposition. The guided sampling approximation can be used in conjunction with the "averaging" method in models with multiple customer classes described previously. Interestingly, the idea of this "sampling" approach was inspired by a testing technique in complex systems with very many possible states, where exhaustive testing might prove prohibitively costly, so that only most likely states are tested (Dimitrijevic & Chen, 1989).

6.9 Node-by-node analysis

We saw in Section 6.1 that the proper selection of the state description may be important in simplifying the solution of a model. Another example of the use of suitably selected state description to model a complex system comes from the performance analysis of optical packet switching bus-based networks with unslotted Carrier Sense Multiple Access with Collision Avoidance (CSMA/CA) protocols (Brandwajn *et al.*, 2009). We concentrate on the transmission bus, which we view as a unidirectional fiber connecting several access nodes (stations) to a so-called point of presence (POP) node. Figure 6.16 illustrates such a unidirectional bus.

In our system, all nodes share a single wavelength to transmit (and therefore compete for access to it). There is no packet-drop

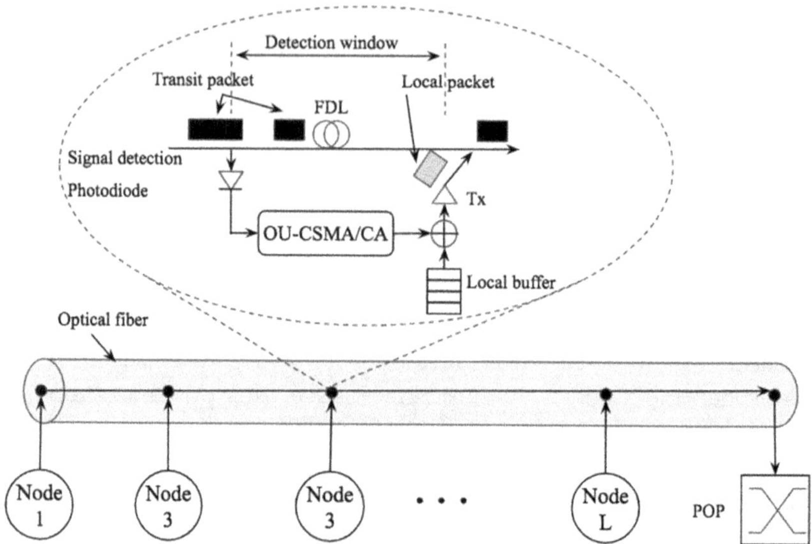

Figure 6.16. Unidirectional bus-based network.

operation at bus nodes, and each node can only insert local traffic into the wavelength but not intercept transit traffic of upstream nodes. Our goal is to analyze the performance of the access nodes on such a transmission bus. With the protocol used, a node begins transmitting a packet only if it detects a void, i.e., a free segment of bandwidth between transit packets from upstream nodes. This void must be at least equal to the size of the packet it wishes to transmit for the transmission to be successful. The most upstream node is always able to transmit since it has available bandwidth at all times, but downstream nodes have only the residual bandwidth left by upstream nodes.

Let's denote by L the number of access nodes sharing the transmission bus. A node with a packet to transmit begins the transmission as soon as it detects a void. The transmission is interrupted if a transit packet from an upstream node is detected before completion of the transmission. The interrupted local packet returns to the queue to wait for the next void. When the next void is detected, the node starts again transmitting the local packet, and this process is repeated until the packet is successfully transmitted. For performance modeling purposes, we elect to view the transmission of a packet in the protocol under consideration as some number of preempted, i.e., interrupted, transmission attempts followed by a successful transmission.

The behavior described above is relatively similar to that of a priority queueing system in which a single server (the shared wavelength) services L queues (the L bus nodes) with Preemptive-Repeat-Identical (PRI) priority discipline. Each access node in the system defines a separate priority level according to its position on the bus. This is the view we will adopt here to model the performance of the bus. We must point out, however, that the queueing system with PRI priority discipline does not match exactly the operation of the real network. In the queueing model, a lower level (downstream node) can start transmitting only if there is no client packet at a higher level (upstream nodes). This implies that the server viewed by a lower-level client remains occupied until all higher-level clients have been successfully served. In the real network, the bandwidth viewed by a downstream node is occupied only during the successful transmission periods of client packets at upstream nodes. This is because when an upstream node detects a void and the void is not long enough (the transmission attempt is interrupted), this void may be used by a downstream node to transmit smaller packets whose length fits this void. Thus, on intuitive grounds, we would expect the real system to be somewhat more egalitarian than the PRI priority model.

In our model, we number the nodes 1 through L, node 1 having the highest priority. We assume that each node has a buffer large enough to hold the client packets to be transmitted, and the packets stored in this buffer are served in First-Come-First-Served (FCFS) order. We also assume that local packets at node i arrive from a Poisson source with rate λ_i and that their service times are mutually independent and have a quasi-general distribution with known finite mean m_i and variance Var_i. The assumption of Poisson arrivals is reasonable if the incoming traffic represents an aggregation of traffic from many sources.

Our goal here is to analyze our model approximately, one priority level at a time. To model the service time distribution for packets at each of the nodes, we use a Coxian distribution (Cox, 1955) with the minimum number of stages required to match the known mean and variance. We denote by k the number of stages, as shown in Figure 6.17.

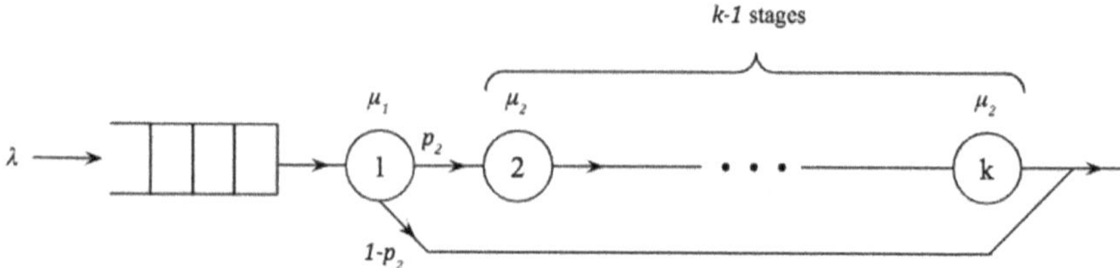

Figure 6.17. k-stage Coxian for service time distribution.

Recall that we encountered Coxian distributions, which are a special case of phase-type distributions, earlier in this text when we were discussing modeling non-exponential service times. The parameters of these Coxian distributions can be readily determined (Bux & Herzog, 1977; Faddy, 2002).

Before we outline the one-node-at-a-time approximate solution to our model, we need to dwell a little on the Preemptive Resume Identical aspect of the priority queue. As discussed above, the service at nodes $i > 1$ may be interrupted by arrivals of client packets at higher-priority nodes. This means that a node may need to attempt the transmission of a packet several times until a long enough void comes along and there is no interruption. Obviously, after each interrupted transmission, the node re-attempts the transmission of the same packet at the detection of next void. It is intuitively clear that smaller packets are less likely to be interrupted than longer packets. This means that unless the packets are all the same length, the distribution of the length of packets at each attempt will be different. Thus, somewhat paradoxically, to describe the fact that it is the same packet whose transmission is re-attempted, we need to represent a potentially different packet length (and service time) distribution at each consecutive attempt.

To explain what is going on, let us examine the case where the original service times are exponentially distributed with parameter μ, and interruptions arrive from a Poisson source with rate α as shown in Figure 6.18(a).

On the first transmission attempt, the service time distribution is the original distribution corresponding to the packet lengths of all client packets at the node considered. Here, this is an exponential distribution with parameter μ. This service may be interrupted by the Poisson source with rate α.

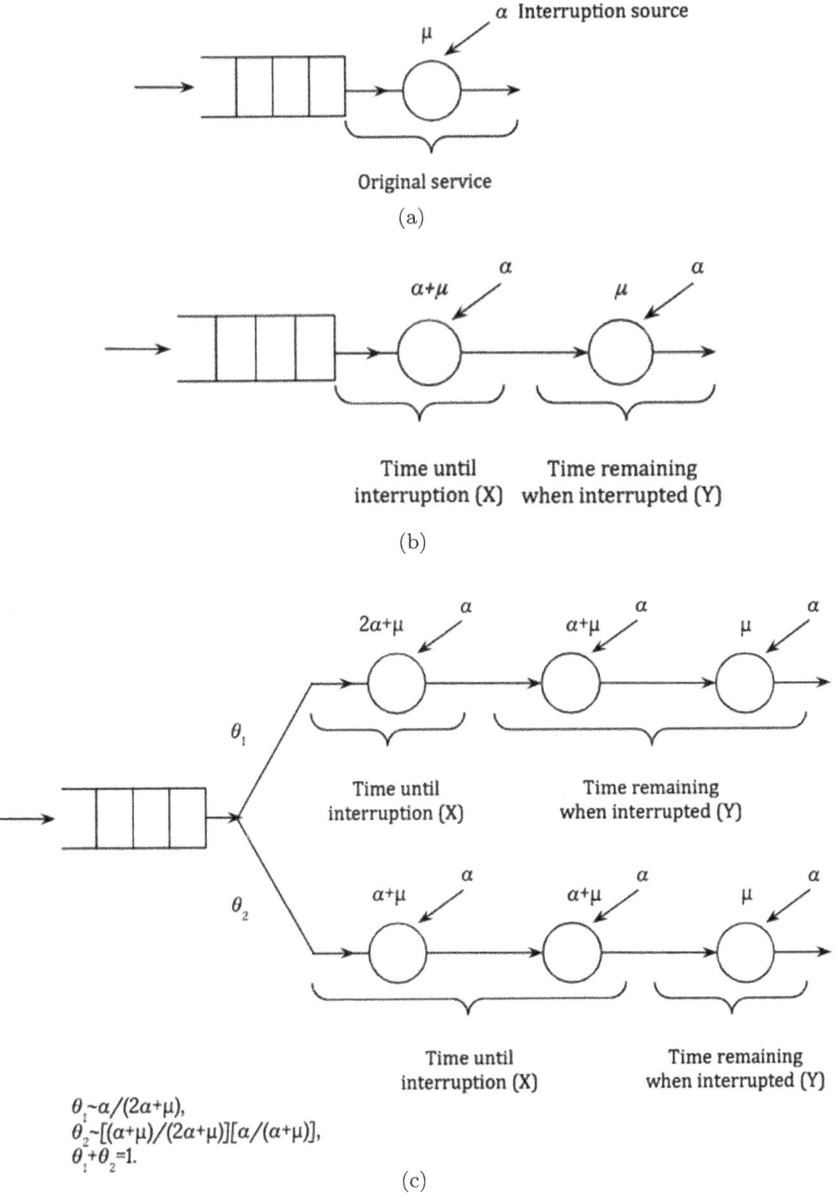

Figure 6.18. Evolution of service time distribution after interruption. (a) System with original service time distribution. (b) Service time distribution on the second transmission attempt. (c) Service time distribution on the third transmission attempt.

On the second attempt, what remains is a subset of client packets whose transmission got interrupted for the first time (in other words, we exclude all packets that managed to be successfully transmitted on the first attempt). To obtain the distribution of the service time of this subset of packets, let's consider the time until the interruption, X, and the time remaining when the interruption occurred, Y. Let's consider a small interval of time $(t, t + \Delta t]$. The probability that a first service interruption will happen during this interval is $\Delta t \alpha e^{-\alpha t} e^{-\mu t} + o(\Delta t)$ where $\lim_{\Delta t \to 0} o(\Delta t)/\Delta t = 0$. The overall probability that the first attempt will be interrupted is simply the probability that an interruption arrives before the service finishes. In our case, this yields $\frac{\alpha}{\alpha+\mu}$. Hence, the conditional density of the time to interruption given that the service is interrupted is $(\alpha+\mu)e^{-(\alpha+\mu)t}$. This in turn means that the time until interruption X is exponentially distributed with parameter $\alpha+\mu$. Due to the memoryless property of the exponential distribution, the remaining service time at the point of interruption Y is exponentially distributed with the original parameter μ. Therefore, the service time distribution of the subset of client packets on the second transmission attempt is the hypoexponential distribution shown in Figure 6.18(b).

On the third attempt, we exclude all client packets that were successfully transmitted on the first and second attempts, so that what remains is the subset of packets whose transmission was interrupted for the second time. This means that we are now dealing with the two-stage hypoexponential distribution discussed in Figure 6.18(b) interrupted by a Poisson process of rate α. The interruption in question might have taken place while the service was in the first or the second stage of the two-stage hypoexponential. Denote by θ_1 the probability that it was during the first stage. In this case, the time to interruption is exponentially distributed with parameter $2\alpha + \mu$. This is followed by an exponentially distributed residual of the first stage (parameter $\alpha + \mu$) and a full second stage (parameter μ). Let $\theta_2 = 1 - \theta_1$ be the probability that the interruption happened during the second stage of the two-stage hypoexponential of Figure 6.18(b). Then, the time before interruption consists of the full first stage (exponential with parameter $\alpha + \mu$), followed by the part of second stage that precedes the interruption, i.e., an exponential with parameter $\alpha + \mu$, and the time after the interruption is exponentially distributed with parameter μ (due to memoryless property of the exponential distribution).

The resulting distribution for packets on third attempt is shown in Figure 6.18(c).

An analogous reasoning can be used for subsequent interruptions. Similarly, we can derive the distribution of service times at each interruption for non-exponential distributions of the packet length. With the obvious exception of packets of constant length, the mean increases while the coefficient of variation decreases on each subsequent transmission attempt. This means that we need to keep track of the current attempt number and model the evolution of the service time for each node other than node 1.

To reduce the complexity of the problem, we try to analyze the nodes (i.e., the priority levels) one by one, starting with node 1 at which there are no interruptions. To keep the notation light, we omit the node subscript to identify the node in our discussion where it should not lead to confusion. Let's denote by n the current number of packets (customers) at this node and by ℓ the current stage of the Coxian service time distribution. Assuming the system reaches equilibrium, we can describe the steady-state behavior by the probability $p(n,\ell)$. Following the approach introduced in Section 2.6, we use the fact that $p(n,\ell) = p(n)p(n|\ell)$, where $p(n|\ell)$ is the conditional probability that the current stage is ℓ given n. With the particular form of Coxian distribution with a total of k stages shown in Figure 6.17, which we selected to match the first two moments, we can express the conditional rate of packet completions at node 1 as

$$u(n) = p(\ell = 1|n)\mu_1(1 - p_2) + p(k|n)\mu_k. \tag{6.64}$$

Since the arrivals come from a Poisson source with rate λ, we have for the steady-state probability for the number of packets at node 1

$$p(n) = \frac{1}{\mathrm{G}} \prod_{i=1}^{n} \frac{\lambda}{u(i)}, \quad n = 0, 1, \ldots. \tag{6.65}$$

We have no trouble deriving the balance equations for $p(n,\ell)$. With the help of formula (6.77), we transform them into the following set of equations for $p(\ell|n)$ when $n = 1$:

$$\begin{aligned} p(2|n)(\lambda + \mu_2) &= p(1|n)\mu_1 p_2, \\ p(\ell|n)(\lambda + \mu_\ell) &= p(\ell - 1|n)\mu_{\ell-1}, \quad \ell = 3, \ldots, k. \end{aligned} \tag{6.66}$$

We note that these equations allow us to express all probabilities $p(\ell|n = 1)$ in terms of $p(1|n = 1)$ as $p(\ell|1) = c(\ell)p(1|1)$ with $c(2) = \mu_1 p_2/(\lambda + \mu_2)$ and $c(\ell) = c(\ell - 1)\mu_{\ell-1}/(\lambda + \mu_\ell)$ for $\ell > 2$. The conditional probabilities must be normalized for each value of $n = 1, 2, \ldots$, i.e., we must have $\sum_{\ell=1}^{k} p(\ell|n) = 1$, $\forall n > 0$. This allows us to determine $p(1|1)$ as $p(1|1) = \left[1 + \sum_{\ell=2}^{k} c(l)\right]^{-1}$.

For $n > 1$, we get the following equations for $p(\ell|n)$.

$$p(2|n)(\lambda + \mu_2) = p(1|n)\mu_1 p_2 + p(2|n-1)u(n),$$
$$p(\ell|n)(\lambda + \mu_\ell) = p(\ell - 1|n)\mu_{\ell-1} + p(\ell|n-1)u(n), \quad \ell = 3, \ldots, k. \tag{6.67}$$

We consider these equations in the order of increasing n. The form of equations (6.67) suggests the following expression for its solution $p(\ell|n) = a(\ell)p(1|n) + b(\ell)u(n)$, where

$$a(2) = \mu_1 p_2/(\lambda + \mu_2); \quad a(\ell) = \mu_{\ell-1}/(\lambda + \mu_\ell) \quad \text{for } \ell > 2 \quad \text{and}$$
$$b(\ell) = p(\ell|n-1)/(\lambda + \mu_\ell) \quad \text{for } \ell > 1.$$

For $n = 2$, the values of $p(\ell|n - 1)$ have already been computed so that the coefficients $a(\ell)$ and $b(\ell)$ are known for $\ell > 1$. Let's define $a(1) = 1$ and $b(1) = 0$. To determine $p(1|n)$ and $u(n)$, we use the normalization condition and formula (6.76), i.e., $p(1|n)\sum_{\ell=1}^{k} a(\ell) + u(n)\sum_{\ell=1}^{k} b(\ell) = 1$ and $u(n) = p(1|n)\mu_1(1 - p_2) + [a(k)p(1|n) + b(k)u(n)]\mu_k$. We then move on to $n = 3$ and repeat the process (of course, the values of the coefficients a and b may change for each value of n). In theory, since there is no upper limit to the values of n in our model, there would be an infinite number of equations to solve. In practice, the conditional probabilities $p(\ell|n)$ and the conditional rate of packet transmission $u(\ell)$ tend to quickly reach limiting values as n increases. Thus, the solution for the highest priority node is reduced to a straightforward recurrence. Clearly, knowing $u(n)$, we can compute $p(n)$ and any derived performance measures.

For downstream nodes, we need to represent the presence of upstream nodes, i.e., preemptions of service caused by them. A convenient approach is to view preemptions as the server disappearing and reappearing. After all, from the perspective of a lower-priority level customer (packet) in service, this is what is happening. Let us denote by α_i and β_i the respective rates of disappearance and reappearance of the server seen by node $i > 1$. Since the service distribution may change at consecutive transmission attempts in our model (as shorter packets tend to be less interrupted than longer ones), to represent the Preemptive-Repeat-Identical discipline at node $i > 1$, we adopt a state description that explicitly accounts for service interruptions and retrials. We denote by k_j the number of exponential stages required to represent the service time distribution at the j-th transmission attempt, $j = 1, 2, \ldots$ We describe the state of node by the triple (n, j, ℓ) in which n is the current number of packets at the node (queued and in service), j is the transmission attempt, and ℓ is the number of the current service stage during this attempt $\ell = 1, \ldots, k_j$ or $\ell = 0$ if the service, i.e., the bandwidth, is unavailable.

In theory, in the model, we could have an infinite number of transmission attempts. To limit the size of the state space for each value of n, we explicitly compute the parameters of the service time distributions at transmission attempts up to a certain value $\hat{j}$. For transmission attempts $\hat{j}$ and above, we can use the network physical limit of the transmission time corresponding to the Maximum Transmission Unit of the protocol used. In other words, as an approximation, all service time distributions at attempts above $\hat{j}$ are replaced by a Coxian distribution with $\hat{k}$ stages (in other words, $\hat{k} = k_{\hat{j}}$). Incidentally, this means that the minimum value of the coefficient of variation of the service time is $1/\sqrt{\hat{k}}$ in our approximation, since this is the lowest value we can achieve with $\hat{k}$ exponential stages.

Let's denote by μ_ℓ^j the parameter of stage ℓ ($\ell = 1, \ldots, k_j$) of the Coxian representation of the service time at transmission attempt j, and by $\overline{q}_\ell^j$ the corresponding probability that stage ℓ will be followed by another service stage (clearly, for the last stage of the Coxian, this probability is 0). We let $q_\ell^j = 1 - \overline{q}_\ell^j$. We consider that attempt numbers increase just after a service interruption by a higher-priority customer. A new service must always start in stage 1 of the corresponding attempt. We obtain the following balance equations for $n > 1$ with the selected state description.

- Server unavailable at node:

$$p(n, j = 1, \ell = 0)(\lambda + \beta) = p(n-1, j = 1, \ell = 0)\lambda,$$

$$p(n, j, \ell = 0)(\lambda + \beta) = \sum_{\ell=1}^{k_{j-1}} p(n, j-1, \ell)\alpha + p(n-1, j.\ell = 0)\lambda,$$

$$j = 2, \ldots, \hat{\jmath} - 1,$$

- Server available at node, attempt $j = 1, \ldots, \hat{\jmath} - 1$:

$$p(n, j, \ell = 1)(\lambda + \mu_1^j + \alpha) = p(n, j, \ell = 0)\beta + p(n-1, j, \ell = 1)\lambda + \sum_{\ell=1}^{k_j} p(n+1, j, \ell)\mu_\ell^j q_\ell^j,$$

$$p(n, j, \ell)(\lambda + \mu_\ell^j + \alpha) = p(n, j, \ell - 1)\mu_{\ell-1}^j \overline{q}_{\ell-1}^j + p(n-1, j, \ell)\lambda, \quad \ell > 1.$$

- For the "catch all" attempt $\hat{\jmath}$:

$$p(n, \hat{\jmath}, \ell = 0)(\lambda + \beta) = \sum_{\ell=1}^{k_{\hat{\jmath}-1}} p(n, \hat{\jmath} - 1, \ell)\alpha + \sum_{\ell=1}^{\hat{k}} p(n, \hat{\jmath}, \ell)\alpha + p(n-1, \hat{\jmath}, \ell = 0)\lambda,$$

$$p(n, \hat{\jmath}, \ell = 1)(\lambda + \mu_1^{\hat{\jmath}} + \alpha) = p(n, \hat{\jmath}, \ell = 0)\beta + p(n-1, \hat{\jmath}, \ell = 1)\lambda + \sum_{\ell=1}^{\hat{k}} p(n+1, \hat{\jmath}, \ell)\mu_\ell^{\hat{\jmath}} q_\ell^{\hat{\jmath}},$$

$$p(n, \hat{\jmath}, \ell)(\lambda + \mu_\ell^{\hat{\jmath}} + \alpha) = p(n, \hat{\jmath}, \ell - 1)\mu_{\ell-1}^{\hat{\jmath}} \overline{q}_{\ell-1}^{\hat{\jmath}} + p(n-1, \hat{\jmath}, \ell)\lambda, \quad \ell > 1.$$

Let's denote by $p(n = 0, \ell = 0)$ and by $p(n = 0, \ell = \overline{0})$ the probability that there are no customers (packets) at the node and the server is unavailable and that it is available, respectively. The term for $n - 1$ vanishes in some of the equations for $n = 1$:

- Server unavailable at node:

$$p(n=1,j=1,\ell=0)(\lambda+\beta)=p(n=0,\ell=0)\lambda,$$

$$p(n=1,j,\ell=0)(\lambda+\beta)=\sum_{\ell=1}^{k_{j-1}}p(n=1,j-1,\ell)\alpha,$$

$$j=2,\ldots,\hat{j}-1.$$

- Server available at node, attempt $j=1,\ldots,\hat{j}-1$:

$$p(n=1,j,\ell=1)(\lambda+\mu_1^j+\alpha)=p(n=1,j,\ell=0)\beta+p(0,\ell=\overline{0})\lambda$$
$$+\sum_{\ell=1}^{k_j}p(n=2,j,\ell)\mu_\ell^j q_\ell^j,$$

$$p(n=1,j,\ell)(\lambda+\mu_\ell^j+\alpha)=p(n=1,j,\ell-1)\mu_{\ell-1}^j\overline{q}_{\ell-1}^j,\quad \ell>1.$$

- For the "catch all" attempt $\hat{j}$:

$$p(n=1,\hat{j},\ell=0)(\lambda+\beta)=\sum_{\ell=1}^{k_{\hat{j}-1}}p(n=1,\hat{j}-1,\ell)\alpha$$
$$+\sum_{\ell=1}^{\hat{k}}p(n=1,\hat{j},\ell)\alpha,$$

$$p(n=1,\hat{j},\ell=1)(\lambda+\mu_1^{\hat{j}}+\alpha)=p(n=1,\hat{j},\ell=0)\beta$$
$$+\sum_{\ell=1}^{\hat{k}}p(n=2,\hat{j},\ell)\mu_\ell^{\hat{j}}q_\ell^{\hat{j}},$$

$$p(n=1,\hat{j},\ell)(\lambda+\mu_\ell^{\hat{j}}+\alpha)=p(n=1,\hat{j},\ell-1)\mu_{\ell-1}^{\hat{j}}\overline{q}_{\ell-1}^{\hat{j}},\ \ell>1.$$

For $n=0$, we are only dealing with the availability of the server

$$p(n=0,\ell=0)(\lambda+\beta)=p(n=0,\ell=\overline{0})\alpha.$$

The conditional rate of customer completions at node $i>1$, given that there are n customers at this node, i.e., the conditional rate of packet transmissions in our model, can be expressed as

$$u(n) = \sum_{j\geq 1} \sum_{\ell=1}^{k_j} p(j,\ell|n)\mu_\ell^j q_\ell^j, \quad n = 1, 2, \ldots. \tag{6.68}$$

$p(j,\ell|n)$ is the conditional probability that the transmission attempt is j and the current service stage is ℓ given n. As was the case for node 1, the steady-state probability distribution for n can be expressed as

$$p(n) = \frac{1}{\mathrm{G}} \prod_{i=1}^{n} \frac{\lambda}{u(n)}. \tag{6.69}$$

G is a normalizing constant such that $\sum_n p(n) = 1$.

Following what we have done several times before, we can transform the balance equations into equations for the conditional probabilities $p(j,\ell|n)$ using the identity $p(n,j,\ell) = p(j,\ell|n)p(n)$ together with formula (6.69). These conditional probabilities are normalized for every value of n, which means that we must have $\sum_{j\geq 1}\sum_{\ell=0}^{k_j} p(j,\ell|n) = 1$ for $n > 0$ and $\sum_\ell p(\ell|n=0) = 1$ for $n = 0$. This independent normalization for each n helps us solve our model. From the balance equation for $n = 0$, we readily get

$$p(\ell = 0|n = 0) = \alpha/(\alpha + \beta + \lambda). \tag{6.70}$$

For $n = 1$, skipping equations for $\ell = 1$, we obtain the following equations for the conditional probabilities:

- Server unavailable at node:

$$p(j = 1, \ell = 0|n = 1)(\lambda + \beta) = p(n = 0, \ell = 0)u(1),$$

$$p(j, \ell = 0|n = 1)(\lambda + \beta) = \sum_{\ell=1}^{k_{j-1}} p(j - 1,\ell|n = 1)\alpha,$$

$$j = 2, \ldots, \hat{\jmath} - 1.$$

- Server available at node, attempt $j = 1, \ldots, \hat{\jmath} - 1$:

$$p(j,\ell|n = 1)(\lambda + \mu_\ell^j + \alpha) = p(j, \ell - 1|n = 1)\mu_{\ell-1}^j \overline{q}_{\ell-1}^j, \quad \ell > 1.$$

- For the "catch all" attempt $\hat{\jmath}$:

$$p(\hat{\jmath}, \ell = 0|n = 1)(\lambda + \beta) = \sum_{\ell=1}^{k_{\hat{\jmath}-1}} p(\hat{\jmath} - 1, \ell|n = 1)\alpha + \sum_{\ell=1}^{\hat{k}} p(\hat{\jmath}, \ell|n = 1)\alpha,$$

$$p(\hat{\jmath}, \ell|n = 1)(\lambda + \mu_\ell^{\hat{\jmath}} + \alpha) = p(\hat{\jmath}, \ell - 1|n = 1)\mu_{\ell-1}^{\hat{\jmath}}\overline{q}_{\ell-1}^{\hat{\jmath}}, \quad \ell > 1.$$

We quickly note that all probabilities $p(j, \ell|n = 1)$ can be expressed in terms of $p(j, \ell = 1|n = 1)$ and $u(1)$. These two unknowns can then be determined from the normalizing condition and from formula (6.68), i.e., from the definition of $u(n)$. We then can move on to the next value of n, for which the probabilities $p(j, \ell|n - 1)$ are already known, and repeat the process. As was the case for node 1, the values of $p(j, \ell|n)$ tend to reach their limiting values rather quickly as n increases, which allows us to obtain the solution of the node considered.

Of course, we need to determine the rates of server disappearance α_i and reappearance β_i for our node-by-node approach to be of practical value. Since in our model, the arrivals to each node come from a Poisson source, the disappearance rate for node $i > 1$, α_i, is given by $\alpha_i = \sum_{j<i} \lambda_j$. We can compute the reappearance rate β_i approximately as follows. *A priori*, the reappearance rate is a function of the retained state description, i.e., $\beta_2(n_2, j, \ell_2 = 0) = p(n_1 = 1|j, \ell_2 = 0) \sum_{i=1}^{k} \mu_i q_i p(i|n_1 = 1, n_2, j, \ell_2 = 0)$. This expression corresponds to the idea that the server becomes available for node 2 if there is only one packet at node 1 and its transmission ends. If we drop the dependence on n_2 and on the attempt number j, the sum in this expression becomes the conditional rate of completions at node 1, $u_1(n_1)$. Since, by the very nature of our node-by-node approach, we don't have joint information of the state of neighboring nodes, we use a simple approximation

$$\beta_2 \approx u_1(n_1 = 1)p_1(1)/[1 - p_1(0)]. \tag{6.71}$$

In formula (6.71), all the terms are known once we have solved node 1.

For nodes $i > 2$, we would need to account for the case where there is just one customer (packet) at the preceding node and its service completes, as well as the case where the preceding node is empty, and the server is unavailable there but becomes available to node i, and hence to node $i+1$.

$$\begin{aligned}\beta_{i+1}(n_{i+1}, j_{i+1}, \ell_{i+1} = 0) &= p(n_i = 1|n_{i+1}, j_{i+1}, \ell_{i+1} = 0)\\ &\times \sum_{j\geq 1}\sum_{\ell=1}^{k_j} \mu_\ell^j q_\ell^j p(j, \ell|n_1 = 1, n_{i+1}, j_{i+1}, \ell_{i+1} = 0)\\ &+ p(n_i = 0|n_{i+1}, j_{i+1}, \ell_{i+1} = 0)\\ &\times p(\ell_i = 0|n_i = 0, n_{i+1}, j_{i+1}, \ell_{i+1} = 0)\beta_i(n_i = 0, \ell_i = 0).\end{aligned}$$

As an approximation, we drop some of the dependencies in the conditionals and we use

$$\beta_{i+1} \approx \frac{p(n_i = 1)u_i(n_i = 1) + p(n_i = 0)p(\ell_i = 0|n_i = 0)\beta_i}{1 - [1 - p(\ell_i = 0|n_i = 0)]p(n_i = 0)}, \quad i = 2, \ldots. \tag{6.72}$$

The term $p(\ell_i = 0|n_i = 0)$ for the preceding node is simply given by formula (6.70).

We note that in our approach, the rates of server reappearance are computed for node $i+1$ based on the solution of node i and the rate of server reappearance at this node. Of course, our node-by-node computation comes at the cost of some loss of accuracy since formulas (6.71) and (6.72) are only approximate. Additional approximation errors are potentially introduced by limiting the number of transmission attempts and by limiting the number of stages in the Coxian distribution of the service times at each attempt. Despite these potential sources of errors, the overall accuracy of this approach appears to be reasonably good (Brandwajn *et al.*, 2009), and the simplification in the analysis of such a PRI priority queue makes it well worth it.

6.10 Hybrid solution methods: A model of live VM migration in cloud computing

As we have seen, different solution methods, such as discrete-event simulation, numerical methods or analytical solutions (exact or approximate), each have their strong points and their weaknesses. So, naturally, when confronted with a model of a complex system, an appealing idea is to leverage the strong points of each method. Typically, such a hybrid solution approach requires a decomposition of the complex model into sub-models, each solved by the most appropriate method.

As an example, consider a study of the effects of server consolidation and live migration in cloud computing with virtualized data centers. In the latter, Virtual Machines (VMs) running on a set of Physical Machines (PMs or servers) can share the available physical resources. This has the potential to improve overall resource utilization and lower energy requirements. VMs are launched (created) and terminated in response to customer requests. At VM launch, one of the objectives of the initial VM placement may be to balance (or, depending on the management emphasis, to maximize) the PM utilization. With live VM migration (Clark *et al.*, 2005), it is possible to reposition a running VM on a different PM. The desire to reduce energy consumption figures among the objectives of live migration. Our main goal in considering live VM migration here is to determine how aggressive PM consolidation should be if we try to take into account the combined effects of energy consumption, user task performance, and equipment wear caused by repeatedly turning PMs on and off.

What do we need to include in our model? Clearly, we need to somehow represent the life cycle of a VM, including its initial placement on a PM and its possible live migration. We also need a model of task performance since it is one of the dimensions along which we wish to measure the effects of live migration. And, of course, we need a model of energy consumption that takes into account the number of VMs running on a PM. It is clear that, in general, VM lifetime and migration, on the one hand, and task executions, on the other hand, occur at very different time scales. This suggests that it may be

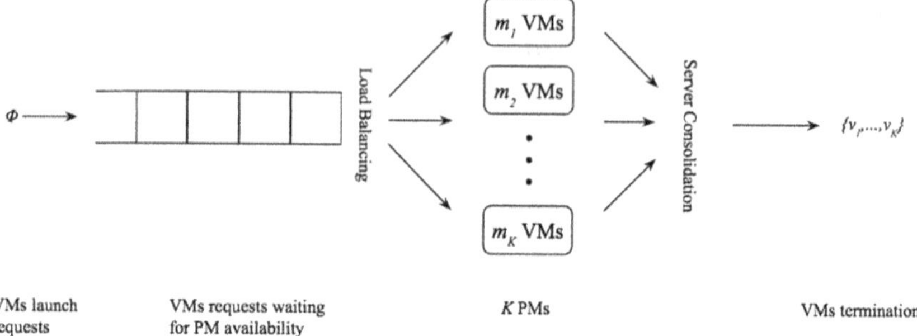

Figure 6.19. Assignment of VMs on PMs (upper-level model for live migration).

advantageous to consider these sets of events separately. Hence, we adopt a two-level view of the system.

At the upper level, as shown in Figure 6.19, we represent the set of Physical Machines (PMs) and the Virtual Machines (VMs) competing for their use. Our lower-level model will represent individual VMs and the tasks executing on them. We use the term "tasks" to denote the end-user jobs executed by the VMs, the latter running on a set of PMs. We denote by K the total number of available PMs and by m_i the current number of VMs at PM number i, $i = 1, \ldots, K$. This number includes VMs in normal operation and VMs being migrated to and from the given PM. Differences in the individual PMs may include their memory capacity, execution speed, and energy consumption characteristics.

For our purpose here, we assume that PM memory (and not the number of CPU cores) is the principal limiting resource (cf. (Lim *et al.*, 2009)). We denote by S_i the available memory capacity of PM number i. The time between consecutive VM launch requests has a distribution with mean $1/\phi$. To represent the fact that individual VMs may require different amounts of memory and may remain in the system for different time durations, we use a distribution of memory requirements as well as a distribution of lifetime durations for VMs. To keep this aspect of our model realistic, we use distributions derived from published measurement data for the times between VM launch requests, VM memory requirements, and VM lifetime (Cortez *et al.*, 2017). Requests for the launch of a new VM are routed to a specific PM (initial VM placement) according to an algorithm that takes into account the current state of the system, including the amount of memory available on the PM. This amount must, of course, be sufficient to accommodate the new VM.

If there are no VMs executing on a PM, the PM becomes inactive and is placed in a low power state. With power savings an important consideration, the initial placement algorithm may attempt to assign the new VM to one of the currently active PMs. In this case, an inactive PM is activated only if no active PM has enough resources to accommodate the new VM. If no PM has enough resources left, the VM launch request is queued. We assume that VM launch requests are considered in FCFS order.

We denote by ν_i the rate with which a VM currently running on PM number i completes its lifetime and leaves the PM. Depending on the VM launch request at the head of the queue (if any), the VMs remaining on PM i may be migrated to another PM number j following such a completion. For the migration to happen, the target PM j must have the capacity to accommodate the remaining VMs from PM i and the migration must be deemed desirable under the migration policy in effect. It may also be possible and preferable to consolidate the VMs from another PM on PM number i following the departure of a VM from the latter.

We denote by $1/\gamma$ the mean time it takes to migrate a single VM. For the purpose of our study, to control how aggressively the system attempts to consolidate, we use a "trigger" fraction f_t so that PM consolidation is attempted if available memory on the PM falls below $f_t S_i$. Thus, with $f_t = 1$, PM consolidation is considered following every VM departure, while with $f_t = 0$ there is no live VM migration.

Clearly, the goal of this upper-level model is to represent server consolidation and initial load balancing in the interactions between VMs and the PMs on which they execute. From this upper level, we can assess various quantities related to resource utilization such as the probability that a VM is initially assigned to a given PM, denoted by q_i $(i = 1, \ldots, K)$, the mean number of VMs at a given PM, denoted by $\overline{m}_i$, the mean time to migration out of PM i when it is active, denoted by $1/\alpha_i$, as well as the probability that, when a migration out of PM i takes place, its target is PM number j, denoted by p_{ij}. As we see shortly, these quantities are needed to evaluate the performance of tasks executed by the VMs.

We can also assess quantities related to energy consumption for each PM, including the number of migrations per time unit, $\Theta_{\text{mig},i}$, the number of times the PM is switched on and off per time unit, $\Theta_{\text{on},i}$ and $\Theta_{\text{off},i}$, as well as the fraction of time the given PM is active denoted by $f_{\text{act},i}$ for $i = 1, \ldots, K$. A model of energy consumption for our purposes needs to account for the number of VMs running on a PM, VM migration, and the switching on and off of the PMs. Thus, we assume that we can express the energy consumed per time unit by PM number i as

$$P_i = P_{\text{base},i} f_{\text{act},i} + \overline{m}_i P_{VM,i} + \Theta_{\text{mig},i} E_{\text{mig}} + \Theta_{\text{on},i} E_{\text{on},i} + \Theta_{\text{off},i} E_{\text{off},i}. \tag{6.73}$$

In formula (6.73), $P_{\text{base},i}$ is the base power consumption of the PM in the absence of VMs running on it, $P_{VM,i}$ is the average consumption of a VM on the PM, E_{mig} is the average amount of energy consumed by a single migration, $E_{\text{on},i}$ and $E_{\text{off},i}$ are the amounts of energy consumed by turning the PM on and off, respectively. The results of the upper-level model directly drive our energy model. Formula (6.73) accounts for the energy use due to the activity of PMs and VM migrations (Callau-Zori *et al.*, 2018) but does not include incompressible data center energy consumption (HVAC, lighting, etc.). This formula assigns the energy use due to a VM migration to the PM that hosted the VM being migrated. It is, of course, possible to design a more sophisticated model of power consumption; in particular, formula (6.73) implies that the energy consumption grows linearly with the number of VMs, while some authors argue that it should be tied non-linearly to the CPU utilization at the PM (Zhang *et al.*, 2013).

Given that we want to use our upper-level model to explore different initial VM placement policies and VM migration strategies, and given that we wish to use in our model empirically derived distributions of times between VM launch requests and of VM lifetime, discrete-event simulation is without a doubt the most appropriate solution method. The main reason for this is the flexibility of this method when it comes to evaluating different algorithms of VM migration and initial VM placement. Of course, as we have seen in Section 4.4, discrete-event simulation produces statistical estimates, so we will need to run it long enough so that the confidence intervals

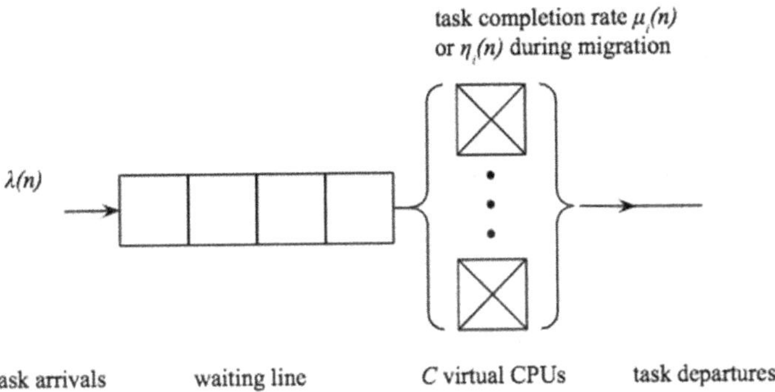

Figure 6.20. Task execution on a VM (lower-level model for live migration).

of its results are narrow enough and we can use their mid-points to safely feed other parts of our study.

Since VM migration and the sharing of a PM by VMs may impact task performance, we need a model of task execution on a VM. This is the objective of our lower-level model, shown in Figure 6.20, representing the tasks on a single VM.

In this model, end-user tasks arrive at instantaneous rate $\lambda(n)$ when the current number of tasks at the VM is n. The maximum number of tasks that can be present on the VM at any given time is limited to N. There are C virtual CPUs available to execute these tasks. To account for initial VM placement and possibly different CPU speeds on different PMs, with probability q_i a VM starts its life on PM number i ($i = 1, \ldots, K$). A VM currently located on PM i can be in its "normal" state, or it can be in "migration" state, i.e., it can be undergoing migration out to another PM. The rate with which a VM in its "normal" state at PM i begins a migration to PM number j is given by $\alpha_i p_{ij}$. Recall that $1/\alpha_i$ is the mean time to migrations out of PM i (when the PM is active), and p_{ij} is the probability that the target of the migration out of PM i is PM number j. Conveniently, both quantities are known from the solution of the upper-level model discussed above. The rate with which a migration from PM i to PM number j ends (i.e., the VM returns to its "normal" state) is given by γ. Recall that $1/\gamma$ is the mean time to complete a VM migration, and it's assumed to be known.

Referring to a VM located on PM number i, we let $\mu_i(n)$ be the rate of task completion for a VM in "normal" state when the current number of tasks on the VM is n. Analogously, when the VM is undergoing migration out of the PM number, its rate of task completion is $\eta_i(n)$. Clearly, we will usually have $\eta_i(n) \ll \mu_i(n)$, and, if task execution was to be totally stopped during migration, we would have $\eta_i(n) = 0$. These completion rates are assumed to be parameters in our study. In practice, their values would have to come from workload measurements or assumptions on the type of workload the VM is (or is supposed to be) running. Although virtualization provides logical isolation between VMs, one can expect that, because of possible competition for PM resources, the task completion rates may depend on the number of VMs packed onto the PM. We certainly can use the average number of VMs on the PM, known from the upper model, to adjust these rates according to how busy the PM is.

Let's assume that our main task performance metrics are the mean task response time and the attained task throughput and that we are comfortable with considering memoryless distributions for the time between task arrivals. If such is the case, then a numerical or semi-analytical approach is probably the most appropriate to solve our lower-level model. This is because it is quite straightforward to include state-dependent rates in the balance equations of a system. By contrast, state-dependent rates in a discrete-event simulation, while feasible, lead to a more involved simulation since events already scheduled have to be canceled and rescheduled as the system state changes. Even so, if the task performance metrics include the percentage of tasks whose response time remains below or exceeds a certain value, then a discrete-event simulation may be a more suitable approach to solve our model. In a solution based on balance equations, in certain cases, we may be able to estimate higher moments of the response time and gain additional knowledge about how meaningful the mean response time is. We will have the opportunity to revisit this subject later in this book.

By semi-analytical (or semi-numerical) solution approach, we mean here a solution of the balance equations of the system in which some part of the solution is given by a formula and some part may have to be obtained numerically. If we do adopt a solution other than discrete-event simulation for our lower-level model, we have what is referred to as a hybrid solution for our overall model of server consolidation in a virtualized data center. In other words, a hybrid approach utilizes more than one solution method to solve a model, typically combining discrete event simulation with an analytical or numerical solution.

Let's go this hybrid route, and let's derive the balance equations for our lower-level model representing the execution of tasks on a VM, the latter potentially subject to migrations. The state of a VM is defined by the couple (n, i) where n is the current number of tasks at this VM, and i refers to the PM on which the VM is located. We use positive values $i = 1, \ldots, K$ if the VM is in its "normal" execution mode, and negative values $(-i)$ if the VM is being migrated from PM number i to another PM. We denote by $p(n, i)$ the steady-state probability that the VM is in state (n, i). For $n = 1, \ldots, N-1$; $i = 1, \ldots, K$, we have the following balance equations:

$$p(n,i)[\lambda(n) + \mu_i(n) + \alpha_i] = p(n-1,i)\lambda(n-1) + p(n+1,i)\mu_i(n+1) + \sum_{j \neq i} p(n,-j)\gamma p_{ji}, \tag{6.74}$$

$$p(n,-i)[\lambda(n) + \eta_i(n) + \gamma] = p(n-1,-i)\lambda(n-1) + p(n+1,-i)\eta_i(n+1) + p(n,i)\alpha_i. \tag{6.75}$$

For $n = N$, these equations become

$$p(N,i)[\mu_i(N) + \alpha_i] = p(N-1,i)\lambda(N-1) + \sum_{j \neq i} p(N,-j)\gamma p_{ji}, \tag{6.76}$$

$$p(N,-i)[\eta_i(N) + \gamma] = p(N-1,-i)\lambda(N-1) + p(N,i)\alpha_i. \tag{6.77}$$

The derivation of the balance equations for $n = 0$ is a bit more involved. We assume that a VM can only end its life at a moment when there are no user tasks at the VM, and the VM is in its normal state, i.e., when $n = 0$ and the VM is not in the process of migrating. Since our lower-level model is to model the execution of tasks on a VM in a steady state, we assume that a new instance of a VM is launched following the end of the life of a VM. To account for the initial placement of VMs on PMs, we assume that the probability that the new instance starts its life on PM number i is given by q_i.

Recall that the values of this probability are known from the solution of the upper-level model. We denote by δ_i the rate with which a VM ends its life given that the current system state is $(0, i)$. This allows us to write the following balance equations for $n = 0$.

$$p(0,i)[\lambda(0) + \alpha_i + \delta_i(1 - q_i)] = p(1,i)\mu_i(1) + \sum_{j \neq i} p(0,-j)\gamma p_{ji} + \sum_{j \neq i} p(0,j)\delta_j q_i \tag{6.78}$$

$$p(0 - i)[\lambda(0) + \gamma] = p(1,-i)\eta_i(1) + p(0,i)\alpha_i. \tag{6.79}$$

We don't know directly the value of the parameter δ_i but we know the rate with which a VM running on PM i ends its lifetime: ν_i. We interpret ν_i as pertaining to normal VM operations, excluding possible migration. This implies that we must have $\sum_{i=1}^{K} \delta_i p(0,i) = \sum_{i=1}^{K} \nu_i \sum_{n=0}^{N} p(n,i)$ to maintain the correct rate of VM end-of-life events. In order to satisfy this relationship, we set

$$\delta_i = \nu_i \sum_{n=0}^{N} p(n,i)/p(0,i) \quad \text{for } i = 1, \ldots, K. \tag{6.80}$$

Thus, to obtain a solution of our lower-level model, we need to obtain the solution of the set of balance equations (6.74–6.79) together with relationship (6.80) and the normalizing condition

$$\sum_{i=1}^{K} \sum_{n=0}^{N} [p(n,i) + p(n,-i)] = 1. \tag{6.81}$$

If we use (6.80) directly in the balance equation (6.78), we obtain a system of linear equations, which can be solved using a standard numerical method. If we keep (6.80) as a separate relationship and use an iterative method, such as the semi-numerical method of conditionals discussed earlier in this book, we can embed equation (6.80) in the iteration process, i.e., at every iteration of the numerical method recompute the value of δ_i for $i = 1, \ldots, K$.

Incidentally, one might be tempted to base the method of conditionals for our model on the relationship $p(n,i) = p(i|n)p_1(n)$, where

$p_1(n)$ is the probability that there are n tasks on the VM. This turns out not to be a good idea in practice because the rates of change in i are generally much lower than the rates of transitions that change n, leading to convergence and numerical problems for the method of conditionals. A better idea if one wants to use the method of conditionals is to apply the relationship $p(n, i) = p(n|i)p_2(i)$, where $p_2(i)$ is the steady-state probability that the VM is located on PM number i and is in its normal or migration state. This allows us to transform the balance equations into equations for $p(n|i)$, the conditional probability that there are n tasks on the VM given i, which avoids the numerical problems mentioned above.

On the surface of it, hybrid solutions mixing discrete-event simulation and analytical or numerical methods, like the one described for the model of live VM migration and server consolidation, leverage the best features of each solution approach. Despite their obvious conceptual appeal, hybrid approximate solutions have been relatively rarely used (Schwetman, 1977; Shanthikumar & Sargent, 1983). This may be due in part to the fact that, in addition to an appropriate decomposition of the original problem, one needs to be comfortable with different solution approaches and, in part, to personal preferences (or biases) of computer performance modelers. We note in closing that, even if we choose to stick with discrete-event simulation throughout, it does not seem like the best of ideas to include VM-task interactions in a single simulation together with PM–VM interactions. Indeed, there can be many orders of magnitude of end-user task executed during a single VM lifetime, and we need a fair number of VM launch and end-of-life events to assess with a reasonable degree of accuracy the effects of live migration policies. This would result in a very large number of task completions to simulate. So, in all cases, an appropriate decomposition of the problem is in order.

References

Balsamo, S., & Iazeolla, G. (1982). An extension of Norton's theorem for queueing networks. *IEEE Transactions on Software Engineering*, SE-8 (4), 298–305.

Bard, Y. (1980). A model of shared DASD and multipathing. *Communications of the ACM*, 23(10), 564–572.

Boxma, O. J., & Konheim, A. G. (1980). Approximate analysis of exponential queueing systems with blocking. *Acta Informatica*, 15, 19–66.

Brandwajn, A. (1975, October). A model of a time-sharing system with two classes of processes. In *GI (German Informatics Society) Annual Conference* (pp. 545–566). Berlin, Heidelberg: Springer.

Brandwajn, A. (1980). Further results on equivalence and decomposition in queueing network models. *ACM SIGMETRICS Performance Evaluation Review*, 9(2), 93–104.

Brandwajn, A. (1982, August). Fast approximate solution of multiprogramming models. In *Proceedings of the 1982 ACM SIGMETRICS Conference on Measurement and Modeling of Computer Systems* (pp. 141–149). Association for Computing Machinery, New York.

Brandwajn, A. (1985). Equivalence and decomposition in queueing systems – a unified approach. *Performance Evaluation*, 5(3), 175–186.

Brandwajn, A. (1998, October). Fast decomposition in large stochastic models. In *SMC'98 Conference Proceedings. 1998 IEEE International Conference on Systems, Man, and Cybernetics (Cat. No. 98CH36218)* (Vol. 4, pp. 3073–3078). IEEE.

Brandwajn, A. (2002). A note on SCSI bus waits. *ACM SIGMETRICS Performance Evaluation Review*, 30(2), 41–47.

Brandwajn, A., & Begin, T. (2016). Breaking the dimensionality curse in multi-server queues. *Computers & Operations Research*, 73, 141–149.

Brandwajn, A., & Begin, T. (2017). Multi-server preemptive priority queue with general arrivals and service times. *Performance Evaluation*, 115, 150–164.

Brandwajn, A., & Begin, T. (2019). First-come-first-served queues with multiple servers and customer classes. *Performance Evaluation*, 130, 51–63.

Brandwajn, A., & Begin, T. (2023). Approximation method for a non-preemptive multiserver queue with quasi-Poisson arrivals. *ACM Transactions on Modeling and Performance Evaluation of Computing Systems*, 9(1), 1–21.

Brandwajn, A., & Jow, Y. L. L. (1988). An approximation method for tandem queues with blocking. *Operations Research*, 36(1), 73–83.

Brandwajn, A., Nguyen, V. H., & Atmaca, T. (2009). A conditional probability approach to performance analysis of optical unslotted bus-based networks. *Current Research Progress of Optical Networks*, 65–94.

Bryant, R. M., Krzesinski, A. E., Lakshmi, M. S., & Chandy, K. M. (1984). The MVA priority approximation. *ACM Transactions on Computer Systems (TOCS)*, 2(4), 335–359.

Bux, W., & Herzog, U. (1977). The phase concept: Approximation of measured data and performance analysis. *Computer Performance*, 23–38.

Callau-Zori, M., Samoila, L., Orgerie, A. C., & Pierre, G. (2018). An experiment-driven energy consumption model for virtual machine management systems. *Sustainable Computing: Informatics and Systems*, 18, 163–174.

Chandy, K. M., Herzog, U., & Woo, L. (1975). Parametric analysis of queuing networks. *IBM Journal of Research and Development*, 19(1), 36–42.

Chandy, K. M., & Sauer, C. H. (1978). Approximate methods for analyzing queueing network models of computing systems. *ACM Computing Surveys (CSUR)*, 10(3), 281–317.

Clark, C., Fraser, K., Hand, S., Hansen, J. G., Jul, E., Limpach, C., ... & Warfield, A. (2005, May). Live migration of virtual machines. In *Proceedings of the 2nd Conference on Symposium on Networked Systems Design & Implementation* (Vol. 2, pp. 273–286), Boston, MA, USA, usenix.org.

Cortez, E., Bonde, A., Muzio, A., Russinovich, M., Fontoura, M., & Bianchini, R. (2017, October). Resource Central: Understanding and predicting workloads for improved resource management in large cloud platforms. In *Proceedings of the 26th Symposium on Operating Systems Principles* (pp. 153–167), Association for Computing Machinery, New York, NY, United States.

Courtois, P. J. (2014). *Decomposability: Queueing and Computer System Applications*. Academic Press.

Cox, D. R. (1955, April). A use of complex probabilities in the theory of stochastic processes. In *Mathematical Proceedings of the Cambridge Philosophical Society* (Vol. 51, No. 2, pp. 313–319). Cambridge University Press.

Dimitrijevic, D. D., & Chen, M. S. (1989, June). Dynamic state exploration in quantitative protocol analysis. In *Proceedings of the IFIP WG6. 1 Ninth International Symposium on Protocol Specification, Testing and Verification IX* (pp. 327–338), North-Holland Publishing Co.Div. of Elsevier Science Publishers B.V. P.O. Box 211 1000 AE Amsterdam Netherlands.

Eager, D. L., & Lipscomb, J. N. (1988). The AMVA priority approximation. *Performance Evaluation*, 8(3), 173–193.

Faddy, M. J. (2002). Penalised maximum likelihood estimation of the parameters in a Coxian phase-type distribution. In *Matrix-Analytic Methods: Theory and Applications* (pp. 107–114), Proceedings of the Fourth International Conference: Adelaide, Australia, 14–16 July 2002. World Scientific, 2002.

Herzog, U. (1974). Some remarks concerning the extended analytic models for system evaluation (X-model). *IBM*, Thomas J. Watson Research Division.

Lazowska, E. D., & Zahorjan, J. (1982). Multiple class memory constrained queueing networks. *ACM SIGMETRICS Performance Evaluation Review*, 11(4), 130–140.

Lim, K., Chang, J., Mudge, T., Ranganathan, P., Reinhardt, S. K., & Wenisch, T. F. (2009). Disaggregated memory for expansion and sharing in blade servers. *ACM SIGARCH Computer Architecture News*, 37(3), 267–278.

Schweitzer, P. (1981). Approximate analysis of multiclass closed networks of queues. *Journal of the ACM*, 29(2), 358–371.

Schwetman, H. D. (1977, April). Hybrid simulation models: A speed-up technique combining analytic and discrete-event modeling. In *Modelle für Rechensysteme: Workshop der GI, Bonn, 31. 3.-1. 4. 1977* (pp. 226–237). Berlin, Heidelberg: Springer Berlin Heidelberg.

Shanthikumar, J. G., & Sargent, R. G. (1983). A unifying view of hybrid simulation/analytic models and modeling. *Operations Research*, 31(6), 1030–1052.

Simon, H. A., & Ando, A. (1961). Aggregation of variables in dynamic systems. *Econometrica: Journal of the Econometric Society*, 111–138.

Zahorjan, J., Eager, D. L., & Sweillam, H. M. (1988). Accuracy, speed, and convergence of approximate mean value analysis. *Performance Evaluation*, 8(4), 255–270.

Zhang, X., Lu, J. J., Qin, X., & Zhao, X. N. (2013). A high-level energy consumption model for heterogeneous data centers. *Simulation Modelling Practice and Theory*, 39, 41–55.

Chapter 7

Multi-server Queues

7.1 M/M/C model and saturation patterns

In the early days of computer performance modeling, much effort was invested into developing models of systems with single-server devices. Since more recent computer systems incorporate a significant degree of parallelism, from multiple CPU cores to multiple I/O access paths, it became important to be able to model systems with multiple-server devices. Luckily, Jacksonian and some BCMP networks can incorporate servers with state-dependent service rates, and, as we have seen on several occasions, such state-dependent service rates can be used to model multiple servers if the service time is exponentially distributed. This is why, for instance, formulas (2.23–2.29) pertain in fact to an M/M/C/K queue.

Similarly, if we want to assume an unrestricted queueing room, it is not difficult to derive the steady-state solution for a system with Poisson arrivals, exponentially distributed service times, and C servers, i.e., the M/M/C queue. Let's denote by $1/\lambda$ the mean time between customer arrivals, by $1/\mu$ the mean service time of a customer, and by $n = 0, 1, \dots,$ the current number of customers in the system (queued and in service). We let $\nu(n) = \min(n, C)\mu$ to obtain the following balance equations for the steady-state probability that there are n customers in the system:

$$p(0)\lambda = p(1)\nu(1),$$
$$p(n)[\lambda + \nu(n)] = p(n-1)\lambda + p(n+1)\nu(n+1), n = 1, \dots.$$

If $\lambda < C\mu$, their solution has the form

$$p(n) = \begin{cases} \dfrac{1}{G}\displaystyle\prod_{i=1}^{n}\frac{\lambda}{i\mu}, & n = 0, \ldots, C-1 \\ \dfrac{1}{G}\left(\displaystyle\prod_{i=1}^{C}\frac{\lambda}{i\mu}\right)\left(\dfrac{\lambda}{C\mu}\right)^{n-C}, & n = C, C+1, \ldots \end{cases} \tag{7.1}$$

G is a normalizing constant given by $G = \sum_{n=0}^{C-1}\prod_{i=1}^{n}\frac{\lambda}{i\mu} + \frac{\prod_{i=1}^{C}\frac{\lambda}{i\mu}}{1-\frac{\lambda}{C\mu}}$. The system is unstable and possesses no steady state if, on average, customers arrive faster than the C servers can serve them, i.e., if $\lambda \geq C\mu$.

If we let $E[N]$ denote the mean number of customers in the system and $E[\Lambda]$ the mean rate of customer arrivals, the mean time a customer spends in the system can be obtained from Little's formula

$$E[W] = \frac{E[N]}{E[\Lambda]}.$$

For the M/M/C model,

$$E[N] = \sum_{n=0}^{\infty} np(n) = \frac{1}{G}\left\{\sum_{n=1}^{C-1} n\prod_{i=1}^{n}\frac{\lambda}{i\mu} + \left(\prod_{i=1}^{C}\frac{\lambda}{i\mu}\right) \times \left[\frac{C}{1-\frac{\lambda}{C\mu}} + \frac{\frac{\lambda}{C\mu}}{\left(1-\frac{\lambda}{C\mu}\right)^2}\right]\right\} \quad \text{and} \quad E[\Lambda] = \lambda.$$

To develop (or improve, as the case may be) our intuition, let us examine how the mean response time in the M/M/C model grows with the server utilization and how the pattern of this growth changes with the number of servers.

Figure 7.1(a) shows the mean response time in an unrestricted M/M/C model with $C = 1$, $C = 2$, and $C = 4$ servers. The service rate is taken to be $\mu = 1$. The x-axis gives the utilization per server. We note that the mean response time curve becomes closer to a straight angle as the number of servers increases. In other words, the response time seems to remain relatively flat for higher server

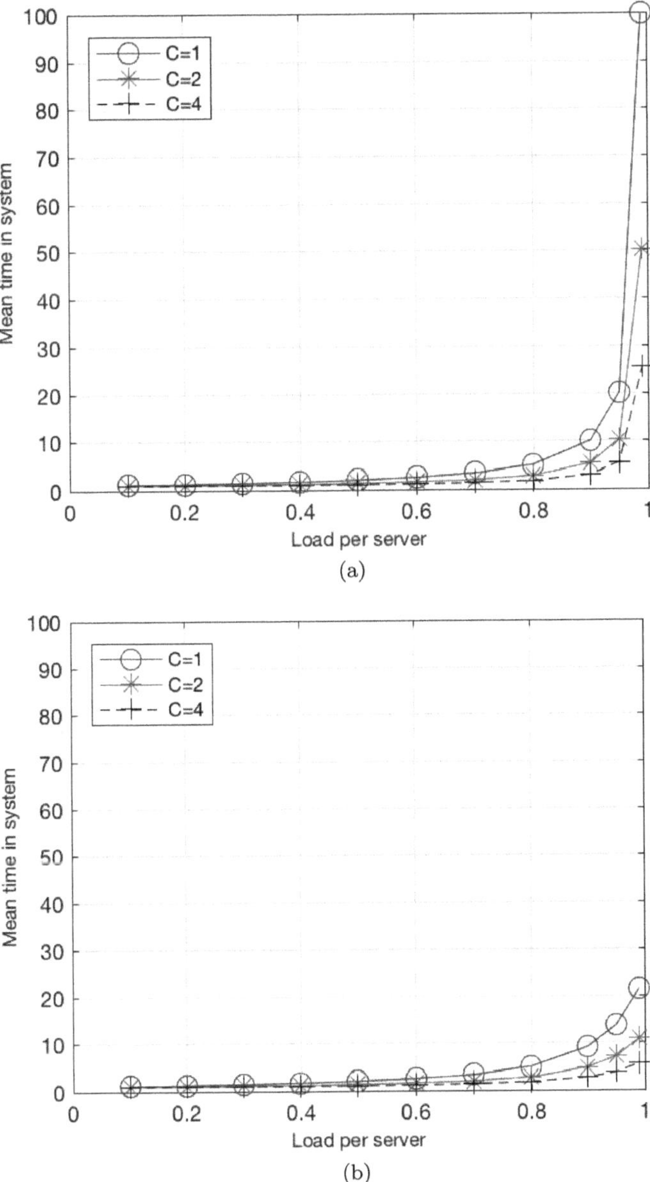

Figure 7.1. (a) Mean response time in M/M/C as a function of server utilization. (b) Mean response time in M/M/C/32 as a function of server utilization.

utilizations and then it "takes off" more suddenly in a system with a higher number of servers. Figure 7.1(b) shows analogous results for a model with a capacity limited to 32 customers, i.e., for the M/M/C/32 queue. It is quite instructive to observe a similar, more sudden increase in the mean response time as the number of servers increases. At the same time, since the system capacity is limited, the mean response time cannot grow without bound.

7.2 Approximations for multiple servers and general service times

In the case of a single server with Poisson arrivals and general service times, the Pollaczek–Khintchine formula (2.35) very elegantly states that, for a given mean service time, the mean waiting time depends on the distribution of the service time only through the coefficient of variation of that distribution. You may recall that we were able to derive this result by considering the expected wait in queue of an arriving customer, which included the residual service time of the customer in service. If we try to carry over this reasoning to a system with $C > 1$ servers, instead of the mean residual service time of a customer, we are now faced with the expected value of the minimum of C residual service times, which is no simple matter, except if the service times are exponential. But in this case, we actually know the full steady-state probability distribution (7.1).

Although there is no equivalent to the Pollaczek–Khintchine formula for multiple servers, several approximate formulas have been proposed to attempt to capture the distributional dependence in a system with multiple servers and general service times. In particular, we focus on easily implementable closed-form formulas to evaluate the mean queueing time for a customer.

Possibly inspired by the distributional dependence factor in the Pollaczek–Khintchine formula for the M/G/1 queue, the majority of existing studies seem to concentrate on the influence of the coefficient of variation of the service time distribution. Most of the existing approximations base their solutions on the interpolation and extrapolation of two special cases of the M/Ph/C queue, viz. the M/D/C and the M/M/C queues, for which the mean queue length $E[N_q]$

can be easily computed (exactly or approximately). We denote by λ the rate of arrivals to our queue, by $1/\mu$ the mean service time, by c_s the coefficient of variation of the service time, and we let $\rho = \lambda/\mu$. Recall that, by Little's law, the mean waiting time in the queue $E[W_q]$ is related to the mean number in the queue: $E[N_q] = \lambda E[W_q]$.

The detailed formulas to compute the mean queueing time $E[W_q]$ for six approximations, often found in textbooks on performance evaluation, are given as follows:

1. Martin's approximation (Martin, 1978):

$$E[W_q] \simeq \frac{P_C/\mu}{1-\rho} \cdot \frac{1+{c_s}^2}{2C},$$

   where P_C is the waiting probability for an arriving customer estimated by the corresponding probability in an M/M/C queue.
2. Cosmetato's (also Björklund and Elldin's) approximation (Björklund & Elldin, 1964; Cosmetatos, 1974):

$$E[W_q] \simeq c_s^2\, E[W_{q,M/M/C}] + (1-c_s^2) E[W_{q,M/D/C}].$$

3. Boxma, Cohen, and Huffles referred to as BCH's approximation (Boxma *et al.*, 1979):

$$E[W_q] \simeq \frac{1+c_s^2}{2} \frac{2E[W_{q,M/D/C}]E[W_{q,M/M/C}]}{2aE[W_{q,M/D/C}] + (1-a)E[W_{q,M/M/C}]},$$

   where $a = 1$ if $C = 1$ and $a = \frac{1}{C-1}\left(\frac{c_s^2+1}{\gamma_1} - C + 1\right)$ if $C > 1$ and $\gamma_1 = \frac{1-c_s^2}{C+1} + \frac{c_s^2}{C}$.
4. Tijms's approximation (Tijms, 1986):

$$E[W_q] \simeq \left((1-\rho)\gamma_1 C + \frac{\rho}{2}(c_s^2+1)\right) E[W_{q,M/M/C}]$$

   where γ_1 is defined as in the BCH approximation.
5. Lee's approximation (Lee & Longton, 1959):

$$E[W_q] \simeq \frac{1+c_s^2}{2} E[W_{q,M/M/C}].$$

6. Kimura's approximation (Kimura, 1986):

$$E[W_q] \simeq \frac{1+c_s^2}{\frac{2c_s^2}{E[W_{q,M/M/C}]} + \frac{1-c_s^2}{E[W_{q,M/D/C}]}}.$$

$W_{q,M/M/C}$ denotes the mean waiting time in an M/M/C queue with the same mean service time and the same arrival rate. Following recommendations in the literature (Bolch *et al.*, 2006), in our study of Kimura's approximation, we used

$$E[W_{q,M/D/C}] \simeq \frac{1}{2}E[W_{q,M/M/C}]\left(1+(1-\rho)(C-1)\frac{\sqrt{4+5C}-2}{16\rho C}\right).$$

Proponents of these approximations indicate that they produce excellent results when the variability of the service time distribution is low, i.e., when c_s ranges from 0 to not much higher than 1. This is quite expected since, by design, these approximations rely on the specific solutions of M/D/C and M/M/C queues. Furthermore, probability distributions with low variability tend to be alike, and thus their expected queueing times tend to be close to those of M/D/C and M/M/C queues. On the other hand, it is known that with increasing values of the coefficient of variation of the service time, the approximation accuracy will gradually decrease, although it is generally not stated at what rate.

We conducted several experiments to assess the accuracy of the above approximate solutions when applied to an M/Ph/C queue (Begin & Brandwajn, 2013). We employed an iterative solution based on the method of conditionals to compute the exact values used to assess the accuracy of the approximations. We explored thousands of examples with different parameters: the number of servers C varied from 2 to 12, the per server utilization ρ from 0.1 to 0.9, the coefficient of variation of the service time c_s from 2 to 10, and the skewness of the service time s_s from 0 to 100. For a random variable X, its skewness is given by $E\left[\left(\frac{X-\mu}{\sigma}\right)^3\right]$, where μ is the mean and σ is the standard deviation of the random variable. It is a measure of the asymmetry of the distribution, and a skewness of 0 occurs if the distribution is nearly symmetric. Here, we denote the skewness of the service time by s_s.

Our results (Begin & Brandwajn, 2013) show that the accuracy of all these approximations can be quite poor (with relative errors above

Table 7.1. Example of inaccuracy for M/G/C simple approximations for the mean number of customers in the system.

	Exact	Martin	Cosmetatos	BCH	Tijms	Lee	Kimura
C = 8, $\rho = 0.7$, $c_s = 8$ and $s_s = 30$	14.82	26.12	23.96	8.46	20.83	23.96	17.58
C = 4, $\rho = 0.5$, $c_s = 5$ and $s_s = 50$	2.45	4.26	3.98	2.59	4.30	3.98	3.54

50%), including for supposedly "easy" examples with a low coefficient of variation of the service time. An example of these results is shown in Table 7.1. Therefore, one needs to be extremely cautious when applying such approximations. Nonetheless, if one is willing to use them, we recommend the following guidelines:

- If C is much larger than ρ (say $C > 33\rho - 4.66$), or conversely much smaller (say $C < 50\rho - 33$), then any approximate solution yields excellent results.
- Otherwise,
  - If both c_s and s_s are low (say less than 4 and 30, respectively), any of the six approximate solutions provides accurate results.
  - If s_s is large (say 10 times greater than c_s) and c_s is low, we recommend the use of Kimura solution.
  - If s_s is low and c_s is large, we recommend the use of the BCH solution.
  - If both c_s and s_s are high, none of the approximations seems to be acceptable.

The use of these guidelines would ensure that, in almost all our numerous experiments, the discrepancy between the exact and the approximate mean number of customers in the queue stays within a range of 25% relative error (except obviously when both c_s and s_s are high and ρ is moderate).

The failure of formulaic approximations discussed above stems from the fact that performance indices, including the simple mean value, related to the number of customers in an M/G/C queue depend on the shape (and thus higher order moments) of the service time distribution and no formula is known to capture and quantify this dependence (Wolff, 1977; Whitt, 1980; Gupta *et al.*, 2007; Begin &

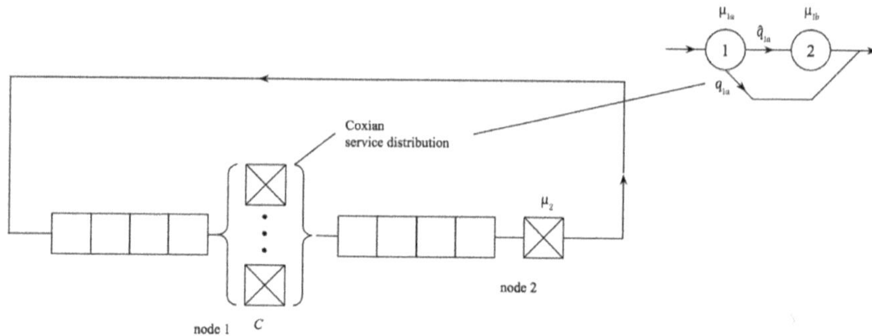

Figure 7.2. Two-node network, Node 1 multiple servers Cox-2 service time, Node 2 exponential.

Brandwajn, 2013). This turns out to be also true for multi-server nodes in closed queueing models (Brandwajn and Begin, 2009). Let's consider the simple two-node network shown in Figure 7.2.

The network consists of a multi-server queue with C servers and a single-server queue, which we refer to as Nodes 1 and 2, respectively. The service time at Node 2 is exponentially distributed with mean $1/\mu_2$, while the service times at Node 1 are assumed to be represented by a two-stage Coxian distribution (Sasaki *et al.*, 2004) with mean $1/\mu_1$. We denote by μ_{1a} and μ_{1b} the respective service rates of the two stages of this distribution, and by $\hat{q}_{1a} = 1 - q_{1a}$ the probability that a customer's service proceeds to the second stage upon completion of the first stage. The coefficient of variation of the service time at Node 1 is denoted by cv_1. We let N be the total number of customers in the network, and we denote by $E[N_1(N)]$ the steady-state mean number of customers at Node 1 in our network with a total of N customers.

Let us consider an example in which the exponential service time distribution at Node 2 has a mean of $1/\mu_2 = 0.35$. To show the dependence on higher-order distributional properties, for Node 1, we use two Cox-2 distributions with mean $1/\mu_1 = 1$, coefficient of variation $cv_1 = 6$ but different higher-order properties of skewness and Kurtosis, related to moments of order three and four of the service time distributions. The parameters of these distributions, labeled D5 and D6, respectively, are shown in Table 7.2.

Table 7.2. Parameters of the different two-stage Coxian (Cox-2) distributions used for the service times at Node 1.

Distribution index	Mean value	cv_1	Skewness	Kurtosis	μ_{1a}	μ_{1b}	q_{1a}
D1	1.0	2.0	19.26	608.91	1.11	6.25E-02	9.938E-01
D2	1.0	2.0	3.07	12.77	1000.0	4.00E-01	6.010E-01
D3	1.0	4.0	54.10	4107.3	1.11	1.32E-02	9.987E-01
D4	1.0	4.0	6.01	48.28	1000.0	1.18E-01	8.830E-01
D5	1.0	6.0	86.00	10087.28	1.11	5.68E-03	9.994E-01
D6	1.0	6.0	9.01	108.30	1000.0	5.40E-02	9.460E-01
D7	1.0	8.0	116.99	18480.19	1.11	3.17E-03	9.997E-01
D8	1.0	8.0	12.01	192.43	1000.0	3.10E-02	9.690E-01
D9	1.0	10.0	147.58	29276.89	1.11	2.02E-03	9.998E-01
D10	1.0	10.0	15.02	300.63	1000.0	1.98E-02	9.802E-01
D11	0.67	6.0	86.04	10097.03	1.67	8.52E-03	9.994E-01
D12	0.67	6.0	9.01	111.30	1500.0	8.10E-02	9.461E-01

As you may recall, skewness is a measure of asymmetry of a distribution. For a random variable X, its skewness is given by $E[(\frac{X-\mu}{\sigma})^3]$, where μ is the mean and σ is the standard deviation of the random variable. Kurtosis is sometimes referred to as a measure of "tailedness" of a distribution (e.g., heavy tail or light tail). For a random variable X, its kurtosis is given by $E[(\frac{X-\mu}{\sigma})^4]$. Both properties relate to the shape of the distribution, but each of them to a different aspect of that shape.

We show in Figure 7.3 the steady-state mean number of customers at Node 1 with 4 servers ($C = 4$) for the two sets of distributional parameters D5 and D6 as a function of the total number of customers in the network N. It is quite striking to see that the values for $E[N_1(N)]$ can be so different just due to properties beyond the mean and the coefficient of variation. This is a clear indicator that the mean number of customers at a server depends on higher-order properties of the service time distributions. It follows that (as is the case for the M/G/C model) approximations attempting to use the distributional factor $(1 + c_s^2)/2$ (where c_s is the coefficient of variation of the service time) of the Pollaczek–Khintchine formula for multi-server queues cannot work in general.

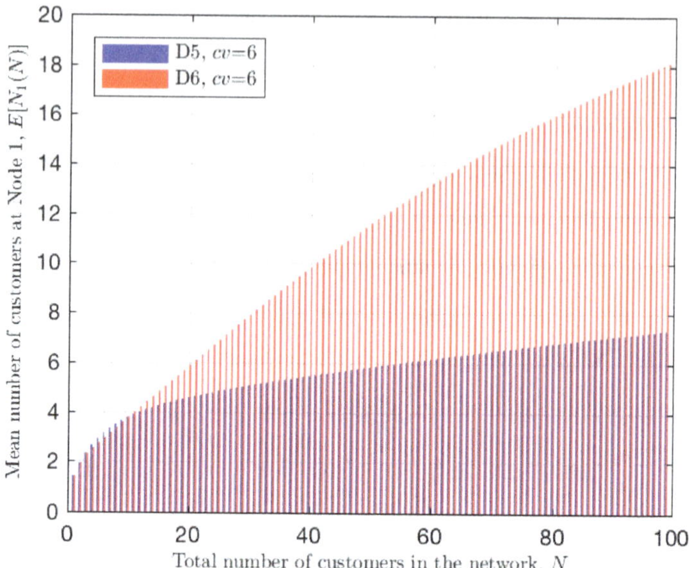

Figure 7.3. Influence of higher-order moments of the service time distribution at Node 1 on the mean number of users at this node.

7.3 State upon arrival with general service times

Before we leave this simple network, we would like to address another aspect of closed models with non-exponential service times at multi-server nodes with First-Come-First-Served queueing discipline. As we mentioned before, such networks generally don't possess a product-form steady-state probability distribution for the number of customers at each node. This means that the theorem of the state upon arrival (arrival theorem, in short) cannot be expected to hold exactly. The interesting question then becomes: is the arrival theorem sufficiently close to being true to serve as a basis for approximate solutions? Nothing like a few simple examples to find out.

Let us denote by $\tilde{n}_1(N)$ the mean number of customers found in steady state by a customer arriving from Node 2 to Node 1 in our simple two-node model of Figure 7.2. If the arrival theorem holds, we should have $\tilde{n}_1(N) = E[N_1(N-1)]$ (the mean number of customers at Node 1 when there are $N-1$ customers in the network). Based on this, we can measure the degree of deviation from the arrival theorem in our network with a total of N customers by the following quantity:

$$\Delta(N) = |1 - E[N_1(N-1)]/\tilde{n}_1(N)|. \tag{7.2}$$

We refer to $\Delta(N)$ as the relative deviation from the arrival theorem.

How can we obtain $\tilde{n}_1(N)$ in our two-node model? A relatively simple approach is to consider the steady-state probability $p(l_2, n)$ that the current number of customers at Node 1 is n ($n = 0, \ldots, N$) and that l_2 of them are in the second stage of their Coxian service time ($l_2 = 0, \ldots, \min(C, n)$). Let us assume that we know the probabilities $p(l_2, n)$. Clearly, we can generate the balance equations for the state description (l_2, n) and use, for example, the method of conditionals to solve these balance equations. To derive the steady-state probabilities that a customer arriving from Node 2 to Node 1 finds n ($n < N$) customers at Node 1, we note that the rate of arrivals from Node 2 is simply the rate of customer completions at Node 2 when there are n customers at Node 1, i.e., $\mu_2 \sum_{l_2=0}^{\min(C,n)} p(l_2, n)$. The overall rate of departures from Node 2, corresponding to all possible system states, can be expressed as $\sum_{l=0}^{N-1} \mu_2 \sum_{l_2=0}^{\min(C,l)} p(l_2, l)$. Following the approach used by Cooper (1984), the probability that a customer leaving Node 2 finds n customers at Node 1, which we denote by $P_A^N(n)$, can be written as the ratio of these two rates:

$$P_A^N(n) = \frac{\sum_{l_2=0}^{\min(C,n)} p(l_2, n)}{\sum_{l=0}^{N-1} \sum_{l_2=0}^{\min(C,l)} p(l_2, l)}. \tag{7.3}$$

From formula (7.3), the expected number of customers found at Node 1 by a customer arriving from Node 2 can be expressed as

$$\tilde{n}_1(N) = \sum_{n=0}^{N-1} n P_A^N(n). \tag{7.4}$$

We are now ready to look at the relative deviation from the arrival theorem. We start by assuming that there is only one non-exponential server at Node 1 ($C = 1$) so that Node 1 may be viewed as an M/G/1-like queue in a closed network.

We use a network with $N = 10$ customers and several Cox-2 distributions at Node 1, all with mean of 1 but different coefficients of variation of $2, 4, 6, 8$, and 10, respectively. The parameter values for the

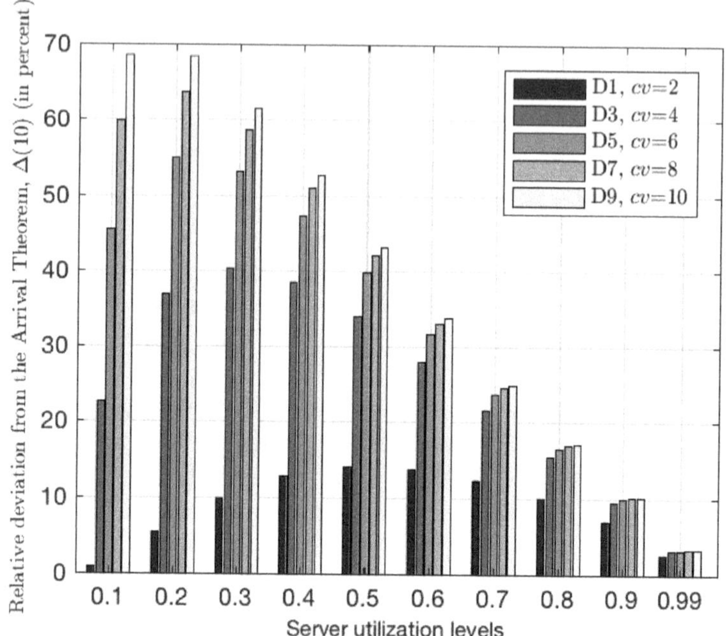

Figure 7.4. Example of deviation from the arrival theorem.

Cox-2 distributions used are given in Table 7.2; the distributions are identified in our figures by their index in Table 7.2 and by their coefficient of variation denoted by *cv*. We have represented in Figure 7.4 the relative deviation from the arrival theorem, expressed in percent, for a range of server utilization levels at Node 1. We observe that, for the distributional parameters considered, the deviation from the arrival theorem depends on both the server utilization level and the coefficient of variation of the service time distribution at Node 1. In this particular example, the deviation ranges from about 10% to around 70% and tends to peak for relatively small levels of server utilization. It is important to keep in mind that the service rate at Node 2 has been adjusted for each value of *cv* in order to maintain the specified utilization levels.

It turns out that things are a bit more complicated and the deviation from the arrival theorem may be affected also by higher-order properties of the service time distributions. To demonstrate this fact, let us use a different set of Cox-2 distributions with the same mean

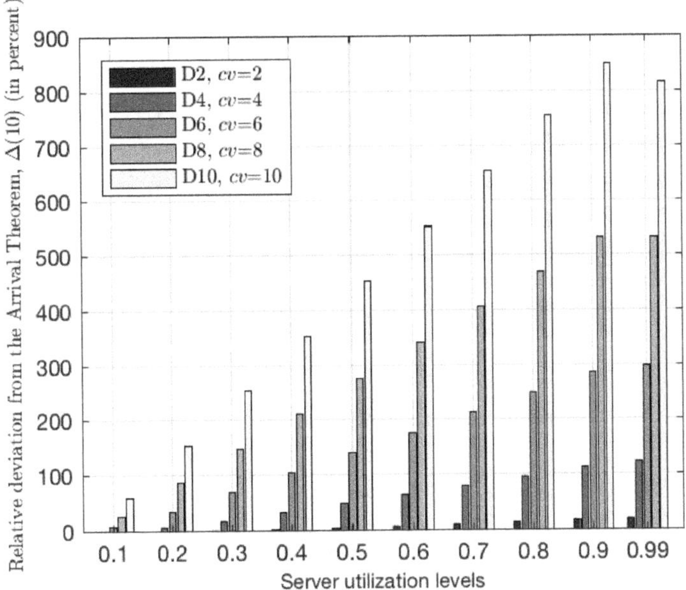

Figure 7.5. Relative deviation from arrival theorem for another subset of distributions.

and coefficients of variation as in Figure 7.4 but with different higher-order properties. We have represented in Figure 7.5 results analogous to those in Figure 7.4, obtained with another subset of distributions given in Table 7.2. As before, the service rate at Node 2 has been adjusted for each distribution so as to maintain the desired utilization level at Node 1.

What a striking difference from the results shown in Figure 7.4! With the new distributions, the deviation from the arrival theorem reaches a whopping 800% and appears to increase as the server utilization level increases. Moreover, large deviations from the arrival theorem are observed for relatively large values of the mean number of customers at Node 1 and thus cannot be viewed as an artifact of relative errors on small quantities. Since the only difference between the service time distributions used in Figures 7.4 and 7.5 lies in their higher-order properties, it is clear that properties of higher order (beyond the mean and the coefficient of variation) of the service time distributions may have an important effect on the mean number of customers found upon arrival. In fact, higher-order properties may influence not only the state found upon arrival but also customary performance measures such as the mean number of customers at a node or the node utilization level.

Let us now turn our attention to the case where there are multiple servers at Node 1. As before, the service time at Node 2 is exponentially distributed. The results shown in Figure 7.6 illustrate how the deviation from the arrival theorem varies with the number of customers in the network for 2, 4 and 8 servers. The parameters of the Cox-2 service time distribution at Node 1 used here correspond to the distribution labeled D4 in Table 7.2 ($cv = 4$). The mean service time at Node 2 is $1/\mu_2 = 1$.

We observe in Figure 7.6 that $\Delta(N)$ peaks for lower values of N and then decreases as the number of customers in the network increases. The relative deviation decreases also as the number of servers increases, which is perhaps not surprising given that there tends to be less queueing with a larger number of servers. Still, even with 8 servers, $\Delta(N)$ can exceed 40%.

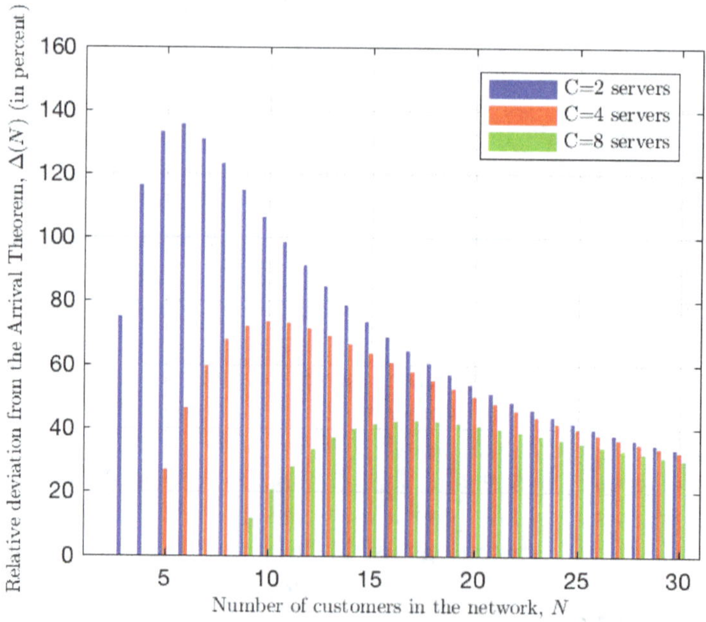

Figure 7.6. Relative deviation from the arrival theorem for distribution D4 ($cv = 4$).

To conclude our study of this subject here, we examine the deviation from the arrival theorem for a fixed population of $N = 16$ customers as a function of the server utilization level for different numbers of servers at Node 1. The same Cox-2 distribution labeled D4 in Table 7.2 is used for the service times at Node 1. We have adjusted the value of the service rate at Node 2 (μ_2) for each value of the number of servers at Node 1 (C) so as to maintain the specified server utilization levels.

It is clear from the results of Figures 7.6 and 7.7 that the arrival theorem is oftentimes violated to a significant degree in networks with non-exponential servers. This, in turn, means that we should approach with caution approximations based on the arrival theorem for such networks.

The fact that the performance of multi-server facility or device may exhibit an important dependence on higher-order properties of the service time distribution has direct implication for system instrumentation. Indeed, it means that collecting just the mean or the variance of the service times may not be enough to characterize the

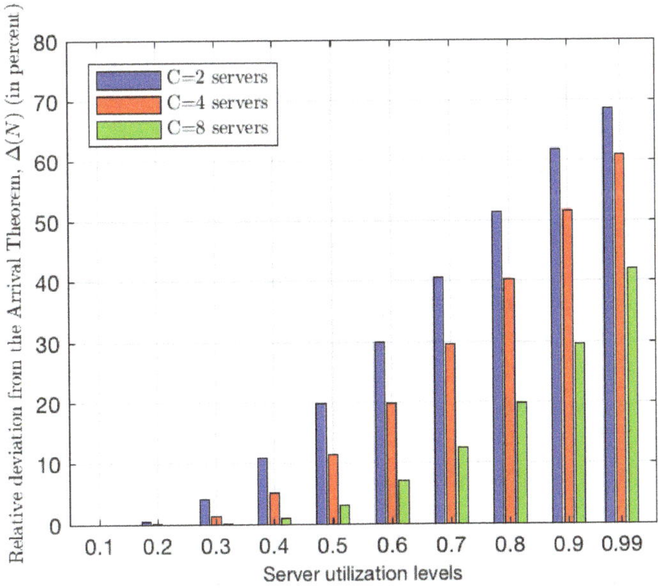

Figure 7.7. Relative deviation from the arrival theorem for distribution D4 ($cv = 4$) for different number of servers at Node 1.

workload if one plans to build a queueing model of the performance of the device or facility being considered.

7.4 Reduced-state approximation for the M/Ph/C/K model

Let us now turn our attention precisely to a multi-server queue in which the service times have a general distribution. If the pattern of customer arrivals is relatively well behaved so that the times between arrivals can be represented as memoryless, a large number of real-life systems, such as multi-core processors or call centers, can be modeled as instances of the M/G/C/K queue, i.e., an M/G/C queue with a maximum number of K customers in the system (queued and in service). The exact analytical solution of this queueing model is not known except in certain special cases. As we have seen earlier in this book for the M/G/1/K model, a frequently used approach when dealing with non-exponential service times is to replace the general service time distribution with a phase-type distribution. This is based on the fact, discussed in Chapter 2, that any distribution can be approximated arbitrarily closely by a phase-type distribution. As noted before, the advantage of this approach is that the resulting M/Ph/C/K model can be described in the steady state by the familiar balance equations.

There are, in general, two possible full state descriptions that can be used for these balance equations. Both involve the current number of customers in the system (or in the waiting line) and a vector to represent the state of the servers. The first description uses the vector of the current number of servers in each phase of the service process. In the second description, the vector used is that of the current phase for each server. (The servers are assumed to be identical but are not synchronized, i.e., each server may be in a different phase of service or idle if not all servers are currently in use.) This latter state description is generally more costly than the first one and thus rarely used. For both descriptions, the number of states increases rapidly as the number of phases in the service time distribution and of servers in the system grows.

When the number of phases and servers is small, the balance equations for the M/Ph/C/K model can be solved numerically by any of a number of methods (Ramaswami & Lucantoni, 1985; Seelen, 1986; Latouche & Ramaswami, 1993; Stewart, 1995; Bini *et al.*, 2005). As mentioned before, the size of the system of equations to be solved suffers from what has been termed the "dimensionality curse", in reference to the combinatorial growth as the number of servers and phases increases. This makes numerical solutions impractical for a larger number of servers. Unfortunately, in current modeling applications the number of servers can be quite high. For example, if our model is used to represent processor cores in a cloud system, we are talking about hundreds of servers. The number of phases in the service time distribution depends on the workload being modeled and can range from just 2 to 15 or more to represent (approximately) a heavy-tailed type of distribution (Crovella & Bestavros, 1997; Willinger *et al.*, 1997; Hsu & Smith, 2003; Willinger *et al.*, 2004; Jiang & Dovrolis, 2005; McNutt, 2005) which may be the result of a mixture of several types of jobs with very different characteristics. We are therefore looking for an approach that can handle larger numbers of servers (say, 100 or more) with a reasonable number of phases in the service time distribution.

It is quite vexing to have to describe the simultaneous state of a large number of servers (as required in the classical full state description) while these servers are identical. This leads us to the idea of using a reduced-state description where the state of only one server is described in detail in order to circumvent the explosion of the number of states discussed above while taking into account the full service time distribution.

Our M/Ph/C/K model is represented in Figure 7.8. We denote by n ($n = 0, \ldots, K$) the current number of customers in the system, including customers queued and in service. There are C homogeneous servers, and the total number of customers is limited to K with $K > C$. We assume that the times between arrivals are memoryless with rate $\lambda(n)$. The service times are represented by a phase-type distribution with b phases (akin to the one shown in Figure 2.3(a)). We denote by σ_i the probability that the service of a customer starts in phase i, $i = 1, \ldots, b$. The intensity, i.e., the completion rate, for phase i is denoted by μ_i. We let q_{ij} be the probability that the service continues in phase j following the completion of phase i, and we denote by $\hat{q}_i$ the probability that the service ends upon completion of phase i.

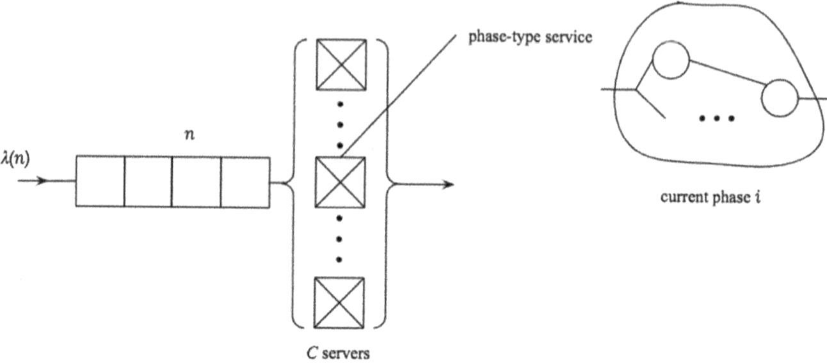

Figure 7.8. M/Ph/C/K queue.

Our discussion follows the spirit of published work (Brandwajn & Begin, 2014). To study our system in the steady state, we select one server among the C servers and we describe the state of our system by the total number of requests and the current phase of service at the selected server (n, i). Obviously, when $n < C$, with probability $(C - n)/C$ the selected server is idle. In this case, we use the value $i = 0$ to denote the idle state of the selected server. We denote by $p(n, i)$ the steady-state probability corresponding to the reduced-state description, i.e., the probability that there are n users in the system and the current phase of service at the selected server is i, $i = 0, 1, \ldots, b$. We note in passing that our reduced-state description can be viewed as a marginal probability compared with the full state description of the second type mentioned above, in which one describes explicitly the current phase of each server. We also denote by $\nu(n, i)$ the conditional rate of customer completions by servers other than the selected server given the current state (n, i). This conditional rate of completions is not given directly and we will address the question of how to assess it shortly.

With this notation, we can write the balance equations for $p(n, i)$ using our familiar "rate of flow out = rate of flow in" approach. We get the following equations. In the case $n = 0$, we have

$$p(0,0)[\lambda(0)] = \sum_{i=1}^{b} p(1,i)\mu_i \hat{q}_i + p(1,0)\nu(1,0). \tag{7.5}$$

For $0 < n < C$, the selected server can be busy, or it can be idle, and we have

$$p(n,i)[\lambda(n)+\mu_i+\nu(n,i)]$$
$$= p(n-1,0)\lambda(n-1)\frac{\sigma_i}{c-n+1} + p(n-1,i)\lambda(n-1)$$
$$+\sum_{j=1}^{b} p(n,j)\mu_j q_{ij} + p(n+1,i)\nu(n+1,i), \quad i=1,\ldots,b, \tag{7.6}$$

$$p(n,0)[\lambda(n)+\nu(n,0)]$$
$$= \sum_{i=1}^{b} p(n+1,i)\mu_i\hat{q}_i + p(n+1,0)\nu(n+1,0)$$
$$+p(n-1,0)\lambda(n-1)(c-n)/(c-n+1). \tag{7.7}$$

For $n = C$, only the values $i = 1,\ldots,b$ are possible and we get

$$p(n,i)[\lambda(n)+\mu_i+\nu(n,i)]$$
$$= p(n-1,0)\lambda(n-1)\sigma_i + p(n-1,i)\lambda(n-1) + \sum_{j=1}^{b} p(n,j)\mu_j q_{ji}$$
$$+\sum_{j=1}^{b} p(n+1,j)\mu_j\hat{q}_j\sigma_i + p(n+1,i)\nu(n+1,i). \tag{7.8}$$

For $n = C+1,\ldots,K-1$ and $i = 1,\ldots,b$, the only i values possible, we have

$$p(n,i)[\lambda(n)+\mu_i+\nu(n,i)]$$
$$= p(n-1,i)\lambda(n-1) + \sum_{j=1}^{b} p(n,j)\mu_j q_{ji} + \sum_{j=1}^{b} p(n+1,j)\mu_j\hat{q}_j\sigma_i$$
$$+p(n+1,i)\nu(n+1,i). \tag{7.9}$$

Finally, for $n = K$, we get

$$p(n,i)[\mu_i + \nu(n,i)] = p(n-1,i)\lambda(n-1) + \sum_{j=1}^{b} p(n,j)q_{ji}. \quad (7.10)$$

If we knew $\nu(n,i)$, the conditional completion rates by servers other than the selected server given the current number of customers in the system and the current service phase at the selected server, the solution of the balance equations (7.5–7.10), together with the obvious normalizing condition, would yield the exact steady-state probabilities $p(n,i)$. We don't know the exact values for the $\nu(n,i)$ but we are able to obtain a pretty reasonable approximation for these completion rates. Let us denote by $\omega(n)$ the rate at which customers depart the system from the selected server, given that the current total number of customers in the system is n and that the selected server is busy. This rate can be expressed as

$$\omega(n) = \sum_{i=1}^{b} p(n,i)\mu_i \hat{q}_i \Big/ \sum_{i=1}^{b} p(n,i). \quad (7.11)$$

Since the C servers are statistically identical, $u(n)$, the overall rate of customer departures when the current number of customers is n must be

$$u(n) = \min(n,C)\omega(n). \quad (7.12)$$

Formula 7.12 reflects the fact that, on average, the conditional completion rate of each server is the same. We assume that the rate of completions for servers other than the selected server exhibits little dependence on the current service phase of the selected server if it is currently active. Based on this assumption, we use the following intuitive approximation:

$$\nu(n,i) \approx u(n) - \omega(n), \quad \text{for } i = 1,\ldots,b, \quad (7.13)$$

$$\nu(n,0) \approx u(n), \quad \text{for } n < C. \quad (7.14)$$

For $n \geq C$, approximation 7.13 amounts simply to assuming that $\nu(n,i) \approx (C-1)\omega(n)$ for all $i = 1,\ldots,b$. Both relationships 7.13 and 7.14 happen to be exact when the service time is exponentially distributed, and it is intuitively clear that, in general, the current phase of service of the selected server should matter less and less as the number of servers increases. Although this may not be quite intuitive, this is not necessarily so with very few servers, so we would expect the approximation to be less accurate for smaller values of the number of servers C.

Looking back at our approximation and formulas (7.11) and (7.12), we note that the conditional rates of departure $\nu(n,i)$ needed to compute $p(n,i)$ are expressed as a function of the reduced-state description $p(n,i)$. If we consider these formulas together with the balance equations (7.5–7.10), we get a system of equations for $p(n,i)$ which can be solved in several different ways. Before we describe in some detail one possible simple solution approach, let us briefly touch on the computation savings afforded by the reduced-state description. It is clear that the number of states in the reduced state approximation grows linearly with the number of servers and phases in the service time distribution, as opposed to combinatorially in the standard full state description.

As an example, we can see in Figure 7.9(a) how the number of states grows with the number of servers in an M/Ph/C/K model with $b = 4$ phases and the total system capacity $K = 5C + 20$. We observe that when the number of servers approaches 100, there are 5 orders of magnitude differences in the number of states between the two state descriptions. Figure 7.9(b) shows the growth of the number of states as a function of the number of phases, b, in the service time distribution for a model with $C = 64$ servers and a system capacity $K = 5C + 20$, i.e., $K = 340$. It is quite impressive how quickly the number of states grows in the full state description. We are sure you noted the logarithmic scale used for the y-axis in Figures 7.9(a) and 7.9(b).

Let us now return to the solution of our M/Ph/C/K model using the reduced-state description. We denote by $p(n)$ the steady-state probability that there are n $(n = 0, \ldots, K)$ customers in the system. We also denote by $p(i|n)$, $i = 0, 1, \ldots b$, the conditional probability that the current service phase at the selected server is i given that there are n customers. Our reduced-state probability can be expressed as

$$p(n,i) = p(i|n)p(n). \tag{7.15}$$

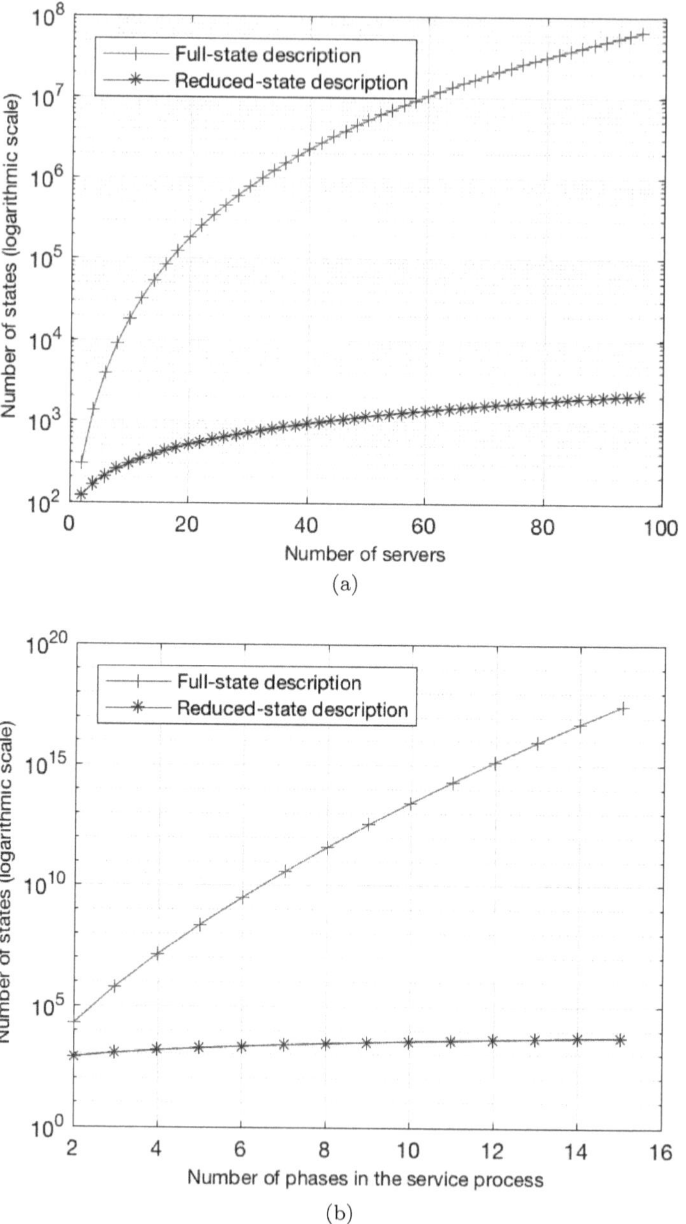

Figure 7.9. (a) Number of states as a function of the number of servers. (b) Number of states as a function of the number of phases.

Using the above relationship in formula (7.10), we get

$$\omega(n) = \sum_{i=1}^{b} p(i|n)\mu_i \hat{q}_i \Bigg/ \sum_{i=1}^{b} p(i|n). \tag{7.16}$$

The conditional probabilities $p(i|n)$ are normalized independently for each value of n. This means that we must have $\sum_{i=0}^{b} p(i|n) = 1\ \forall n$, and, since $p(i = 0|n) = 0$ for $n \geq C$, we have $\sum_{i=1}^{b} p(i|n) = 1$ for $n = C, \ldots, K$. Obviously, when $n = 0$, the selected server is idle, and we have $p(0|0) = 1$, $p(i|0) = 0$ for $i = 1, \ldots, b$. The conditional rate of customer arrivals given the current number of customers in the system is, of course, $\lambda(n)$ and the corresponding conditional rate of customer departures is $u(n)$, expressed in terms of $\omega(n)$ in formula (7.12). Hence, we can readily derive the form of the steady-state probability $p(n)$ as

$$p(n) = \frac{1}{G} \prod_{i}^{n} \frac{\lambda(i-1)}{u(i)}, \quad n = 0, \ldots, K. \tag{7.17}$$

As usual, G is a normalizing constant such that $\sum_{n=0}^{K} p(n) = 1$. It follows from (7.17) that

$$\frac{p(n-1)}{p(n)} = u(n)/\lambda(n-1), \quad n = 1, \ldots, K, \quad \text{and}$$

$$\frac{p(n+1)}{p(n)} = \lambda(n)/u(n+1), \quad n = 0, \ldots, K-1. \tag{7.18}$$

We use the probability formula (7.15) together with (7.16), (7.12), (7.13), (7.15), and (7.18) in the balance equations for our M/Ph/C/K queue to transform these balance equations into equations for the conditional probabilities $p(i|n)$. This transformation yields the following set of equations:

For $0 < n < C$, we have

$$\begin{aligned} &p(0|n)[\lambda(n) + \nu(n,0)] \\ &\quad = \left[\sum_{j=1}^{b} p(i|n+1)\mu_i \hat{q}_i + p(0|n+1)\nu(n+1,0)\right] \lambda(n)/u(n+1) \\ &\qquad + p(0|n-1)u(n)(c-n)/(c-n+1), \end{aligned} \tag{7.19}$$

$$
\begin{aligned}
&p(i|n)[\lambda(n) + \mu_i + \nu(n,i)] \\
&\quad = \left[p(0|n-1)\frac{\sigma_i}{c-n+1} + p(i|n-1)\right] u(n) + \sum_{j=1}^{b} p(j|n)\mu_j q_{ij} \\
&\qquad + p(i|n+1)\nu(n+1,i)\lambda(n)/u(n+1), i = 1, \ldots, b. \qquad (7.20)
\end{aligned}
$$

For $n = C$ and $i = 1, \ldots, b$, we get

$$
\begin{aligned}
&p(i|n)[\lambda(n) + \mu_i + \nu(n,i)] \\
&\quad = [p(0|n-1)\sigma_i + p(i|n-1)]u(n) + \sum_{j=1}^{b} p(j|n)\mu_j q_{ji} \\
&\qquad + \left[\sum_{j=1}^{b} p(j|n+1)\mu_j \hat{q}_j \sigma_i + p(i|n+1)\nu(n+1,i)\right] \lambda(n)/u(n+1).
\end{aligned}
\tag{7.21}
$$

For $n = C+1, \ldots, K-1$, we have for $i = 1, \ldots, b$

$$
\begin{aligned}
&p(i|n)[\lambda(n) + \mu_i + \nu(n,i)] \\
&\quad = p(i|n-1)u(n) + \sum_{j=1}^{b} p(j|n)\mu_j q_{ji} + \left[\sum_{j=1}^{b} p(j|n+1)\mu_j \hat{q}_j \sigma_i \right. \\
&\qquad \left. + p(i|n+1)\nu(n+1,i)\right] \lambda(n)/u(n+1). \qquad (7.22)
\end{aligned}
$$

For $n = K$, we get for $i = 1, \ldots, b$

$$
p(i|n)[\mu_i + \nu(n,i)] = p(i|n-1)u(n) + \sum_{j=1}^{b} p(j|n)q_{ji}. \tag{7.23}
$$

There is no equation for $n = 0$ since $p(0|0)$ is known, and, as you may have noted, transforming the balance equation (7.5) results in $p(0|0)[\lambda(0)] = \lambda(0)[\sum_{i=1}^{b} p(i|1)\mu_i \hat{q}_i + p(0|1)\nu(1,0)]/u(1)$, i.e., an identity: $\lambda(0) = \lambda(0)$.

Since the (approximate) conditional rates of customer completions by servers other than the selected server are expressed in terms of the conditional probabilities $p(i|n)$, while at the same time, these rates are needed to compute the latter probabilities, an obvious possible approach is to try to use a fixed-point iteration. We can devise a straightforward scheme as follows. As usual, a superscript denotes the iteration number, the number 0 referring to initial values. We use a single array to store the values of the conditional probabilities $p^k(i|n)$ so that any newly computed value immediately replaces the value from the previous iteration. We also keep a single array for the rates of customer completions at the selected server $\omega^k(n)$. In our description of the computation scheme, we use the following short hands: $u^k(n) = \min(n, C)\omega^k(n)$, $\nu^k(n, 0) = u^k(n)$ and $\nu^k(n, i) = u^k(n) - \omega^k(n)$. Our computation proceeds in the following steps.

Algorithm 7.1.

Step 1. Select an initial distribution $p^0(i|n)$ and compute the corresponding values for the conditional rates of customer completions at the selected server $\omega^0(n)$ from formula (7.16). A simple choice is to make the $p^0(i|n)$ equally likely in each active service phase: for $\geq C$, $p^0(i|n) = 1/b$ for $i = 1, \ldots, b$; for each $n < C$, $p^0(0|n) = (C - n)/C$ and $p^0(i|n) = n/(Cb)$, for $i = 1, \ldots, b$.

Step 2. At iteration k, $k = 1, 2, \ldots$, enumerate the states in the order of increasing values of $n = 1, \ldots, K$.

For each n, consider the states in the order $i = 0, 1, \ldots, b$ and compute non-normalized values $\tilde{p}^k(i|n)$ directly from the corresponding equations. Specifically, for $n = 1, \ldots, C - 1$,

$$\tilde{p}^k(0|n) = \frac{1}{\lambda(n) + \nu^{k-1}(n, 0)} \left\{ \left[\sum_{j=1}^{b} p^{k-1}(i|n+1)\mu_i \hat{q}_i \right.\right.$$
$$\left. + p^{k-1}(0|n+1)\nu^{k-1}(n+1, 0) \right] \lambda(n)/u^{k-1}(n+1)$$
$$\left. + p^k(0|n-1)u^{k-1}(n)(c-n)/(c-n+1) \right\}, \qquad (7.24)$$

$$\tilde{p}^k(i|n) = \frac{1}{\lambda(n) + \mu_i + \nu^{k-1}(n,i)}$$
$$\times \left\{ \left[p^k(0|n-1)\frac{\sigma_i}{c-n+1} + p^k(i|n-1) \right] u^{k-1}(n) \right.$$
$$+ \sum_{j=1}^{i-1} \tilde{p}^k(j|n)\mu_j q_{ij} + \sum_{j=i}^{b} p^{k-1}(j|n)\mu_j q_{ij}$$
$$\left. + p^{k-1}(i|n+1)\nu^{k-1}(n+1,i)\lambda(n)/u^{k-1}(n+1) \right\},$$
$$i = 1, \ldots, b. \tag{7.25}$$

For $n = C$,

$$\tilde{p}^k(i|n) = \frac{1}{\lambda(n) + \mu_i + \nu^{k-1}(n,i)} \left\{ [p^k(0|n-1)\sigma_i + p^k(i|n-1)] \right.$$
$$\times u^{k-1}(n) + \sum_{j=1}^{i-1} \tilde{p}^k(j|n)\mu_j q_{ji} + \sum_{j=i}^{b} p^{k-1}(j|n)\mu_j q_{ji}$$
$$+ \left[\sum_{j=1}^{b} p^{k-1}(j|n+1)\mu_j \hat{q}_j \sigma_i + p^{k-1}(i|n+1)\nu^{k-1}(n+1,i) \right]$$
$$\left. \times \lambda(n)/u^{k-1}(n+1) \right\}, \quad i = 1, \ldots, b. \tag{7.26}$$

For $n = C+1, \ldots, K-1$ and $i = 1, \ldots, b$,

$$\tilde{p}^k(i|n) = \frac{1}{\lambda(n) + \mu_i + \nu^{k-1}(n,i)} \left\{ p^k(i|n-1)u^{k-1}(n) \right.$$
$$+ \sum_{j=1}^{i-1} \tilde{p}^k(j|n)\mu_j q_{ji} + \sum_{j=i}^{b} p^{k-1}(j|n)\mu_j q_{ji}$$

$$+\left[\sum_{j=1}^{b} p^{k-1}(j|n+1)\mu_j \hat{q}_j \sigma_i + p^{k-1}(i|n+1)\nu(n+1,i)\right]$$
$$\left.\times \lambda(n)/u^{k-1}(n+1)\right\}. \tag{7.27}$$

And for $n = K$ and $i = 1, \ldots, b$,

$$\tilde{p}^k(i|n) = \frac{1}{\mu_i + \nu^{k-1}(n,i)}\left[p^k(i|n-1)u^{k-1}(n) + \sum_{j=1}^{i-1} \tilde{p}^k(j|n)q_{ji}\right.$$
$$\left.+\sum_{j=1}^{b} p^{k-1}(j|n)q_{ji}\right]. \tag{7.28}$$

Step 3. As soon as all the values of i have been enumerated for the value of n considered, compute normalized values $p^k(i|n)$:

$$p^k(i|n) = \tilde{p}^k(i|n) \Big/ \sum_j \tilde{p}^k(j|n). \tag{7.29}$$

This normalization happens before moving on to the next value of n.

Step 4. Compute new values of the conditional customer completion rates at the selected server given n for $n = 1, \ldots, K$ as

$$\omega^k(n) = \sum_{i=1}^{b} p^k(i|n)\mu_i \hat{q}_i \Big/ \sum_{i=1}^{b} p^k(i|n). \tag{7.30}$$

Check for convergence: if $\max_n \left|1 - \frac{\omega^{k-1}(n)}{\omega^k(n)}\right| < \varepsilon$ where ε is the desired convergence stringency, continue with Step 5, else go back to Step 2 for the next iteration of the scheme. The ratio $\frac{\omega^{k-1}(n)}{\omega^k(n)}$ can be computed as each new value of $\omega^k(n)$ is obtained so that a single array suffices to store the $\omega^k(n)$. Reasonable values for ε may range from 10^{-5} to 10^{-8}.

Step 5. Use the values for $u(n) = \min(n, C)\omega(n)$ obtained from the last iteration to compute the steady-state distribution $p(n)$ from formula (7.17), as well as any derived performance metrics for our M/Ph/C/K model.

The fixed-point iteration described above is in essence a variant of our semi-numerical method of conditionals, where we have embedded the computation of the conditional completion rates by the selected server, $\omega(n)$, into the fixed-point iteration. Although we have no theoretical proof of convergence for this scheme, it is clear that, if it converges, it must converge to the solution of our model, and in practice it seems to converge in most if not all cases (Brandwajn & Begin, 2014). The number of iterations required to attain convergence varies from a few hundred to several thousand depending on the service time distribution and the number of servers. Figure 7.10 shows how the number of iterations increases as a function of the coefficient of variation of the service time distribution and the number of servers. Nonetheless, since the computational effort at each

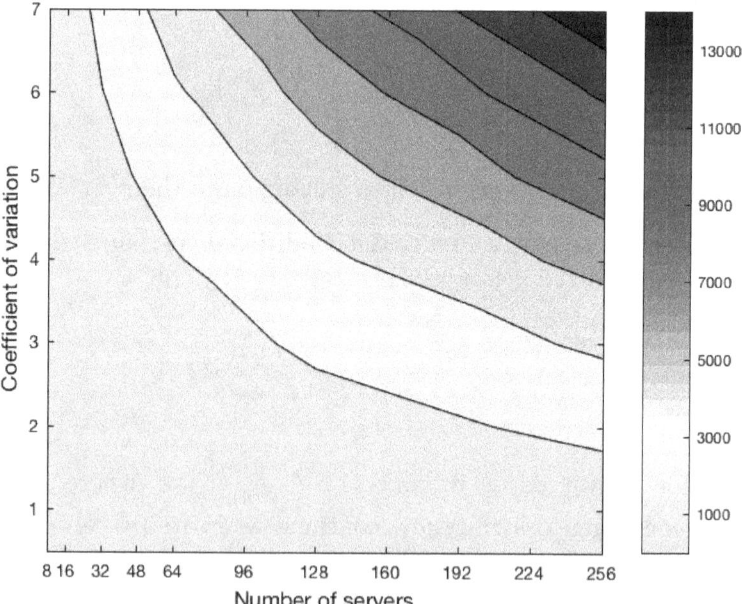

Figure 7.10. Number of iterations for the fixed-point iteration in reduced state approximation.

iteration is quite limited, the overall execution speed remains very fast even when the system has hundreds of servers.

It is clear that the reduced-state approximation breaks the "dimensionality curse" of the M/Ph/C/K model, but this fact is only of value if the results obtained using this approach are sufficiently accurate. An in-depth study of the accuracy of the reduced state approximation was conducted by the authors (Brandwajn & Begin, 2014). We considered several sets of values for the number of servers C and the maximum number of customers in the system K, as well as the offered load λ (for simplicity, we limit our discussion here to Poisson arrivals with rate λ). Since performance measures such as the mean number of customers in the system in an M/Ph/C/K queue are known to depend on the shape of the service time distribution (and not only its first two moments), we also considered several sets of values for the coefficient of variation of the service time c_s and we build several distributions with different higher-order properties, viz. skewness, denoted by s_s. To build such distributions we kept the mean service time, denoted by m_s here, at 1, and we used the algorithm by Bobbio *et al.* (2005). The values of skewness s_s we explored ranged from c_s to 100. In general, larger values of s_s correspond to longer-tailed distributions. The algorithm used aims to produce a phase-type distribution with the lowest number of phases to match the given first three moments of the distribution. The resulting number of phases in our study varied between 2 and 13, most frequently around 4.

We considered several performance metrics including the mean number of customers in the system (relative error), the loss probability (relative error), the delay probability (relative error) and the steady-state probability distribution $p(n)$ (mean relative error). The relative error, expressed as a percentage, is defined as the ratio 100(approximate − actual)/actual. The mean relative error used for $p(n)$ is defined as the weighted average of the absolute values of the corresponding relative errors. We used the state probabilities as weights so that the mean relative error amounts to $100\sum_{n=0}^{K}|p(n)_{\text{approximate}} - p(n)_{\text{actual}}|$. In our study, the actual values were obtained from a numerical solution of the full-description balance equations for the number of servers $C \leq 64$, and by discrete-event simulation for larger values of C. For the simulation runs we used the independent replications method with 7 replications of

50,000,000 completions each. We only considered relative errors for probabilities exceeding 0.001 when the actual values were obtained from a numerical solution and 0.01 when the values came from simulation. The main reason for this choice is that, for small quantities, it is easy to have large relative errors that don't seem especially meaningful as a measure of the practical accuracy of an approximation. In describing our results, we use the notion of offered load per server, which we define as the ratio λ/C. Since the system has a finite buffer capacity, the offered load can be greater than 1.

In order to get a somewhat comprehensive view of the accuracy of the reduced state description approximation, we ran close to 10,000 experiments in which the number of servers C ranged from 4 to 256, the buffer size K varied between 8 and 1024, the coefficient of variation of the service time distribution c_s spanned the range 0.4 to 7, and the offered per server load λ/C varied from 0.2 to 2.

Table 7.3 and 7.4 summarizes the accuracy of the approximate solution for the mean number of customers in the system.

The overall relative errors for the mean number in the system for a large number of experiments shown in Table 7.3 indicate that in some 95% of cases, the errors remain below 5% and exceed 10% in about 2% of cases. Our study indicates that the approximation produces essentially flawless results when the coefficient of variation of the

Table 7.3. Accuracy of the reduced-state approximation for the mean number of customers in M/Ph/C/K queue. Overall accuracy of the approximate solution for the mean number of customers in M/Ph/c/N queue

Average	Median	<5%	5–10%	10–15%	>15%
0.87%	0.03%	94.98%	3.05%	0.97%	1.00%

Table 7.4. Accuracy of the approximate solution for the mean number of customers in M/Ph/C/K queue with $C = 32$, $K = 64$, and $\lambda = 0.8\times\text{C}$ as a function of the coefficient of variation c_s.

Coefficient of variation c_s	0.5	1	2	3	4	5	6	7
Average error	0.10%	0%	0.04%	0.20%	0.46%	0.75%	1.08%	1.40%

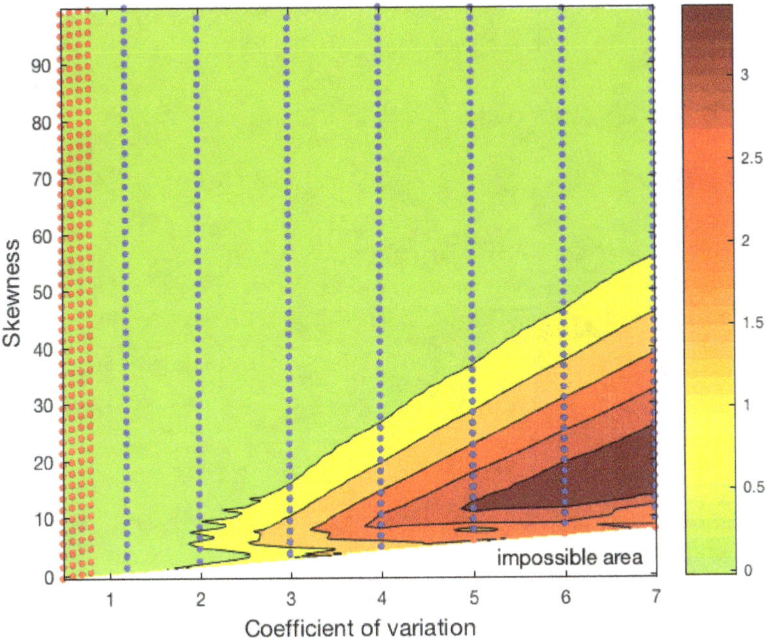

Figure 7.11. Relative error (%) for the mean number of customers.

service time distribution in the M/Ph/C/K model does not exceed 1. Approximation errors tend to appear for larger values of c_s and their extent may depend on the shape of the service time distribution, represented by its skewness s_s. Figure 7.11 illustrates this dependence for the relative error in the mean number of customers in the system with $C = 32$ server, the buffer size $K = 64$, and the offered load per server kept at $\lambda/C = 0.8$.

The relative error tends to increase as the coefficient of variation of the service time c_s increases, especially for smaller values of the skewness. Keep in mind that for a non-negative distribution (as is the case for the service time distribution) with a given coefficient of variation c_s the skewness s_s must be greater than a certain value. This is the reason behind the white "impossible area" band in Figure 7.11. We note that the relative error for the mean number in the system remains below 5% even for larger values of the coefficient of variation of the service time, such as 6 or 7.

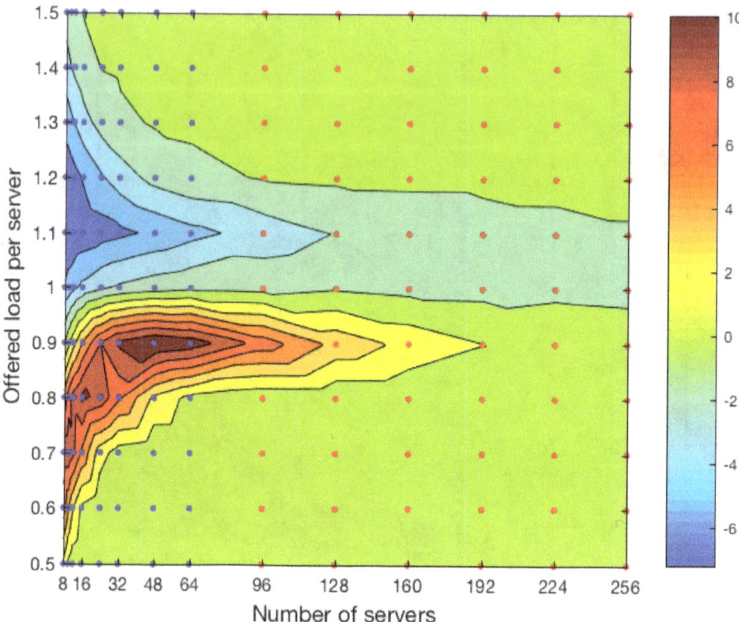

Figure 7.12. Accuracy as a function of the number of servers for a range of load levels.

When discussing the basic idea of the approximation expressed in formulas (7.13) and (7.14), we mentioned that, on intuitive grounds, we would expect the accuracy of the approximation to improve as the number of servers grows. This turns out to be indeed the case. Figure 7.12 illustrates this for the mean number of customers in an M/Ph/C/K model with a capacity of $K = 4C$ and a large coefficient of variation $c_s = 4$.

We see that the relative error may depend on the offered load per server. It appears to be the highest when the offered load is in the neighborhood of 0.9, and, as expected, it decreases relatively rapidly as the number of servers increases.

Another performance metric of interest is the loss probability, i.e., the probability that an arriving request finds the system at capacity and thus cannot be queued. For Poisson arrivals, as considered in the results described here, the loss probability is simply the steady-state probability $p(K)$ because of the PASTA property. As a reminder, the PASTA property states that Poisson arrivals see

Table 7.5. Overall accuracy of the approximate solution for the loss probability in M/Ph/C/K queue.

Average	Median	<5%	5–10%	10–15%	>15%
8.6%	0.24%	87.19%	2.32%	1.66%	8.83%

the regular steady state of the system. Table 7.5 summarizes the distribution of relative errors in the loss probability over a large number of experiments.

As we can see, in about 87% of cases, the relative error remains below 5%, although it exceeds 10% in close to 9% of cases. Apparently, these larger relative errors tend to occur when the loss probability is near or below 0.01. The influence of these larger relative errors is quite clear if we compare the mean error of 8.6% with the median error of less than 0.25%.

While the loss probability pertains to customers that find no room to queue, the delay probability measures how likely it is that an arriving customer has to queue. In other words, it is the probability that an arriving customer finds all servers busy serving other customers. With Poisson arrivals, again, the PASTA property is in effect, and the delay probability is simply the steady-state probability that all servers are busy. Just as a note in passing, if the arrivals were state dependent with rate $\lambda(n)$, i.e., quasi-Poisson, the probability of a delay or loss would have to be computed as

$$\text{Prob}_A\{n \geq C\} = \frac{\sum_{n=C}^{K} \lambda(n)p(n)}{\sum_{k=0}^{K} \lambda(k)p(k)}, \tag{7.31}$$

where $p(n)$ is the steady-state probability that there are n customers in the system. As we have seen before, formula (7.31) is the ratio of arrivals that happen when all servers are busy to all customer arrivals during a very long period of time. Similarly, with state-dependent arrivals, the loss probability would be expressed as the fraction of arrivals that find the system at capacity, i.e.,

$$\text{Prob}_A\{n = K\} = \frac{\lambda(K)p(K)}{\sum_{k=0}^{K} \lambda(k)p(k)}. \tag{7.32}$$

The subscript "A" is there to remind us that these are probabilities seen by an arriving customer as opposed to regular steady-state probabilities.

Table 7.6 shows the distribution of relative errors in the delay probability observed in a large number of experiments.

The reduced-state approximation seems to perform quite well for this metric since in 93% of the cases considered the relative error is below 5% and it exceeds 10% in less than 4% of cases, while the mean relative error for the delay probability is below 2%.

It is quite likely that the good accuracy of the approximation for the performance metrics discussed above would be sufficient in the majority of computer modeling applications. However, occasionally, we may be interested in the shape of the probability distribution for $p(n)$, and it is not rare that, for a number of reasons, an approximation may produce reasonable results for mean numbers but dismal results for the shape of the distribution. Table 7.7 illustrates the ability of the reduced-state approximation to reproduce the shape of the steady-state probability distribution $p(n)$.

This table was obtained as follows. For each set of parameters, i.e., for each steady-state distribution $p(n)$, we considered the absolute values of the relative errors in the $p(n)$ for all $n = 0, \ldots, K$, and we used the weighted average of these absolute values, where the state probabilities $p(n)$ were the weight, as the relevant metric. We refer to this metric of the deviation between the exact and approximate values of $p(n)$ as the mean relative error since

Table 7.6. Overall accuracy of the approximate solution for the delay probability in M/Ph/C/K queue.

Average	Median	<5%	5–10%	10–15%	>15%
1.89%	0.20	92.95%	3.60%	1.22%	2.23%

Table 7.7. Overall accuracy of the approximate solution for the steady-state probability $p(n)$ in an M/Ph/C/K queue.

Average	Median	<5%	5–10%	10–15%	>15%
3.76%	0.43%	80.42%	7.64%	3.17%	8.77%

it is the expected value of the absolute values of the relative errors for each $n = 0, \ldots, K$. As mentioned earlier, this amounts to using $100 \sum_{n=0}^{K} |p(n)_{\text{approximate}} - p(n)_{\text{actual}}$ as the mean relative error. Since it is difficult to estimate accurately very small values in discrete-event simulation, the data summarized in Table 7.7 pertains to systems with no more than $C = 64$ servers, for which numerical results for the state probabilities could be obtained through a numerical solution of the exact balance equations for the model. It seems worthwhile to note that, in general, the reduced-state approximation tends to correctly reproduce the shape of the steady-state distribution $p(n)$ as the mean of the error metric is below 4%, and in about 80% of cases the errors remain below 5%. Clearly, the error metric used, i.e., the mean relative error, emphasizes the most likely states at the expense of states that are less likely.

7.5 Reduced-state approximation for the M/Ph/C model

While all human-made systems have a finite queueing capacity, an M/Ph/C model with an infinite queueing room may be of interest to represent a system that is not too heavily loaded and in which the loss probability is minimal. As it turns out, the extension of reduced-state approximation to handle an open M/Ph/C queue is relatively simple if we consider the solution in terms of the conditional probabilities $p(i|n)$. As before, n is the current number of customers in the system, except that $n = 0, 1, \ldots,$ and i denotes the current service phase at the selected server. In essence, if we refer to the equations for $p(i|n)$, equations (7.19–7.21) remain unchanged, equation (7.22) is now for all values of $n > C$, and equation (7.23) is simply absent since there is no upper limit on the number of customers in the system. Specifically, if we let λ be the rate of Poisson arrivals to our system, the equations for $p(i|n)$ are as follows:

For $0 < n < C$, we have

$$\begin{aligned}
&p(0|n)[\lambda + \nu(n,0)] \\
&\quad = \left[\sum_{j=1}^{b} p(i|n+1)\mu_i \hat{q}_i + p(0|n+1)\nu(n+1,0)\right] \lambda/u(n+1) \\
&\qquad + p(0|n-1)u(n)(c-n)/(c-n+1), \qquad (7.33)
\end{aligned}$$

$$p(i|n)[\lambda + \mu_i + \nu(n,i)]$$
$$= \left[p(0|n-1)\frac{\sigma_i}{c-n+1} + p(i|n-1)\right] u(n) + \sum_{j=1}^{b} p(j|n)\mu_j q_{ij}$$
$$+ p(i|n+1)\nu(n+1,i)\lambda/u(n+1), \quad i = 1,\ldots,b. \tag{7.34}$$

For $n = C$ and $i = 1,\ldots,b$, we get

$$p(i|n)[\lambda + \mu_i + \nu(n,i)]$$
$$= [p(0|n-1)\sigma_i + p(i|n-1)]u(n) + \sum_{j=1}^{b} p(j|n)\mu_j q_{ji}$$
$$+ \left[\sum_{j=1}^{b} p(j|n+1)\mu_j \hat{q}_j \sigma_i + p(i|n+1)\nu(n+1,i)\right] \lambda/u(n+1). \tag{7.35}$$

For $n > C$, we have for $i = 1,\ldots,b$

$$p(i|n)[\lambda + \mu_i + \nu(n,i)]$$
$$= p(i|n-1)u(n) + \sum_{j=1}^{b} p(j|n)\mu_j q_{ji}$$
$$+ \left[\sum_{j=1}^{b} p(j|n+1)\mu_j \hat{q}_j \sigma_i + p(i|n+1)\nu(n+1,i)\right] \lambda/u(n+1). \tag{7.36}$$

As before, we have $\omega(n) = \sum_{i=1}^{b} p(i|n)\mu_i \hat{q}_i / \sum_{i=1}^{b} p(i|n)$ and $u(n) = \min(n,c)\omega(n)$. Formulas (7.13) and (7.14) expressing the reduced-state approximation remain valid so that $\nu(n,i) \approx u(n) - \omega(n)$, for $i = 1,\ldots,b$ and $\nu(n,0) \approx u(n)$, for $n < C$. The form of the steady-state probability $p(n)$ remains unchanged

$$p(n) = \frac{1}{\mathrm{G}} \prod_{i}^{n} \frac{\lambda}{u(i)} = \frac{\lambda^n}{\mathrm{G}} \prod_{i=1}^{n} \frac{1}{u(i)}, \quad n = 0,1,\ldots. \tag{7.37}$$

Let us now, somewhat informally, figure out what happens as the number of customers n tends to infinity. We assume that the system possesses a steady state, and we also assume that the conditional probability $p(i|n)$ reaches a limit as $n \to \infty$. We denote by $\overline{\overline{p}}(i)$ this limit so that

$$\overline{\overline{p}}(i) = \lim_{n\to\infty} p(i|n), \quad i = 1, \ldots, b \quad \text{and} \quad \sum_{i=1}^{b} \overline{\overline{p}}(i) = 1. \tag{7.38}$$

If the conditional probabilities $p(i|n)$ reach a limit for $n \to \infty$, so must the customer completion rate by the selected server $\omega(n)$, the overall rate of completions $u(n)$, and the rate of completions by servers other than the selected server $\nu(n, i)$. Thus,

$$\overline{\overline{\omega}} = \lim_{n\to\infty} \omega(n) = \sum_{i=1}^{b} \overline{\overline{p}}(i)\mu_i \hat{q}_i, \tag{7.39}$$

$$\overline{\overline{u}} = \lim_{n\to\infty} u(n) = C\overline{\overline{\omega}}, \tag{7.40}$$

$$\overline{\overline{\nu}}(i) = \lim_{n\to\infty} \nu(n, i) \approx (C-1)\overline{\overline{\omega}}, \quad \text{for } i = 1, \ldots, b. \tag{7.41}$$

Since $\overline{\overline{u}}$ is the limit of the conditional rate of customer completions $u(n)$, it follows that the steady-state probability $p(n)$ must be asymptotically geometric (with factor $\lambda/\overline{\overline{u}}$) as $n \to \infty$, which is in agreement with known related results (Takahashi, 1981). It also means that, for the steady state to exist, the factor in the asymptotic geometric distribution must be less than 1, i.e., we must have

$$\lambda < \overline{\overline{u}} \quad \text{or, equivalently,} \quad \lambda < C\overline{\overline{\omega}}. \tag{7.42}$$

It is not difficult to find the (approximate) value of the asymptotic rate of completion at the selected server by letting $n \to \infty$ in equation (7.36) for $p(i|n)$. We obtain for $i = 1, \ldots, b$

$$\overline{\overline{p}}(i)[\lambda + \mu_i + \overline{\overline{\nu}}(i)] = \overline{\overline{p}}(i)\overline{\overline{u}} + \sum_{j=1}^{b} \overline{\overline{p}}(j)\mu_j q_{ji} + \left[\sum_{j=1}^{b} \overline{\overline{p}}(j)\mu_j \hat{q}_j \sigma_i + \overline{\overline{p}}(i)\overline{\overline{\nu}}(i)\right] \lambda/\overline{\overline{u}}. \tag{7.43}$$

Taking into account formulas (7.39–7.41), this equation becomes

$$\bar{\bar{p}}(i)[\lambda + \mu_i + (C-1)\bar{\bar{\omega}}] = \bar{\bar{p}}(i)C\bar{\bar{\omega}} + \sum_{j=1}^{b} \bar{\bar{p}}(j)\mu_j q_{ji}$$

$$+ \left[\sum_{j=1}^{b} \bar{\bar{p}}(j)\mu_j \hat{q}_j \sigma_i + \bar{\bar{p}}(i)(C-1)\bar{\bar{\omega}}\right] \lambda/(C\bar{\bar{\omega}}),$$

$$i = 1, \ldots, b. \tag{7.44}$$

For a given set of model parameters, the solution of this set of equations, with the added normalization condition $\sum_{i=1}^{b} \bar{\bar{p}}(i) = 1$, yields the asymptotic probabilities $\bar{\bar{p}}(i)$ and the asymptotic completion rate at the selected server $\bar{\bar{\omega}}$. Such a solution can be obtained using a straightforward fixed-point iteration, whose design we leave to the reader.

The net of this is that we can readily obtain the asymptotic value for the conditional probabilities $p(i|n)$ as $n \to \infty$, and, as luck would have it, in practice, the convergence of the $p(i|n)$ to $\bar{\bar{p}}(i)$ tends to occur relatively fast as n increases. This fact allows us to design the following fixed-point iterative scheme (Algorithm 7.2) to obtain the steady-state solution of an M/Ph/C model.

Algorithm 7.2.

Step 1. Compute the asymptotic probabilities $\bar{\bar{p}}(i)$ by solving the set of equations (7.44) together with $\sum_{i=1}^{b} \bar{\bar{p}}(i) = 1$. Set a reasonable limit for the enumeration of the values of n in your implementation, e.g., $\hat{K} = 2000$ or $\hat{K} = 20C$, whichever is greater. Select an initial distribution $p^0(i|n)$ for $n < C$ and use $p^0(i|n) = \bar{\bar{p}}(i)$ for $n \geq C$. For $n < C$, a simple choice is to make the $p^0(i|n)$ equally likely in each active service phase: $p^0(0|n) = (C-n)/C$ and $p^0(i|n) = n/(Cb)$, for $i = 1, \ldots, b$. Compute the corresponding values for the conditional rates of customer completions at the selected server $\omega^0(n)$ from formula (7.16).

Step 2. At iteration k, $k = 1, 2, \ldots$, enumerate the states in the order of increasing values of $n = 1, 2,$. For each n, in the order $i = 0, 1, \ldots, b$ start by computing non-normalized values $\tilde{p}^k(i|n)$ directly from the corresponding equations. Specifically, for $n = 1, \ldots, C-1$,

$$\tilde{p}^k(0|n) = \frac{1}{\lambda(n) + \nu^{k-1}(n,0)} \left\{ \left[\sum_{j=1}^{b} p^{k-1}(i|n+1)\mu_i \hat{q}_i \right.\right.$$

$$\left. + p^{k-1}(0|n+1)\nu^{k-1}(n+1,0) \right] \lambda(n)/u^{k-1}(n+1)$$

$$\left. + p^k(0|n-1)u^{k-1}(n)(c-n)/(c-n+1) \right\}, \tag{7.45}$$

$$\tilde{p}^k(i|n) = \frac{1}{\lambda(n) + \mu_i + \nu^{k-1}(n,i)} \left\{ \left[p^k(0|n-1)\frac{\sigma_i}{c-n+1} \right.\right.$$

$$\left. + p^k(i|n-1) \right] u^{k-1}(n) + \sum_{j=1}^{i-1} \tilde{p}^k(j|n)\mu_j q_{ij}$$

$$+ \sum_{j=i}^{b} p^{k-1}(j|n)\mu_j q_{ij} + p^{k-1}(i|n+1)\nu^{k-1}(n+1,i)$$

$$\left. \times \lambda(n)/u^{k-1}(n+1) \right\}, \quad i = 1, \ldots, b. \tag{7.46}$$

For $n = C$,

$$\tilde{p}^k(i|n) = \frac{1}{\lambda(n) + \mu_i + \nu^{k-1}(n,i)} \left\{ [p^k(0|n-1)\sigma_i \right.$$

$$+ p^k(i|n-1)]u^{k-1}(n) + \sum_{j=1}^{i-1} \tilde{p}^k(j|n)\mu_j q_{ji}$$

$$+ \sum_{j=i}^{b} p^{k-1}(j|n)\mu_j q_{ji} + \left[\sum_{j=1}^{b} p^{k-1}(j|n+1)\mu_j \hat{q}_j \sigma_i \right.$$

$$\left. \left. + p^{k-1}(i|n+1)\nu^{k-1}(n+1,i) \right] \lambda(n)/u^{k-1}(n+1) \right\},$$

$$i = 1, \ldots, b. \tag{7.47}$$

For $n = C+1, C+2, \ldots$ and $i = 1, \ldots, b$,

$$\tilde{p}^k(i|n) = \frac{1}{\lambda(n) + \mu_i + \nu^{k-1}(n,i)} \Bigg\{ p^k(i|n-1)u^{k-1}(n) + \sum_{j=1}^{i-1} \tilde{p}^k(j|n)\mu_j q_{ji} + \sum_{j=i}^{b} p^{k-1}(j|n)\mu_j q_{ji} + \left[\sum_{j=1}^{b} p^{k-1}(j|n+1)\mu_j \hat{q}_j \sigma_i + p^{k-1}(i|n+1)\nu(n+1,i) \right] \times \lambda(n)/u^{k-1}(n+1) \Bigg\}. \quad (7.48)$$

Step 3. Compute normalized values $p^k(i|n)$

$$p^k(i|n) = \tilde{p}^k(i|n) \Big/ \sum_j \tilde{p}^k(j|n). \quad (7.49)$$

This normalization happens as soon as all values of i have been enumerated for a given value of n, before moving on to the next value of n.

Step 4. Compute new values of the conditional customer completion rates at the selected server for the value of n currently considered

$$\omega^k(n) = \sum_{i=1}^{b} p^k(i|n)\mu_i \hat{q}_i \Big/ \sum_{i=1}^{b} p^k(i|n). \quad (7.50)$$

Starting with some value of $n > C$ (for example, $n = C + 10$), check for asymptotic convergence: if $\left|1 - \frac{\omega^k(n-1)}{\omega^k(n)}\right| < \delta$, where δ is the desired asymptotic convergence stringency, set $\overline{\overline{n}} = n$. For all values of $n > \overline{\overline{n}}$, set $p^k(i|n) = \overline{\overline{p}}(i)$ and $\omega^k(n) = \overline{\overline{\omega}}$, and proceed to Step 5. As an example, reasonable values for δ range from 10^{-6} to 10^{-8}. If asymptotic convergence was not reached before all values of n have been enumerated, increase the value of $\hat{K}$, set $p^k(i|n) = \overline{\overline{p}}(i)$

and $\omega^k(n) = \overline{\overline{\omega}}$ for all n greater than the previous value of $\hat{K}$, and return to Step 2 for the next iteration.

Step 5. If $\max_n \left|1 - \frac{\omega^{k-1}(n)}{\omega^k(n)}\right| < \varepsilon$ where ε is the desired overall convergence stringency, continue with Step 6, else go back to Step 2 for the next iteration. The ratio $\frac{\omega^{k-1}(n)}{\omega^k(n)}$ can be computed as each new value of $\omega^k(n)$ is obtained so that a single array suffices to store the $\omega^k(n)$. Reasonable values for ε may range from 10^{-5} to 10^{-8}.

Step 6. Use the values for $u(n) = \min(n, C)\omega(n)$ obtained from the iterative scheme, as well as the asymptotic value $\overline{\overline{u}} = C\overline{\overline{\omega}}$ obtained in Step 1, to compute the steady-state distribution $p(n)$ from formula (7.37), as well as any derived performance metrics for our M/Ph/C model. Specifically, for $p(n)$ we compute

$$p(n) = \begin{cases} \dfrac{1}{\mathrm{G}} \displaystyle\prod_{i=1}^{n} \frac{\lambda}{u(i)}, & \text{for } n < \overline{\overline{n}} \\ \dfrac{1}{\mathrm{G}} \displaystyle\prod_{i=1}^{\overline{\overline{n}}} \frac{\lambda}{u(i)} \left(\frac{\lambda}{\overline{\overline{u}}}\right)^{n-\overline{\overline{n}}}, & \text{for } n \geq \overline{\overline{n}} \end{cases}. \tag{7.51}$$

The normalizing constant G is given by

$$\mathrm{G} = \sum_{n=0}^{\overline{\overline{n}}-1} \prod_{i=1}^{n} \frac{\lambda}{u(i)} + \frac{1}{1 - \lambda/\overline{\overline{u}}} \prod_{i=1}^{\overline{\overline{n}}} \frac{\lambda}{u(i)}. \tag{7.52}$$

As you may have noted, the iterative scheme described above for solution of the M/Ph/C queue avoids arbitrary truncation, which might be used as one alternative for dealing with an infinite buffer size or explicit reliance on the repetitive structure of matrices, which is the avenue used in the matrix geometric method. Instead, the approach presented above uses the known simple form of the steady-state probabilities $p(n)$ and the asymptotic convergence of the conditional probabilities $p(i|n)$. In practice, the interest of this approach lies in the fact that asymptotic convergence tends to be reached for relatively small values $\overline{\overline{n}}$. The latter may be significantly smaller than the actual value of the system capacity K in an M/Ph/C/K queue, resulting in big computational savings compared to the solution of the M/Ph/C/K model. However, keep in mind that, as mentioned before, the system load must not be too high for the open M/Ph/C queue to be a good alternative to the M/Ph/C/K model.

The solution approach we just described can also be applied if the rate of customer arrivals is a function of the current number of customers in the system, as long as this rate becomes constant starting with some value of n, i.e., $\lambda(n) = \lambda$ for all $n \geq \hat{n}$. Clearly, we must then have $\overline{\overline{n}} > \hat{n}$ in our iterative scheme.

The value of the asymptotic service rate $\overline{\overline{\omega}}$ varies with the values of model parameters and, in particular, with the rate of customer arrivals λ. We must have $\lambda < C\overline{\overline{\omega}}$ for the M/Ph/C system to be stable. Thus, if we let $\lambda = C\overline{\overline{\omega}}$ in equation (7.44), we can determine the stability condition for our system from the resulting simpler set of equations

$$\overline{\overline{p}}(i)\mu_i = \sum_{j=1}^{b} \overline{\overline{p}}(j)\mu_j q_{ji} + \sum_{j=1}^{b} \overline{\overline{p}}(j)\mu_j \hat{q}_j \sigma_i, \quad \text{for } i = 1, \ldots, b, \tag{7.53}$$

with

$$\sum_{i=1}^{b} \overline{\overline{p}}(i) = 1.$$

This is a simple, easy-to-solve set of linear equations. As an example, consider the particular case of a Coxian distribution with b stages; μ_i is the intensity of stage $i = 1, \ldots, b$, α_i denotes the probability that the service proceeds to stage $i+1$ following the completion of stage i, for $i = 1, \ldots, b-1$ and $\beta_i = 1 - \alpha_i$ is the probability that the service ends after stage i. We set $\beta_b = 1$ for the last stage in the distribution. From equation (7.44), we get the very simple relationship

$$\overline{\overline{p}}(i)\mu_i = \overline{\overline{p}}(i-1)\mu_{i-1}\alpha_{i-1}, \quad i = 2, \ldots, b. \tag{7.54}$$

From here, all $\overline{\overline{p}}(i)$ can be expressed in terms of $\overline{\overline{p}}(1)$, which can then be determined from the normalization condition $\sum_{i=1}^{b} \overline{\overline{p}}(i) = 1$. The limiting service rate $\overline{\overline{\omega}} = \sum_{j=1}^{b} \overline{\overline{p}}(j)\mu_j\beta_j$ is then readily obtained and, perhaps not surprisingly, it turns out to be equal to the inverse of the mean service time of the Coxian distribution

$$\frac{1}{\overline{\overline{\omega}}} = \frac{1}{\mu_1} + \frac{\alpha_1}{\mu_2} + \frac{\alpha_1\alpha_2}{\mu_3} + \cdots + \frac{\alpha_1 \ldots \alpha_{b-1}}{\mu_b}. \tag{7.55}$$

Thus, the stability condition becomes $\lambda < \frac{C}{E[S]}$, where $E[S]$ is the mean service time of a customer. This correlates with the observation that in an M/Ph/C system, as the servers approach saturation, i.e., as they become continuously busy, the precise nature of the service distribution stops mattering.

7.6 Quasi-general time between arrivals and service times (Ph/Ph/C/K model)

With the reduced-state approximation, it is possible to handle large numbers of servers in M/Ph/C queues. Of course, the arrival process is still assumed to be memoryless, although the arrival rate can be state dependent. It is known from workload measurements (Hsu & Smith, 2003; Willinger *et al.*, 2004; Jiang & Dovrolis, 2005; McNutt, 2005) that the pattern of customer arrivals may be significantly different from the "well behaved" Poisson arrivals. This has been observed in a number of systems, including the pattern of I/O requests in larger systems or that of VM activation requests in cloud systems. For such systems, it is important to consider a model in which not only the service distribution can be quasi-general, but also the arrivals can be non-Poisson. We use a phase-type distribution for the times between consecutive arrivals so that, with a finite capacity of K customers, our model becomes a Ph/Ph/C/K queue, as shown in Figure 7.13.

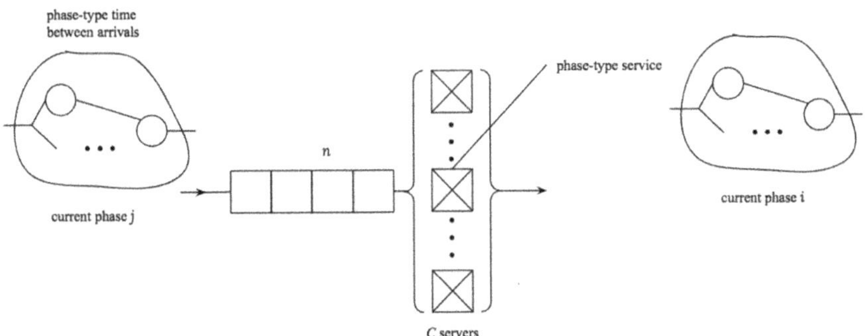

Figure 7.13. A Ph/Ph/C/K queue.

Specifically, we assume that the times between arrivals are represented as an acyclic phase-type distribution with a phases. We denote by τ_j the probability that the arrival process starts in phase j, $j = 1, \ldots, a$, by λ_j the completion rate of phase j, by r_{jl} $(l > j)$ probability that phase j is followed by phase l, and by $\hat{r}_j$ the probability that the arrival process terminates after phase j. Recall that any distribution may be represented arbitrarily closely by such a phase-type distribution (O'Cinneide, 1990; Neuts, 1994; Bolch *et al.*, 2006).

As in the M/Ph/C/K model discussed earlier, the service times are represented by an acyclic phase-type distribution with b phases. We denote by σ_i the probability that the service of a customer starts in phase i, $i = 1, \ldots, b$. The intensity, i.e., the completion rate, for phase i is denoted by μ_i. We let q_{ij} be the probability that the service continues in phase j following the completion of phase i, and we denote by $\hat{q}_i$ the probability that the service ends upon completion of phase i.

Clearly, the issue of rapidly increasing computational complexity discussed earlier for the M/Ph/C/K queue is only compounded by the more involved arrival process. Therefore, we will use the same idea of reduced-state description, in which we describe in detail only the state of one of the C identical but non-synchronized servers. With this in mind, we can describe the state of the system by the vector (n, i, j) where $n = 0, \ldots, K$ is the current number of customers in the system (queued and in service), i is the current phase of service at the selected server $(i = 0, 1, ..b)$, and $j = 1, \ldots, a$ is the current phase of the arrival process. The service phase $i = 0$ corresponds to an idle server and is only possible if $n < C$.

We have two possible avenues from here. Our first possibility is to leverage the fact that we have an exact recurrent solution for Ph/M/C/K queues with state-dependent service rates, as described in Section 2 (formulas (2.62–2.65) together with (2.58)), as well as a pretty accurate approximate solution for the M/Ph/C/K queue with state-dependent arrival rates summarized in Algorithm 7.1. We can proceed as follows (Atmaca *et al.*, 2015). Instead of looking directly at the probability $p(n, i, j)$, let us first consider only the probability that there are n customers in the system and that the arrival process is in its phase j, which we denote by $p_1(n, j)$, $j = 1, \ldots, a; n = 0, \ldots, K$. We let $p(i|n, j)$ denote the conditional probability that the service at the selected server is in phase i, $i = 0, 1, \ldots, b$, given (n, j), $n > 0$ (phase $i = 0$ is only possible for $n < C$). The conditional rate of customer service at the selected server can be expressed as

$$\omega_1(n,j) = \sum_{i=1}^{b} p(i|n,j)\mu_i\hat{q}_i \Big/ \sum_{i=1}^{b} p(i|n,j) \quad \text{for } n > 0, \quad j = 1, \ldots, a. \tag{7.56}$$

As an approximation, we assume that the current phase of service depends primarily on n, the number of customers in the system, and not on the phase of the arrival process, j. Consequently, we get

$$\omega_1(n,j) \approx \omega(n) = \sum_{i=1}^{b} p(i|n)\mu_i\hat{q}_i \Big/ \sum_{i=1}^{b} p(i|n), \quad \text{for } n > 0, \quad j = 1, \ldots, a. \tag{7.57}$$

In formula (7.57), $p(i|n)$ is the conditional probability that the service phase is i given that there are n customers in the system. The overall rate of customer completions, given the current number of customers in the system, becomes

$$u(n) \approx \min(n, C)\omega(n) = \min(n, C) \sum_{i=1}^{b} p(i|n)\mu_i\hat{q}_i \Big/ \sum_{i=1}^{b} p(i|n). \tag{7.58}$$

With this approximation, provided we know $u(n)$, we can solve for the probability $p_1(j, n)$ using the recurrence formulas (2.63) and (2.64). In addition to the $p_1(j, n)$, this recurrence produces the conditional rate of customer arrivals, denoted by $\alpha(n)$, as shown in the formula (2.65) for the Ph/M/1/K queue. To summarize, if we know the conditional rate of customer departures $u(n)$, we can obtain the conditional rate of customer arrivals given the current number of customers in the system from the recurrent solution of the Ph/M/C/K queue.

On the other hand, let us consider only the probability that the current service phase at the selected server is i and that there are n customers in the system. We denote this probability by $p_2(n, i)$, $i = 0, 1, \ldots, b$; $n = 0, 1, \ldots, K$ (i is meaningless for $n = 0$, and $i = 0$ is possible only if $n < C$). Let us denote by $p(j|n, i)$ the conditional probability that the phase of the arrival process is j given the system state (n, i). The rate with which arrivals happen (and n increases if $n < K$) can be expressed as

$$\alpha_2(i,n) = \sum_{i=1}^{a} p(j|i,n)\lambda_j \hat{r}_j. \tag{7.59}$$

As an approximation, assume that the current phase of the arrival process depends more on the number of customers in the system than on the phase of service of the user being served at the selected server so that

$$\alpha_2(i,n) \approx \alpha(n) = \sum_{i=1}^{a} p(j|n)\lambda_j \hat{r}_j, \quad \text{for } n = 0, \ldots, K. \tag{7.60}$$

With this approximation, the system considered becomes an M/Ph/C/K queue with state-dependent arrival with rate $\alpha(n)$. If we know this rate, we can solve for the probability $p_2(n,i)$ using the procedure described as Algorithm 7.1 for the M/Ph/C/K queue.

By now, you certainly guessed where this is going. Similar to the iteration between the Ph/M/1/K and the M/Ph/1/K models shown in Figure 6.10, we can iterate between the solution of the Ph/M/C/K and the M/Ph/C/K models. This is summarized in Figure 7.14.

As mentioned in Section 2, the solution of the Ph/M/C/K model uses the same simple recurrence as the Ph/M/1/K queue. The approximate solution of the M/Ph/C/K model is described in Algorithm 7.1. This iteration between two simpler models is our first avenue.

In our second avenue, we consider the probabilities $p(n,i,j)$ directly. Similar to what we did for the M/Ph/C/K queue, we derive balance equations for the steady-state probability $p(n,i,j)$. These equations involve the rate of customer departures from servers other than the selected server given the current system state, which we denote by $\nu(n,i,j)$. We let $p(n)$ be the steady-state probability that there are n customers in the system and $p(i,j|n)$ the conditional probability that the current service phase is i and the arrival process is in phase j given n. From the definition of conditional probability, we have

$$p(n,i,j) = p(n)p(i,j|n) \quad \text{with} \sum_{i,j} p(i,j|n) = 1, \quad \forall n. \tag{7.61}$$

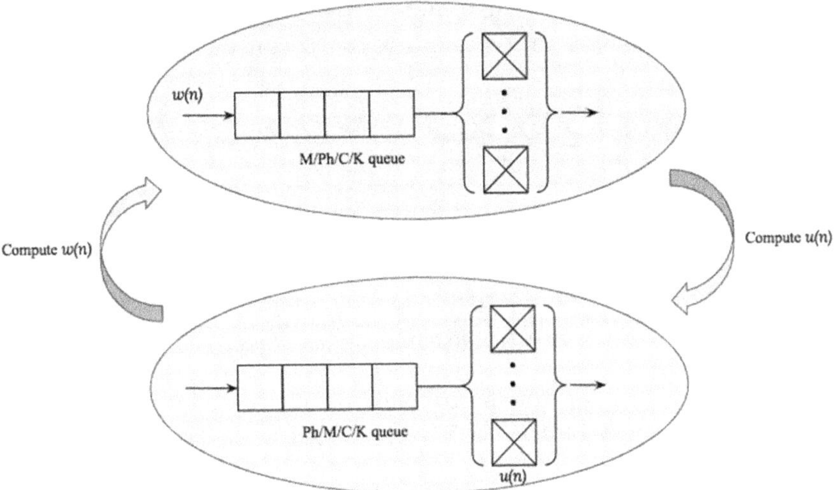

Figure 7.14. Fixed-point iteration Ph/M/C/K–M/Ph/C/K to solve approximately the Ph/Ph/C/K model.

The conditional rate of customer completions, given that there are n customers and that the selected server is active, can be expressed as

$$\omega(n) = \sum_{j=1}^{a}\sum_{i=1}^{b} p(i,j|n)\mu_i\hat{q}_i \Bigg/ \sum_{j=1}^{a}\sum_{i=1}^{b} p(i,j|n), \quad n > 0. \tag{7.62}$$

The overall conditional rate of completions when there are n customers is given by

$$u(n) = \min(n, C)\omega(n), \quad n = 1, \ldots, K. \tag{7.63}$$

As an approximation, we assume that rates of customer departures from servers other than the selected server don't depend on the current phase of service at the selected server, if it is active, nor the current phase of the arrival process so that

$$\nu(n,i,j) \approx u(n) - \omega(n), \quad i = 1, \ldots, b; \quad j = 1, \ldots, a, \quad \text{for all } n > 0, \tag{7.64}$$

$$\nu(n,0,j) \approx u(n), \quad j = 1, \ldots, a, \quad \text{for } n < C. \tag{7.65}$$

The conditional rate of customer arrivals, given that there are n customers in the system, denoted by $\alpha(n)$, can be expressed as

$$\alpha(n) = \sum_{i=0}^{b} \sum_{j=1}^{a} p(i, j|n) \lambda_j \hat{r}_j, \quad n = 0, \ldots, K. \tag{7.66}$$

Thus, the steady-state probability $p(n)$ has a well-known product form

$$p(n) = \frac{1}{\mathrm{G}} \prod_{k=1}^{n} \frac{\alpha(k-1)}{u(k)}, \quad n = 0, 1, \ldots, K. \tag{7.67}$$

As usual, empty products are set to 1, and G is a normalizing constant such that $\frac{1}{\mathrm{G}} \sum_{n=0}^{K} \prod_{k=1}^{n} \frac{\alpha(k-1)}{u(k)} = 1$, i.e., $\mathrm{G} = \sum_{n=0}^{K} \prod_{k=1}^{n} \frac{\alpha(k-1)}{u(k)}$.

With the help of the known product form of $p(n)$ given by (7.67), we can transform the balance equations into equations for the conditional probabilities $p(i, j|n)$, which can be solved using a fixed-point iteration analogous to Algorithm 7.1 described for the M/Ph/C/K queue earlier in this chapter. That's our second possibility.

Clearly, compared to the direct solution for the state description $p(n, i, j)$ (or, equivalently, the conditional probabilities $p(i, j|n)$ and the probability that there are n customers, $p(n)$) our first avenue introduces an additional approximation, which is likely to impact the resulting accuracy. It has the big advantage, however, of using existing solution procedures as elements, so to speak, off the shelf. Although it is difficult to see how (except if the number of servers C is small) the knowledge of the current phase of service at the selected server influences our knowledge of the current phase of the arrival process, and vice versa, in practice, the second approach seems to produce more accurate results. Table 7.8 illustrates this for a few examples.

We give in this table the relative errors, in percent, in the server utilization and the mean number of customers in the system for the two approaches discussed above. Lines 1 through 4 in this table correspond to a model with $C = 10$ servers, system capacity $= 50$, a distribution of the times between arrivals with a coefficient of variation $c_a = \sqrt{3}$ and a service times distribution with a coefficient of variation $c_s = 2$. The last line in this table corresponds to a similar system with a larger coefficient of the times between arrivals $c_a = 2$.

Table 7.8. Examples of relative errors for the two approaches for the approximate solution of the Ph/Ph/C/K model (last line marked with * is for a system with a larger coefficient of variation of the time between arrivals: 2 vs. 1.73).

Arrival rate	Server utilization		Mean number in system	
	Approach 1 (%)	Approach 2 (%)	Approach 1 (%)	Approach 2 (%)
2.5	0.00	−0.94	0.08	−0.94
5	0.00	−0.64	2.94	−0.31
7.5	−0.06	−0.32	7.18	1.09
9	−0.31	−0.23	4.14	0.62
7.5*	−0.16	−0.51	8.58	1.06

The second approach tends to underestimate the server utilization but, on average, is better when it comes to the mean number of customers in the system. This is particularly visible for moderately high server utilization and larger values of the coefficient of variation of the time between arrivals.

Using the asymptotic convergence approach mentioned earlier for the M/Ph/C model, both the direct solution for $p(n, i, j)$ and the iteration between simpler models discussed above can be extended to models with infinite buffer spaces, i.e., to the Ph/Ph/C queue. The details of such an extension are outside the scope of our book.

With the help of the reduced-state approximation (Brandwajn & Begin, 2016), we have at our disposal a relatively simple, fast, and accurate solution for the Ph/Ph/C/K queue, which is a model of a multi-server facility with quasi-general arrivals and quasi-general service times. This queue can be viewed as a high-level model of a number of systems or system components ranging from virtualized cloud servers (Xiong & Perros, 2009; Khazaei *et al.*, 2011; Yang *et al.*, 2013; Atmaca *et al.*, 2015; Patch, 2018; Wang, 2021) to I/O devices in large mainframes such as IBM PAVs (Kuzler *et al.*, 1999; Begin & Brandwajn, 2013).

7.7 Model with multiple servers, classes of customers, and FCFS service

As noted earlier, our approximate solution for the Ph/Ph/C/K model scales rather nicely with the number of servers and the size of the buffer. One limitation of the model considered is that all customers are assumed to be statistically identical, i.e., there is a single class of customers. This is fine as long as there is no need to distinguish between customers with different workload characteristics. Granted, the queueing discipline in the Ph/Ph/C/K model is First-Come-First-Served (FCFS), which many of us may think as intrinsically "fair". Nonetheless, for customers with small service requirements being "stuck" waiting for customers with much longer service times certainly does not seem particularly appealing or fair, for that matter. As an example, if we consider a model of a computer system with parallelized I/O devices (such as the IBM Parallel Access Volumes), no I/O priority queueing, and several types of customers (jobs, processes, tasks, transactions, etc.), the performance of an FCFS multi-server queue with multiple customer classes comes to the forefront to evaluate the expected response time of a specific class of jobs.

We note that the exact analytical solution of such queueing models is not known in general, and, except if the service times are not only the same for all classes but also exponentially distributed at an FCFS center, the BCMP solution cannot be used. Additionally, as we have discussed earlier in this chapter, the accuracy of approximations based on the arrival theorem, which one may be tempted to use to deal with an FCFS system with job classes, should be viewed with caution. Interestingly, there seem to be relatively few studies of systems with multiple servers and multiple customer classes. This may be due to the intrinsic complexity of the classical state description in FCFS queues, which requires a vector whose elements are the classes of customers at each queue position (Baskett *et al.*, 1975). Of course, a state description of this type leads to a combinatorial explosion of the number of states as the number of customer classes increases. And, unless the service times are exponentially distributed, this complexity is in addition to the complexity inherent in the description of the state of the servers themselves.

Following the idea of simplified state descriptions to circumvent the complexity mentioned above, we will develop a simple approach to obtain an accurate approximate solution, in which the number of equations to solve grows linearly with the number of servers and the number of classes. We focus our discussion on systems with memoryless albeit state-dependent arrivals. Our discussion follows the spirit of published research (Brandwajn & Begin, 2019).

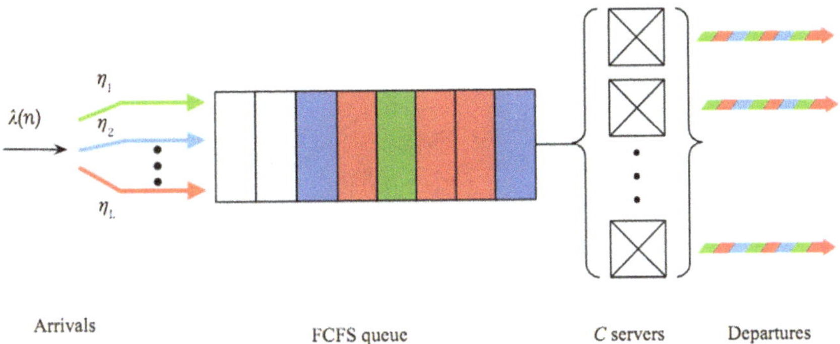

Figure 7.15. Multi-server FCFS queue with multiple classes of customers.

As shown in Figure 7.15, the multi-server FCFS queue with multiple classes of customers considered has C homogeneous servers serving a single queue of customers. Times between customer arrivals are distributed according to a memoryless distribution with rate $\lambda(n)$, where n is the current total number of customers in the system, queued and in service. For systems with finite buffer capacity, we let K be the maximum total number of customers in the system. There are L customer classes and η_ℓ is the probability that an arriving customer is of class $\ell = 1, \ldots, L$. This probability is assumed to be independent of the current state of the system. The queueing discipline is FCFS, i.e., customers queue for service in the order of their arrival and any available server starts serving the customer at the head of the queue.

We assume that each customer class has its own phase-type distribution. As we have mentioned several times in this text, any distribution can be represented arbitrarily closely by a phase-type distribution (O'Cinneide, 1990; Neuts, 1994; Bolch *et al.*, 2006). Figure 7.16 shows an illustration of such a phase-type distribution for customers of class ℓ.

Referring to class ℓ, there are a total of b_ℓ exponential phases, each with intensity (rate) $\mu_{\ell,i}$ ($i = 1, \ldots, b_\ell$). We denote by $\sigma_{\ell,i}$ the probability that the service starts in phase i and by $q_{\ell,ij}$ the probability that the service continues in phase j following the completion of phase i. $\hat{q}_{\ell,i}$ denotes the probability that the service ends and the customer leaves the system after the completion of phase i. T_ℓ is the mean service time for class ℓ customers.

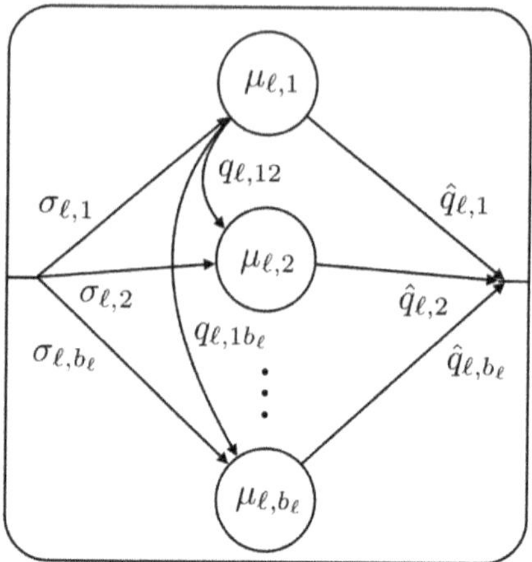

Figure 7.16. Phase-type distribution for the service times of class ℓ.

7.7.1 *Two classes of customers*

We start the derivation of our solution by considering the case where there are only $L = 2$ classes of customers. We let $m = \min(n, C)$ be the current number of busy servers. To deal with non-exponential service times, we borrow the idea of the reduced-state description, which we applied to M/Ph/C models earlier in this book (Section 7.4). That's the easy part. The more challenging part is how to deal with the FCFS order in the queue. As we mentioned earlier, we certainly don't want to describe the state of the queue by the class of each customer in the queue. After some (OK, a lot of) thought, we find that it is sufficient to describe the state of the system by the vector (m_1, ℓ, i, n) where $m_1 = 0, \ldots, m-1$ is the current number of class 1 customers being served by servers other than the selected server, ℓ is the class of the current customer at the selected server, i describes the current service phase of the selected server, and n is the current number of customers in the system. We use the class number 0 to denote an idle server so that the possible values for ℓ are as follows:

$$\ell = \begin{cases} 0 & \text{if the selected server is idle} \\ 1 & \text{if the selected server is serving a customer of class 1} \\ 2 & \text{if the selected server is serving a customer of class 2,} \end{cases}$$

and $i = 1, \ldots, b_\ell$. By convention, we set $b_0 = 1$.

As usual, we consider the system in its steady state. In the case of an unrestricted buffer size, we assume that the steady state exists for the given values of model parameters; the steady state always exists in the queue capacity is finite. We denote by $p(m_1, \ell, i, n)$ the probability that the current state of the system is $(m_1, \ \ell, i, n)$. We also denote by $p(m_1, i|n)$ the probability that there are m_1 class 1 customers in service at servers other than the selected server and that the state of the selected server is (ℓ, i) given that the total number of customers in the system is n. From the definition of conditional probability, we have $p(m_1, \ell, i, n) = p(m_1, \ell, i|n)p(n)$ where $p(n)$ is the marginal probability for n, i.e., the probability that there are n customers in the system, regardless of the values of other state variables. The conditional probabilities $p(m_1, \ell, i|n)$ must be normalized for each value of n so that $\sum_{m_1=0}^{m-1} \sum_{\ell=0}^{2} \sum_{i=1}^{b_\ell} p(m_1, \ell, i|n) = 1 \ \forall n$. Clearly, the value $\ell = 0$ is possible only for $n < C$, and it is the only value possible when $n = 0$.

With the state description defined above, it is not difficult to derive the balance equations for our model. Let us denote by $\nu_k(m_1, \ell, i, n)$ the rate of departures of customers of class k $(k = 1, 2)$ from servers other than the selected server given that the current system state is (m_1, ℓ, i, n). These quantities are not given *a priori* and will be approximated in our solution. This is similar to what we have done earlier for the M/Ph/C/K queue with a single class of customers. As an example, for $n > C$, $0 < m_1 < C - 1$, $\ell = 1, 2$, and $i = 1, \ldots, b_\ell$, we get

$$
\begin{aligned}
&p(m_1, \ell, i, n)[\lambda(n) + \mu_{\ell,i} + \nu_1(m_1, \ell.i, n) + \nu_2(m_1, \ell, i.n)] \\
&\quad = p(m_1, \ell, i, n-1)\lambda(n-1) + \sum_{k=1}^{2} \sum_{j=1}^{b_k} p(m_1, k, j, n+1) \\
&\qquad \times \mu_{k,j}\hat{q}_{k,j}\eta_\ell\sigma_{\ell.i} + \sum_{j=1}^{b_\ell} p(m_1, \ell, j, n)\mu_{\ell,j}q_{\ell,ji} \\
&\qquad + \sum_{k=1}^{2} p(m_1, \ell, i, n+1)\nu_k(m_1, \ell, i, n+1)\eta_k \\
&\qquad + p(m_1 + 1, \ell, i, n+1)\nu_1(m_1 + 1, \ell, i, n+1)\eta_2 \\
&\qquad + p(m_1 - 1, \ell, i, n+1)\nu_2(m_1 - 1, \ell, i, n+1)\eta_1. \qquad (7.68)
\end{aligned}
$$

Note that in our balance equations, the probability that a departing customer is replaced by a customer of class ℓ is given simply by η_ℓ, which is the probability that a customer entering the queue is of class ℓ (here $\ell = 1, 2$). On the left-hand side of equation (7.68), we have the rate of flow out of the state (m_1, ℓ, i, n): arrival of a new customer, end of phase i at the selected server, and departures of customers of each class from other servers. Terms on the right-hand side of equation (7.68) represent the rate of flow into the state considered. The first term corresponds to the arrival of a new customer. The second term corresponds to a completion by a customer of either class from any phase of service, the customer at the head of the queue being of class ℓ and starting its service in phase i. The third term on the right-hand side of equation (7.68) represents a change in the phase of service at the selected server. The fourth term corresponds to the departures of a customer of either class from servers other than the selected server and the customer at the head of the queue being of the same class as the departing customer. The fifth term represents the departure of a customer of class 1 from servers other than the selected server, replaced by a customer of class 2 from the head of the queue. This reduces the number of customers of class 1 at the other servers. Finally, the sixth term corresponds to the reverse situation where a customer of class 2 completes service at the other servers and is replaced by a customer of class 1, thus increasing the number of customers of this class at the other servers.

Equations for other states can be derived in a similar manner. For these balance equations to be useful, we need a way to assess the conditional rates of customer completions at servers other than the given server. We start by computing the conditional rate of customer departures from the selected server given that it is busy serving a customer of class ℓ ($\ell = 1, 2$) and there is a total of n customers in the system. We denote this rate by $\xi_\ell(n)$. Recalling that $m = \min(n, C)$ denotes the number of active servers (including the selected server) when the number of users in the system is n, we get

$$\xi_\ell(n) = \frac{\sum_{m_1=0}^{m-1} \sum_{i=1}^{b_\ell} p(m_1, \ell, i|n)\mu_{\ell,i}\hat{q}_{\ell,i}}{\sum_{m_1=0}^{m-1} \sum_{i=1}^{b_\ell} p(m_1, \ell, i|n)}. \tag{7.69}$$

We approximate the unknown rates of departures from other servers $\nu_k(m_1, \ell, i, n)$ as follows:

$$\nu_k(m_1, \ell, i, n) \approx m_k \xi_k(n) \quad \text{for } k = 1, 2 \quad \text{where}$$
$$m_2 = \begin{cases} m - 1 - m_1 & \text{if } \ell > 0 \\ m - m_1 & \text{if } \ell = 0 \end{cases}. \tag{7.70}$$

Apparently, this approximate computation of the departure rates from servers other than the selected server is the only approximation in this solution of an FCFS multi-server model with two classes of customers. There are some minor adjustments possible to the values obtained from formulas (7.69) and (7.70) that we describe later in this section.

Let us now turn our attention to the rate with which customers leave the selected server, i.e., complete their service there, given that the total number of customers in the system is n. This rate, denoted by $\gamma(n)$, can be expressed as

$$\gamma(n) = \sum_{m_1=0}^{m-1} \sum_{\ell=1}^{2} \sum_{i=1}^{b_\ell} p(m_1, \ell, i|n) \mu_{\ell,i} \hat{q}_{\ell,i}. \tag{7.71}$$

Formula (7.71) is simply the sum over all states in which the selected server is not idle of the completion rates out of each phase of service for both classes of customers. We reason that since all the servers are homogeneous and thus statistically identical, the total rate of customer departures when there are n customers in the system, denoted by $u(n)$, must be given by $u(n) = C\gamma(n)$. It follows that, if it exists, the steady-state probability $p(n)$ can be written as

$$p(n) = \frac{1}{\mathrm{G}} \prod_{j=1}^{n} \frac{\lambda(j-1)}{u(j)}, \quad \text{for } n = 0, 1, \ldots, \tag{7.72}$$

where G is a normalizing constant such that $\sum_{n \geq 0} p(n) = 1$. We have seen this general product form for the probability of the total number of customers in the system in a large number of models.

We can apply the principle of the method of conditionals (Section 5.2) to obtain a solution for our FCFS queue with two classes of customers, viz. we can use formula (7.72) relating the probabilities $p(n)$ to the conditional arrival and completion rates $\lambda(n)$ and $u(n)$ together with the identity $p(m_1, \ell, i, n) = p(m_1, \ell, i|n)p(n)$ to transform the balance equations for our model into equations for the conditional probabilities $p(m_1, \ell, i|n)$. The resulting system of equations for $p(m_1, \ell, i|n)$ can be solved using a straightforward fixed-point iteration, which proceeds as follows.

Let's consider system states in the order of increasing values of $n = 1, 2, \ldots,$ and for each value of n let's enumerate the states in the order of increasing m_1 (recall that m_1 is the number of class 1 customers at servers other than the selected server), for each value of m_1, in the order of increasing values of ℓ, and for each value of ℓ, in the order of increasing $i = 1, \ldots, b_\ell$. We use a superscript t to denote the current iteration number and let $\tilde{p}^t(m_1, \ell, i|n)$ denote non-normalized values obtained during the computation. As usual, we start with a feasible initial distribution $p^0(m_1, \ell, i|n)$ and the corresponding set of $u^0(n)$, $\xi^0_\ell(n)$ and from there $\nu^0_k(m_1, \ell, i, n)$ for $n = 1, 2, \ldots.$ As an example, in the case of equation (7.68) given above for $n > C$, $0 < m_1 < C - 1$, $\ell = 1, 2$, and $i = 1, \ldots, b_\ell$, we can compute values at iteration $t = 1, 2, \ldots,$

$$\begin{aligned}
\tilde{p}^t(m_1, \ell, i|n) = {} & \frac{1}{[\lambda(n) + \mu_{\ell,i} + \nu_1^{t-1}(m_1, \ell.i, n) + \nu_2^{t-1}(m_1, \ell, i.n)]} \\
& \times \left\{ p^t(m_1, \ell, i|n-1)u^{t-1}(n) + \sum_{k=1}^{2} \sum_{j=1}^{b_k} \right. \\
& \times \frac{p^{t-1}(m_1, k, j|n+1)\mu_{k,j}\hat{q}_{k,j}\eta_\ell \sigma_{\ell.i}\lambda(n)}{u^{t-1}(n+1)} \\
& + \sum_{j=1}^{i-1} \tilde{p}^t(m_1, \ell, j|n)\mu_{\ell,j}q_{\ell,ji} + \sum_{j=i}^{b_\ell} p^{t-1}(m_1, \ell, j|n) \\
& \times \mu_{\ell,j}q_{\ell,ji} + \sum_{k=1}^{2} p^{t-1}(m_1, \ell, i|n+1)\nu_k^{t-1} \\
& \times (m_1, \ell, i|n+1)\eta_k \lambda(n)/u^{t-1}(n+1) \\
& + [p^{t-1}(m_1 + 1, \ell, i|n+1)\nu_1^{t-1}(m_1 + 1, \ell, i, n+1)\eta_2
\end{aligned}$$

$$+p^t(m_1-1,\ell,i|n+1)\nu_2^{t-1}(m_1-1,\ell,i,n+1)\eta_1]$$

$$\left.\times\lambda(n)/u^{t-1}(n+1)\right\}. \qquad (7.73)$$

As we mentioned earlier, the conditional probabilities $p(m_1,\ell,i|n)$ must be normalized for each value of n, viz. we must have $\sum_{m_1=0}^{m-1}\sum_{\ell=0}^{2}\sum_{i=1}^{b_\ell} p(m_1,\ell,i|n) = 1$. This allows us to obtain the actual normalized values for a given n using the formula

$$p^t(m_1,\ell,i|n) = \tilde{p}^t(m_1,\ell,i|n) \Bigg/ \sum_{m_1=0}^{m-1}\sum_{\ell=0}^{2}\sum_{i=1}^{b_\ell} \tilde{p}^t(m_1,\ell,i|n). \qquad (7.74)$$

From here, we compute new values for $u^t(n)$ and $\xi_\ell^t(n)$ with the help of formulas (7.71) and (7.69), respectively. This also yields new values for $\nu_k^t(m_{1,},\ell,i,n)$ from formula (7.70). Then, we move on to the next value of n. For values of $n=1,\ldots,C$, the fixed-point iteration proceeds in an analogous way. We stop the iteration when a convergence criterion such as $\left|\frac{u^{t-1}(n)}{u^t(n)}-1\right|<\varepsilon$ for all $n>0$, where ε is the desired convergence stringency, is satisfied (e.g., $\varepsilon=10^{-6}$). Although there is no theoretical proof that this iterative scheme converges to a unique solution, in practice, there seem to be no particular convergence problems.

For systems with a finite buffer capacity, in which the total number of customers is limited to K, the number of equations solved at each iteration with the state description chosen is given by

$$KC(b_1+b_2)+(C-1)-(C-1)C(b_1+b_2-1)/2, \qquad (7.75)$$

where b_1 is the number of phases in the service time of the first class, and b_2 the corresponding number of phases for the second class in our two-class model. For systems with infinite buffer, the effective number of equations to solve at each iteration is determined by the speed of convergence of the conditional probabilities $p(m_1,\ell,i|n)$ to their asymptotic values as n increases. Indeed, the approach we can use for systems with infinite buffers is analogous to the one described for the M/Ph/C model in Section 7.5, i.e., we can use the convergence of the probabilities $p(m_1,\ell,i|n)$ to their limiting values $\overline{\overline{p}}(m_1,\ell,i) = \lim_{n\to\infty} p(m_1,\ell,i|n)$. In other words, in the case of infinite buffer capacity, K is replaced by the value of the number of customers n at which the conditional probabilities become sufficiently close to their asymptotic values. This seems to be typically between 100 and 1000. In all cases, we note that the number of equations to solve at each iteration increases linearly with the number of servers and no more than linearly with the number of service phases.

With the solution of the equations for the conditional probabilities $p(m_1, \ell, i|n)$ we easily get the conditional rate of customer departures from the selected server $\gamma(n)$ from formula (7.71), the overall conditional rate of customer completions $u(n)$, and hence the steady-state probabilities $p(n)$ from formula (7.72). (Clearly, in the case of a model with infinite buffer size, for the steady state to exist, the limiting rate of customer arrivals must be less than the limiting rate of customer departures $\lim_{n\to\infty} u(n)$.) From these quantities, we can compute several performance indices for our FCFS model with two customer classes:

- Attained customer throughput, $\theta = \sum_{n>0} u(n)p(n)$. In steady state, this must be equal to the mean rate of customers entering the system $E[\Lambda]$.
- Mean number of busy servers, $U = \sum_{n>0} \min(n, C)p(n)$.
- Mean number of customers in the system, $E[N] = \sum_{n>0} np(n)$.
- Mean response time, $E[W] = E[N]/\theta$ from Little's formula.
- Mean time in the queue waiting for service, $E[W_q] = (E[N] - U)/\theta$, again from Little's formula since $E[N]-U$ is simply the mean number of customers waiting for service.

Specifically for customers of class 1, we can obtain several quantities through simple reasoning:

- Attained throughput for class 1, $\theta_1 = \theta\eta_1$, since η_1 may be viewed as the fraction of arriving customers that are of class 1.
- Mean number of class 1 customers in service, $U_1 = \theta_1 T_1$. This result follows from Little's law applied to the servers (recall that T_1 is the mean service time for class 1).
- Mean number of class 1 customers in the system (queued and receiving service), $E[N_1] = E[W_q]\theta_1 + U_1$. The first term in this result comes from Little's law applied to the waiting line, and it uses the logical assumption that the mean waiting time in the queue is the same for customers of all classes ($E[W_q]$). The mean

number of customers of class 1 in the system is then the sum of customers of this class in the waiting line and in service (U_1).

- Mean response time for class 1 customers, $E[W_1] = E[N_1]/\theta_1$, from Little's law applied to the system.

We can, of course, derive similar relationships for class 2 customers. What other quantities of interest can we obtain from the state description used? From the probabilities of the total number of customers in the system, $p(n)$, we can get directly the steady-state distribution of the overall queue length. Indeed, the probability that there are k customers waiting for service is simply $p(k+C)$ for $k = 1, 2, \ldots$. The probability that there are no customers waiting is the probability that the total number of customers in the system does not exceed the number of servers, i.e., $\sum_{n=0}^{C} p(n)$. We have also the steady-state distribution for the number of class 1 (and hence class 2) customers in service. With some additional effort, we can derive the probability that there are k_1 ($k_1 = 1, 2, \ldots$) customers of class 1 queued for service. We reason that to have k_1 class 1 customers waiting, we need the overall number of customers waiting to be at least k_1, and, within the resulting number of customers waiting, exactly k_1 must be of class 1. This yields

$$\text{Prob}\{k_1\} = \sum_{n \geq C+k_1} p(n)\eta_1^{k_1}\eta_2^{n-C-k_1}\frac{(n-C)!}{k_1!(n-C-k_1)!}, \quad \text{for} \quad k_1 = 1, 2, \ldots. \tag{7.76}$$

The probability that there are no customers of class 1 queued for service is the sum of the probability that there are no waiting customers at all and of the probability that there are waiting customers but none of class 1, i.e., $\text{Prob}\{k_1 = 0\} = \sum_{n=0}^{C} p(n) + \sum_{n>C} p(n)\eta_2^{n-C}$. In a similar way, we can compute the steady-state distribution of the number of customers of class 2 waiting for service. If desired it is easy to obtain higher order moments of the number of customers queued from the above distributions.

7.7.2 *More than two classes of customers*

With the help of our somewhat unusual state description, we were able to solve an FCFS model with multiple homogeneous servers and two classes of customers. Let us now move on to an FCFS queue with an arbitrary number of classes, i.e., the system represented in Figure 7.15 with $L > 2$ classes of customers. Our approach will be to select any class ℓ, for example, $\ell = 1$. We keep the selected class separate, and we aggregate the remaining $L - 1$ classes of customers into a single class. The service time distribution of this class is the result of a merger of the customer classes being aggregated. In practice, it suffices to combine the phase-type distributions of the $L - 1$ classes as branches of the resulting phase-type distribution. The initial phase selection probabilities in this resulting distribution need to be modified to account for the merger of distributions. Specifically, for class $k \neq \ell$, the initial selection probability $\sigma_{k,i}$ is replaced by $\sigma_{k,i}\eta_k / \sum_{j\neq\ell} \eta_j$ as illustrated in Figure 7.17.

With the aggregation, a multi-class model is reduced to a two-class model. The task now is to obtain meaningful performance measures for individual customer classes. As we saw earlier, the performance indices obtained from the solution of the two-class model include the overall attained throughput for the system θ and the mean time a customer spends in the queue waiting for service $E[W_q]$. We can derive several performance indices of interest for any customer class from these two quantities, viz. the following:

- Attained throughput for customers of class k, $\theta_k = \theta\eta_k$, since η_k is the fraction of arriving customers that belong to class k.
- Mean number of class k customers in service, $U_k = \theta_k T_k$. This follows from Little's law applied to servers for class k. Recall that T_k is the mean service time for class k customers, i.e., the mean time a customer of that class spends at the servers.
- Mean number of class k customers in the system (queued and in service), $E[N_k] = E[W_q]\theta_k + U_k$. This result uses the fact that the mean time in the queue waiting for service is the same for all customer classes, as well as Little's law applied to the waiting line.
- Mean response time for class k, $E[W_k] = E[N_k]/\theta_k$ directly from Little's law applied to the system.

For the arbitrarily selected class ℓ, the results available are those described for class 1 in the two-class model and include, in particular, the distributions of the number of customers of that class queued for service, as well as receiving service. If such distributional results are

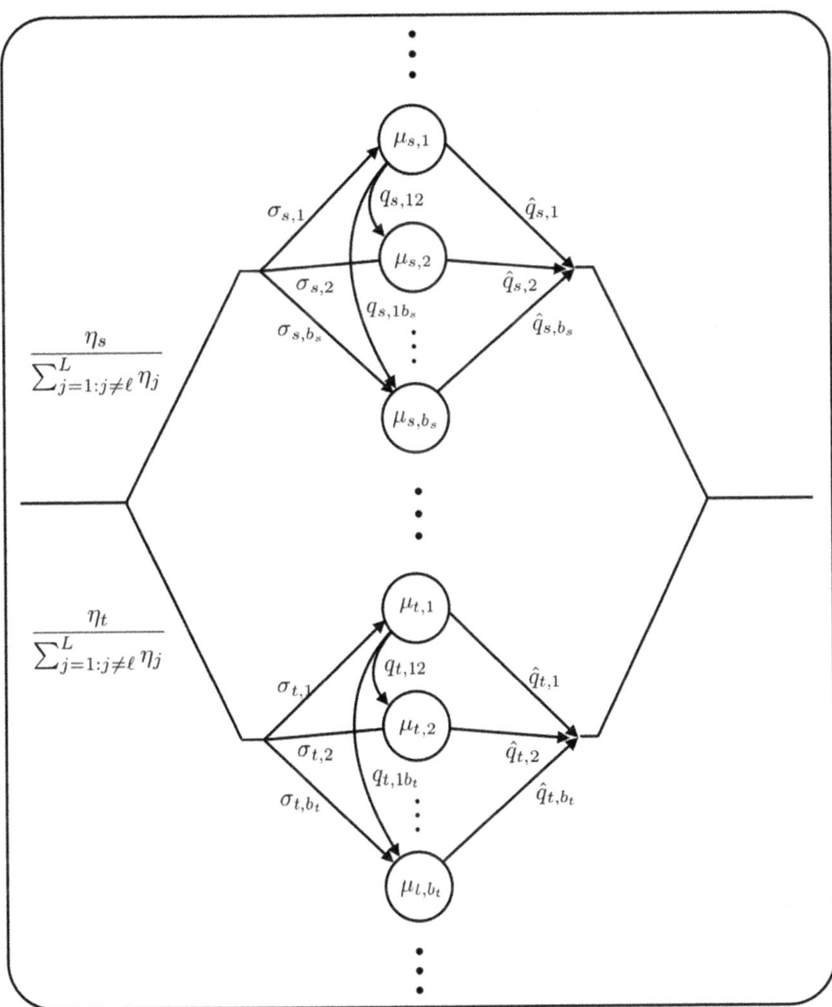

Figure 7.17. Service time distribution resulting from the aggregation of classes other than class ℓ.

needed for other classes, with the state description chosen, we would need to solve the model more than once, each time selecting the desired class as the non-aggregated "class 1" in the two-class model.

The customer classes other than the arbitrarily selected class ℓ are aggregated exactly in the approach described above so that this aggregation introduces no additional approximation. Of course, as there are no free lunches, the aggregation comes at a cost since it increases the number of phases in the service time distribution for the second class in the aggregated model and, thus, the number of equations to solve.

At this point, it is time to look at some numerical results from this approach that we shall refer to as "exact aggregation", remembering that it involves an approximation in formula (7.70).

In our first example, let's consider a system with $L = 4$ classes of customers, $C = 10$ servers, and the total system capacity limited to $K = 70$ customers. The service times are exponentially distributed with respective means $T_1 = 1/\mu_{1,1} = 1/3$, $T_2 = 1/\mu_{2,1} = 1/1$, $T_3 = 1/\mu_{3,1} = 1/9$, and $T_4 = 1/\mu_{4,1} = 1/30$. Arrivals come from a Poisson source with rate λ and arriving customers belong to a specific class of customers with the following class probabilities: $\eta_1 = 6/19$, $\eta_2 = 3/19$, $\eta_3 = 9/19$, and $\eta_4 = 1/19$. Due to the properties of Poisson arrivals, this is equivalent to each class of customers having its own stream of Poisson arrivals with rate $\lambda\eta_k$ for class $k = 1, 2, 3, 4$. In our example, the arrival rate is independent of the current number of customers in the system, n.

We explore the behavior of this system for a range of offered load levels ranging from $\lambda = 1.9$ to $\lambda = 57$. These values of load cover a spectrum of server utilization values from less than 10% to nearly 100%. Per the exact aggregation approach described above, for a given load level, we solve a two-class model in which class $\ell = 1$ is kept separate and unmodified while the remaining three customer classes are combined into a single one. Consequently, the first class in the two-class model has an exponentially distributed service time with mean 1/3. The service time distribution for the second class is hyperexponential with three phases (H-3) whose intensities are given by 1, 9, and 30, respectively. The corresponding phase selection probabilities in this H-3 distribution are 3/13, 9/13, and 1/13. Indeed, the original selection probabilities for the exponential service times were given by $\sigma_{k,1} = 1$ for all customer classes. They are now replaced by $\sigma_{k,1}\eta_k / \sum_{j \neq \ell} \eta_j$ in the H-3 service time distribution. Since a picture is worth a thousand words, we show in Figure 7.18 the results obtained using our state description for the attained throughput and the mean number of customers for each class. For comparison, we have also included the results of discrete-event simulation of this same FCFS model with four customer classes. The simulation

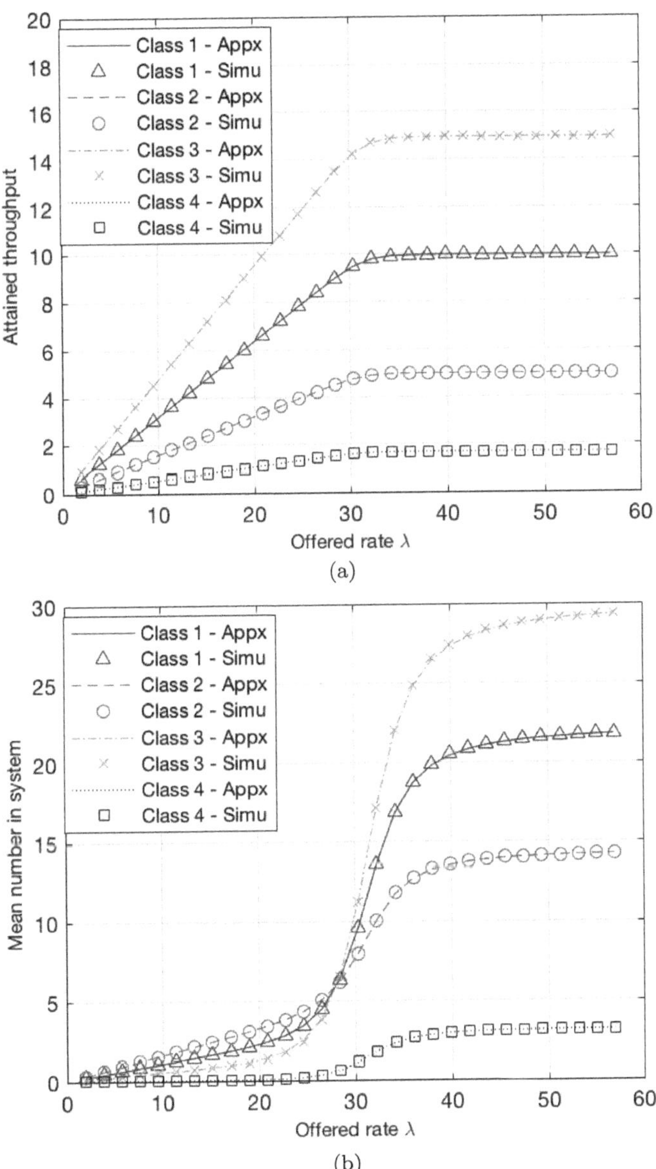

Figure 7.18. Accuracy of the approximate solution using the "exact class aggregation" for the first example. (a) Accuracy for the attained throughput for each class. (b) Accuracy for the mean number of customers for each class.

Table 7.9. Distribution of the relative errors for the mean number of customers per class using the "exact class aggregation" for the first example.

Average	Median	<1%	<5%	<10%	<25%	≥25%
0.59%	0.28%	84.17%	99.17%	100.00%	100.00%	0.00%

results were obtained using the independent replications method with 14 replications of 2,000,000 completions each. The estimated confidence intervals at 95% confidence level turn out to be quite narrow, so we show only their middle points in our figure.

Considering the values obtained from the discrete-event simulation as "exact" reference points, we have derived the relative error distribution for the mean number of customers of each class in the system $E[N_k]$. Table 7.9 summarizes the distribution of these approximation errors.

With mean relative error for the mean number of customers of each class in the system below 0.6% and median relative error below 0.3%, it is clear that the approximation introduces only small errors in our first example. The relative errors for the attained throughputs for each class are even smaller.

In our second example, we wish to see how the approximation fares when the service times are not exponential. Consequently, we consider a similar FCFS model but with service time distributions for each class no longer exponential but given by H-2 distributions. Recall that an H-2 distribution has two exponential branches and can be used to match the first two moments of any distribution whose coefficient of variation is greater than 1. As in our first example, the mean service times are given by 1/3, 1, 1/9, and 1/30. The squared coefficients of variation (defined as the ratio of the variance to the square of the mean) of the service times for the four customer classes are 16, 9, 4, and 2, respectively. For completeness, we show the parameters of the corresponding H-2 distributions in Table 7.10.

Following the "exact class aggregation" approach, the two-class model we need to solve comprises one class with the original H-2 service time distribution and one class with an H-6 distribution, i.e., six exponential branches, which represent the combination of the

Table 7.10. Parameters of the H-2 distributions.

Class 1	$\sigma_{1,1}$	0.115
	$\sigma_{1,2}$	0.885
	$\mu_{1,1}$	0.349
	$\mu_{1,2}$	267.706
Class 2	$\sigma_{2,1}$	0.392
	$\sigma_{2,2}$	0.608
	$\mu_{2,1}$	0.396
	$\mu_{2,2}$	61.0
Class 3	$\sigma_{3,1}$	0.196
	$\sigma_{3,2}$	0.804
	$\mu_{3,1}$	1.782
	$\mu_{3,2}$	729.0
Class 4	$\sigma_{4,1}$	0.653
	$\sigma_{4,2}$	0.347
	$\mu_{4,1}$	19.802
	$\mu_{4,2}$	1030.0

three remaining classes in the original four-class model. Figure 7.19 illustrates the numerical results obtained for the four-class model with H-2 service time distribution.

As before, class $\ell = 1$ was kept separate and the remaining three classes aggregated into a single class. The offered load levels were the same as in our first example with exponential service times. Discrete-event simulation results are included for comparison. Using the simulation results as "exact" values, we again derived the distribution of relative errors. The relative error distribution obtained for the mean numbers of customers of each class in the system is summarized in Table 7.11.

Here, the mean relative error is 3% and the corresponding median value is below 2%. The relative errors for the attained throughput per class (not reported here) are below 1%.

If we compare Figures 7.18 and 7.19, we can see an example of the effect of the service time distributions in an FCFS queue with multiple servers. Looking at the mean numbers of customers of each class in the system, we observe that, in the example considered, the relative difference in the mean number of customers, caused by different service time distributions, may exceed 30%, which is not negligible.

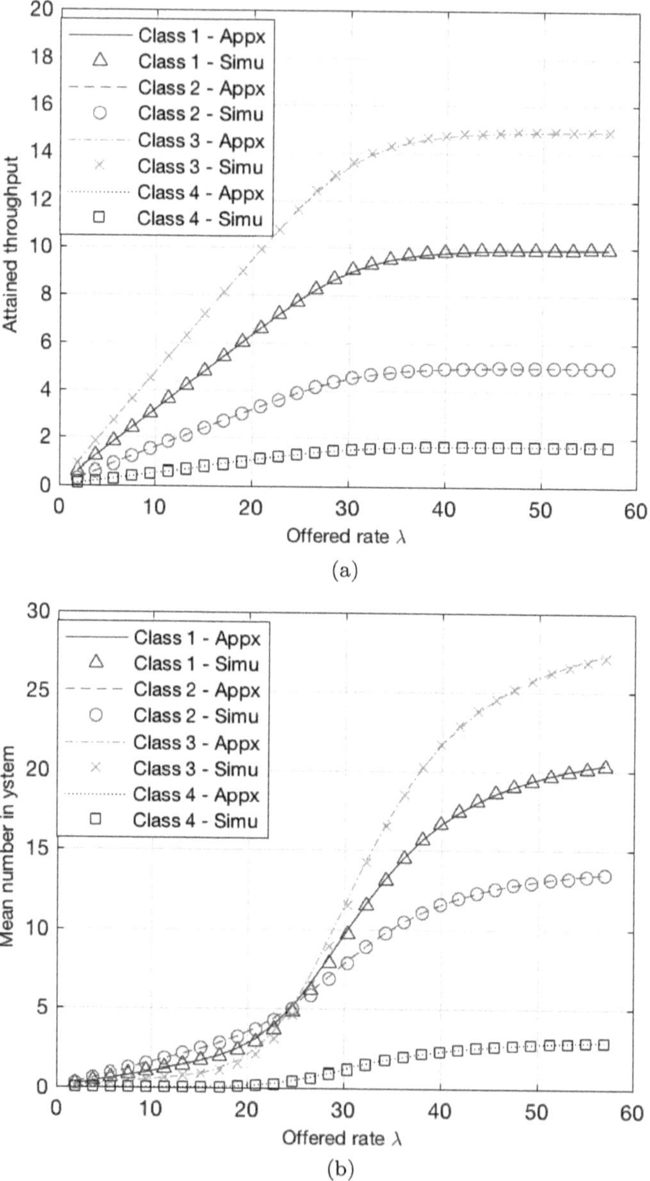

Figure 7.19. (a) Accuracy for the attained throughput for each class. (b) Accuracy for the mean number of customers for each class.

Table 7.11. Distribution of relative errors for the mean number of customers per class using the "exact class aggregation" for the second example.

Average	Median	<1%	<5%	<10%	<25%	≥25%
3.02%	1.75%	28.33%	84.17%	91.67%	100.00%	0.00%

As you may have noted, the relative difference appears most pronounced for medium loads.

Given its good accuracy, the "exact class aggregation" is a good approach to solve an FCFS model with multiple servers and several customer classes unless the number of phases in the phase-type distribution for the aggregated class becomes intractably large. If this is the case, we can simplify the aggregated class and replace its service time distribution with a simpler distribution with the same first two moments. Similar ideas have been proposed for a number of models (Kurinckx & Pujolle, 1979; Puigjaner & Potier, 2012). In practice, if the coefficient of variation of the service time distribution of the aggregated second class is greater than $1/\sqrt{2}$, two phases are sufficient to match the first two moments. We can then solve the resulting simplified two-class model and use the attained throughput for the whole system θ, as well as the mean time in the queue waiting for service $E[W_q]$ to derive several performance indices, as discussed earlier. We refer to this approach as "approximate class aggregation". Obviously, since in this approach, we are matching only the first two moments of the service time distribution for the merged classes and the performance of the M/Ph/C model is known to potentially be sensitive to higher-order properties, we expect some loss of accuracy compared to the "exact class aggregation". For instance, Table 7.12 shows the relative errors for the mean number of customers of each class in the system in the case of our second example considered earlier.

While the approximation results deviate somewhat more from simulation midpoints, they remain sufficiently close to simulation results to be a useful approximation. The complexity of the resulting two-class model is greatly reduced for a larger number of customers of classes.

Table 7.12. Distribution of the relative errors for the mean number of customers per class using the "approximate class aggregation" for the second example.

Average	Median	<1%	<5%	<10%	<25%	≥25%
3.66%	3.00%	25.00%	84.17%	92.50%	100.00%	0.00%

When discussing the approximate formula (7.70) for $\nu_k(m_1, \ell, i, n)$, the conditional rate of completions for customers of class k ($k = 1, 2$) from servers other than the selected server given that the current state of the system is (m_1, ℓ, i, n), we mentioned that some adjustments are possible to the approximation. Indeed, from the very definition of the $\nu_k(m_1, \ell, i, n)$ and from the fact that all servers are statistically identical, it is intuitively clear that the following relationships would be expected to hold

$$\sum_{m_1}\sum_{\ell>0}\sum_{i} p(m_1, \ell, i|n)[\nu_1(m_1, \ell, i, n) + \nu_2(m_1, \ell, i, n)]$$
$$= [\min(n, C) - 1]\gamma(n), \quad \text{if } \ell > 0, \tag{7.77}$$

and

$$\sum_{m_1}\sum_{i} p(m_1, \ell, i|n)[\nu_1(m_1, \ell, i, n) + \nu_2(m_1, \ell, i, n)]$$
$$= \min(n, C)\gamma(n), \quad \text{if } \ell = 0.$$

Formula (7.77) relates the conditional rates $\nu_k(m_1, \ell, i, n)$ to the overall rate of departures. If we use formula (7.70), it turns out that the desired relationships hold exactly when the service times of both classes in the two-class model are exponentially distributed. With non-exponentially distributed service times, formula (7.70) introduces a slight violation of these relationships (it is an approximation, after all). Therefore, we can introduce an adjustment by simply scaling the values of $\nu_k(m_1, \ell, i, n)$ computed from formulas (7.69) and (7.70) so as to have the proper values for

$$\sum_{m_1}\sum_{\ell}\sum_{i} p(m_1, \ell, i|n)[\nu_1(m_1, \ell, i, n) + \nu_2(m_1, \ell, i, n)].$$

Another point related to approximations and their errors pertains to the "approximate class aggregation" approach in which we represented the service time distribution of the aggregated class by its first two moments. The choice of the first two moments is, of course, arbitrary. One might be tempted to think that representing the distribution by matching its first three moments (Feldmann & Whitt, 1998; Khayari *et al.*, 2003; Bobbio *et al.*, 2005; Osogami & Harchol-Balter, 2006; Thummler *et al.*, 2006) would improve the approximation. Interestingly, for our examples, the results of matching the first three moments tend to be only marginally better than using only two moments, but the improvement is not always uniform. For instance, the mean and the median of the relative errors may be smaller, but the maximum errors might be larger. The precise reasons for this behavior are not clear although it is possible that, coincidentally, errors introduced in limiting the approximate aggregation to the first two moments happen to better compensate for errors introduced by the reduced-state approximation for the examples considered.

7.8 Model with multiple servers, quasi-general service times, and preemptive priorities: Level-by-level approximation

Having considered multi-server facilities with FCFS queueing discipline, we now turn our attention to modeling multiple servers in the presence of priority queueing. Systems with multiple servers in which customers are served according to different priorities can be found in many areas of computing, such as cloud computing systems (Ellens *et al.*, 2012) or processor management in Operating Systems (Stallings, 1998; Andrew & Herbert, 2015). Of course, priority-based systems abound in other areas as well, e.g., airport security checkpoints, hospital emergency rooms, etc.

Among the many possible types of priorities, we focus here on preemptive-resume priorities, where a higher-priority customer can interrupt the service of a lower-priority customer, and the interrupted service resumes from the point of interruption when a server becomes available. As was the case for the FCFS queueing discipline with multiple classes of customers, unfortunately, such preemptive-resume priority is not included in the BCMP theorem, i.e., its presence seems to preclude, in general, a product-form solution. For the isolated queue, a simple analytical solution is known in the particular case when there is only a single server for all customers (Allen, 1990; Takagi & Takahashi, 1991), but few exact results seem to exist in the case of multiple servers. This is why we will develop a simple approximate solution for preemptive queues with multiple servers, quasi-general service, and inter-arrival times. Our presentation closely follows published work (Brandwajn & Begin, 2017). As usual, we pay attention to the scalability of the approximate solution in terms of the number of servers and the number of priority levels it may be able to handle.

We start by the case of memoryless (Poisson or quasi-Poisson) arrivals. As shown in Figure 7.20, our model comprises C servers. There are L customer classes; each customer class corresponds to a single distinct priority level. Priority levels (and customer classes) are numbered $1, \ldots, L$, where level 1 is the highest priority. Customers of class ℓ arrive according to a quasi-Poisson process with rate $\lambda_\ell(n_\ell)$, where n_ℓ is the current number of level ℓ customers in

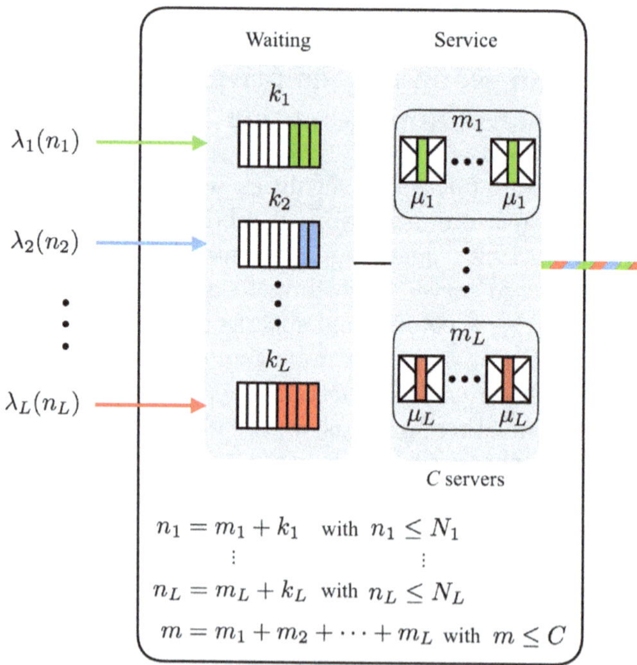

Figure 7.20. The priority queueing system with L priority levels and C servers.

the system. If all servers are currently busy, an arriving customer that finds the waiting line for its level empty preempts, i.e., interrupts the service of a lower-priority customer, if any lower-priority customers are currently using the servers. The interrupted customer resumes its service at the point of interruption (preemptive-resume priorities) when a server becomes available and there are no higher-priority customers waiting. With minor modifications, our approximate solution can accommodate the case where the interrupted customer restarts a whole new service period (preemptive-restart priorities) following an interruption.

The total number of customers at priority level ℓ in the system is limited to K_ℓ. Customers arriving to find their respective queue at capacity are simply lost. Customers at a given priority level are assumed to be statistically identical, and the queueing discipline is FCFS within each class. The service time distribution at each priority level can be different and quasi-general. Specifically, the service times at level ℓ ($\ell = 1, \ldots, L$) are distributed according to a phase-type distribution with a total of b_ℓ phases, as shown in Figure 7.16. Referring to level ℓ, we denote by $\sigma_{\ell,i}$ the probability that service starts in phase i and by $\mu_{\ell,i}$ the intensity of this phase. The probability that the service proceeds in phase j following the completion of phase i is given by $q_{\ell,ij}$ ($i, j = 1, \ldots, b_\ell$) and $\hat{q}_{\ell,i}$ denotes the probability that the service ends with the completion of phase i.

As we mentioned several times before, any distribution can be represented arbitrarily closely by a phase-type distribution (O'Cinneide, 1990; Neuts, 1994; Bolch *et al.*, 2006). If only the first two moments of distribution are known, and its squared coefficient of variation is greater than 0.5, a phase-type distribution with only two phases is enough to match the known first two moments. If more moments are known or if we wish to match a whole distribution (empirical or theoretical), we may typically need many more than two phases. Also, distributions with a low coefficient of variation (below $1/\sqrt{2}$) require more than two phases. There are readily available tools to perform such distribution fitting (Horváth & Telek, 2002).

Let's now return to our model of Figure 7.20. Our goal is to obtain (at the very least) such customary steady-state performance metrics as the mean number of customers at each priority level, the mean sojourn (i.e., response) time, loss probabilities, server utilization, and customer throughput for each class.

A standard state description for our model with priorities might be the vector of the number of customers at each priority level and the number of customers in each phase of service, including customers whose service has been suspended due to preemption by higher-priority classes. This state description would allow us to generate directly the full balance equations for the corresponding steady-state probabilities. As one could expect, the number of states in this description grows combinatorially with the number of servers and the number of priority levels, making such a direct description quickly unmanageable. Since, with the preemptive priority considered in our model, a given priority level is only affected by higher-priority levels, it makes sense to try and look at a single level at a time, starting from the highest priority level ($\ell = 1$).

As we wish to look at a single priority level at a time, we find it convenient to view each level as having at its disposal C service positions. A total of C servers are there to serve level customers currently in the service positions, and any server can serve any service position, provided that the server is not busy serving higher-priority customers. At the highest priority level, all servers are always available to serve service positions occupied by customers. At priority levels other than the top level, not all servers are always available since the servers may disappear and reappear depending on the activity at higher-priority levels. As there is no affinity between service positions and servers, a customer interrupted by higher-priority customers will in general resume service with a different server than before preemption.

We immediately note that the top priority level is simply an instance of the M/Ph/C/K queue, and we already know how to solve it approximately using the reduced-state description (n_1, i_1), where n_1 is the current number of customers at this level and i_1 is the current service phase of a selected service position. Since one of the C servers is always available to serve any of the occupied service positions, at the top level, there is no real difference between the state of a selected server and the state of a selected service position. We discussed the reduced-state approach in detail earlier and we described the computational approach in Algorithm 7.1.

For lower-priority levels, in the case of preemptive-resume priority, we describe the state of priority level ℓ ($\ell = 2, \ldots, L$) by the triple (n_ℓ, m_ℓ, i_ℓ) where n_ℓ is the current number of customers at this level,

m_ℓ is the number of servers unavailable at this level, i.e., serving higher-priority customers, and i_ℓ describes the progress of the service at the current level. We follow the idea of the reduced-state description introduced for the M/Ph/C/K queue and we describe explicitly the progress of the service of only one arbitrarily chosen service position. Thus, $i_\ell = 1, \dots, b_\ell$ if the selected service position is currently active at the given level and the value of i_ℓ then corresponds to the phase of service of the customer. We use negative values $i_\ell = -1, \dots, -b_\ell$ if the level ℓ customer at this position is currently suspended and it was preempted by a higher-priority customer while in its service phase i_ℓ. Finally, we use the value $i_\ell = 0$ to describe a service position without a level ℓ customer. Possible values for the number of unavailable servers are $m_\ell = 0, \dots, C$ (depending on the maximum number of customers at higher levels) and for the current number of level ℓ customers in the system $n_\ell = 0, \dots, K_\ell$.

The number of states in our state description for level ℓ is at most $(K_\ell + 1)(2b_\ell + 1)(C + 1)$ since not all of i_ℓ are possible for all sets of n_ℓ and m_ℓ. The set of values we just defined for i_ℓ corresponds to preemptive-resume priority discipline. If the discipline is preemptive-restart, there is no need to keep track of the service phase in which the customer at the selected service position was preempted so that a single "suspended" state suffices and the total number of possible values for i_ℓ is limited to $(b_\ell + 2)$. As an example, consider a preemptive-resume system with $C = 6$ servers. The state $(n_\ell = 4, m_\ell = 4, i_l = 2)$ means that there are currently 4 customers at level ℓ, the number of servers available at this level is $C - m_\ell = 2$, and the customer at the selected service position is currently active in phase 2 of its service. If $i_\ell = 0$, this would mean that the selected position is one of the $\max(C - n_\ell, 0) = 2$ unoccupied service positions. If $i_\ell < 0$, this would mean in our example that the customer at the selected service position is one of the $\min(n_\ell - C + m_\ell, m_\ell) = 2$ customers currently suspended due to the unavailability of servers, i.e., preempted by higher-priority levels.

It is clear that generally this state description results in a greatly reduced number of states to consider compared to the classical full-state description for a preemptive-resume queue.

We denote by $p(n_\ell, m_\ell, i_\ell)$ the steady-state probability of our state description at level ℓ. As we alluded earlier, from the perspective of the given priority level, the influence of higher-priority levels can be viewed simply as servers disappearing and reappearing with some rates, which correspond to preemptions and higher-priority levels becoming idle. We let $\alpha_\ell(n_\ell, m_\ell, i_\ell)$ be the rate with which servers disappear from level ℓ and $\beta_\ell(n_\ell, m_\ell, i_\ell)$ the rate with which servers reappear at this priority level, given that the current state is (n_ℓ, m_ℓ, i_ℓ). Also, we let $\nu_\ell(n_\ell, m_\ell, i_\ell)$ denote the rate of service completions, i.e., departures, of customers at service positions other than the chosen one given the current state (n_ℓ, m_ℓ, i_ℓ).

We can derive the balance equations for $p(n_\ell, m_\ell, i_\ell)$ in the usual way by equating the rate of flow out of each state to the rate of flow into the state. For example, in the case where $n_\ell > C$ and $0 < m_\ell < C - 1$ (i.e., at least two servers are available to serve customers at this level) with $i_\ell = 1, \ldots, b_\ell$ we get the following balance equation:

$$\begin{aligned}
&p(n_\ell, m_\ell, i_\ell)[\lambda_\ell(n_\ell) + \mu_{\ell,i} + \nu_\ell(n_\ell, m_\ell, i_\ell) + \alpha_\ell(n_\ell, m_\ell, i_\ell) \\
&\quad + \beta_\ell(n_\ell, m_\ell, i_\ell)] = p(n_\ell - 1, m_\ell, i_\ell)\lambda_\ell(n_\ell - 1) \\
&\quad + \sum_{j=1}^{b_\ell} p(n_\ell, m_\ell, j)\mu_{\ell,j} q_{\ell,j i_\ell} + \sum_{j=1}^{b_\ell} p(n_\ell + 1, m_\ell, j)\mu_{\ell,j}\hat{q}_{\ell,j}\sigma_{\ell,i_\ell} \\
&\quad + p(n_\ell, m_\ell + 1, i_\ell)\beta_\ell(n_\ell, m_\ell + 1, i_\ell) \\
&\quad + p(n_\ell, m_\ell + 1, -i_\ell)\beta_\ell(n_\ell, m_\ell + 1, -i_\ell)\frac{1}{m_\ell + 1} \\
&\quad + p(n_\ell, m_\ell - 1, i_\ell)\alpha_\ell(n_\ell, m_\ell - 1, i_\ell)\frac{C - m_\ell}{C - m_\ell + 1} \\
&\quad + p(n_\ell + 1, m_\ell, i_\ell)\nu_\ell(n_\ell + 1, m_\ell, i_\ell).
\end{aligned} \tag{7.78}$$

The terms in the square brackets on the left-hand side of equation (7.78) correspond respectively to the arrival of a customer, completion of the current phase of service at the selected service position, the disappearance of a server (preemption by a higher-priority customer), and the reappearance of a server (release by a higher-priority level). Let's now look at the right-hand side of this balance equation. The first term corresponds to the arrival of a new customer to a state with one less customer at level ℓ. The second term represents transitions in the service time at the selected service position without the customer at this position completing its service. The third term corresponds to the departure of a customer from the selected service position followed by the start of service of a new customer, the service starting in phase i_ℓ. The fourth term on the right-hand side represents the reappearance of a server freed by a higher-priority level to restart serving a suspended customer at service position other than the selected position. Note that we know that the reappearing server will serve another service position since the customer at the selected service position is currently not suspended ($i_\ell > 0$). The fifth term corresponds to the case where the reappearing server is directed to resume service on the customer suspended at the selected service position, which is one of the $m_\ell + 1$ customers currently suspended at level ℓ. The next term corresponds to the disappearance of a server from a position other than the selected position, i.e., one of the $C - m_\ell$ other service positions currently being served. The last term corresponds to the completion of a customer at another service position (other than the selected one).

We can obtain analogous equations for all other values of n_ℓ, m_ℓ and i_ℓ. Clearly, the derivation of these balance equations is a bit tedious as it requires careful consideration of possible transitions in different cases, although, in principle, there is no conceptual difficulty. In addition to the balance equations, we must have $\sum_{n_\ell=0}^{K_\ell} \sum_{m_\ell=0}^{C} \sum_{i_\ell=-b_\ell}^{b_\ell} p(n_\ell, m_\ell, i_\ell) = 1$. This is the usual normalizing condition. We note that, with the state description chosen, the number of equations for a single priority level is moderate and grows only linearly with the number of servers C. The point is that such a set of a moderate number of balance equations can be solved using any of a number of methods, including an adaptation of the semi-analytical method of conditionals introduced in Section 5.3.

Of course, for our level-by-level approach to be useful in practice, we need to know the conditional rates $\nu_\ell(n_\ell, m_\ell, i_\ell)$, $\alpha_\ell(n_\ell, m_\ell, i_\ell)$ and $\beta_\ell(n_\ell, m_\ell, i_\ell)$. If we knew the exact values for these quantities, the solution of our set of equations would give us the exact steady-state probabilities $p(n_\ell, m_\ell, i_\ell)$. We are not able to obtain the exact values for these unknown conditional rates, but we can obtain good approximations by assuming that some variables in each of these rates are more important than others.

Let's start with $\alpha_\ell(n_\ell, m_\ell, i_\ell)$, the rate at which servers disappear given (n_ℓ, m_ℓ, i_ℓ). It seems logical to assume that the current number of users at the given level and the service progress at the selected service position have much less influence on the rate α_ℓ than the number of servers already unavailable at the given level. This implies that

$$\alpha_\ell(n_\ell, m_\ell, i_\ell) \approx \alpha_\ell(m_\ell), \quad m_\ell = 0, \ldots, C-1. \tag{7.79}$$

On similar grounds, we assume for the rate at which servers reappear at level ℓ

$$\beta_\ell(n_\ell, m_\ell, i_\ell) \approx \beta_\ell(m_\ell), \quad m_\ell = 1, \ldots, C. \tag{7.80}$$

At all priority levels, we have $\alpha_\ell(C) = \beta_\ell(0) = 0$.

Let us now see how our level-by-level solution would proceed. We start with the top priority level. The solution of this level produces the steady-state probability $p(n_1, i_1)$ since there are no unavailable servers at this level. The probability that there are n_1 customers at level 1 is $p_1(n_1) = \sum_{i_1=0}^{b_1} p(n_1, i_1)$ and the overall rate of completions at the top level given that the current number of customers is n_1, denoted by $u_1(n_1)$ can be written as

$$u_1(n_1) = C \sum_{i_1=1}^{b_1} p(n_1, i_1)\mu_{1,i_1}\hat{q}_{1,i_1}/p_1(n_1). \tag{7.81}$$

Formula (7.81) relies on the fact that all servers are statistically identical, and it assumes that the selected service position might be idle, i.e., it is possible to have $i_1 = 0$.

Server disappearance at priority level 2 is caused by arrivals of customers at level 1. The rate with which servers disappear at level 2 is simply

$$\alpha_2(m_2) = \lambda_1(n_1 = m_2), \quad \text{for } m_2 = 0, \ldots, C-1. \tag{7.82}$$

Servers reappear at level 2 when a customer at level 1 completes service, and there are no customers queued at level 1 waiting for service. Consequently, the rate at which servers reappear at level 2 can be expressed as

$$\beta_2(m_2) = \begin{cases} u_1(n_1 = m_2), & m_2 = 1, \ldots, C-1 \\ u_1(n_1 = C)p_1(n_1 = C) \Big/ \sum_{n_1 \geq C} p_1(n_1)\,, & m_2 = C. \end{cases} \tag{7.83}$$

The factor $p_1(n_1 = C)/\sum_{n_1 \geq C} p_1(n_1)$ in the expression for $\beta_2(m_2 = C)$ corresponds to the event that there are exactly C customers of class 1, given that all servers are busy at this top level.

We now move on to level 2. The rate of completion of the selected service position when there are n_2 customers and m_2 servers unavailable, given that the position is not idle can be expressed as

$$\xi_2(n_2, m_2) = \sum_{i_2=1}^{b_2} p(n_2, m_2, i_2)\mu_{2,i_2}\hat{q}_{2,i_2} \Big/ \sum_{i_2=1}^{b_2} p(n_2, m_2, i_2),$$

$$n_2 = 1, \ldots, K_2; \quad m_2 = 0, \ldots, C-1. \tag{7.84}$$

The denominator in formula (7.84) corresponds simply to the condition that the service position is not idle for the given n_2 and m_2. We approximate the rate of completions by service position other than the selected one in terms of the quantity $\xi_2(n_2, m_2)$. Specifically, for $\nu_2(n_2, m_2, i_2)$ we limit the dependence on i_2 to whether the service position is active or not as follows:

$$\nu_2(n_2, m_2, i_2) \approx \begin{cases} \min(n_2, C - m_2)\xi_2(n_2, m_2), & i_2 \leq 0 \\ [\min(n_2, C - m_2) - 1]\xi_2(n_2, m_2), & i_2 > 0. \end{cases} \tag{7.85}$$

At this stage, we can solve the balance equations for priority level 2 to obtain the steady-state probabilities $p(n_2, m_2, i_2)$. You may have noted that the rates $\nu_2(n_2, m_2, i_2)$ appearing in the balance equations for $p(n_2, m_2, i_2)$ are effectively expressed in terms of these same probabilities. This makes the system of equations to solve nonlinear, but, as we have seen previously, this is no problem for a fixed-point iterative approach (such as the method of conditionals, for example).

Having obtained the probabilities $p(n_2, m_2, i_2)$ we can compute the steady-state probability that there are n_2 customers at level 2 $p_2(n_2) = \sum_{m_2=0}^{C} \sum_{i_2=-b_2}^{b_2} p(n_2, m_2, i_2)$. Before leaving this level, we compute the rates of server disappearance and reappearance for the immediately following priority level. We will need the probability that a total of m_3 servers are unavailable for level 3, i.e., busy serving customers at levels 1 and 2. We can compute this quantity, which we denote by $P_2(m_3)$, from the probabilities $p(n_2, m_2, i_2)$ as follows:

$$P_2(m_3) = \begin{cases} \sum_{m_2=0}^{m_3} \sum_{i_2=-b_2}^{b_2} p(n_2 = m_3 - m_2, m_2, i_2), & m_3 = 0, \ldots, C-1 \\ \sum_{m_2=0}^{C} \sum_{n_2=C-m_2}^{K_2} \sum_{i_2=-b_2}^{b_2} p(n_2, m_2, i_2), & m_3 = C \end{cases}. \tag{7.86}$$

For $m_3 < C$, the sum is over all states such that m_2 severs are unavailable to level 2, and there are $m_3 - m_2$ customers at level 2. For $m_3 = C$, the sum includes all states where there are at least $C - m_2$ customers at level 2 to occupy the servers, if any, left available by the m_2 servers unavailable to this level.

Servers disappear at level 3 as a result of the combined effect of customer arrivals at level 2 and servers disappearing at this level. Based on this, we can express the rate of server disappearance for level 3 as

$$\alpha_3(m_3) = \frac{\sum_{m_2=0}^{m_3} \sum_{i_2=-b_2}^{b_2} p(n_2 = m_3 - m_2, m_2, i_2)[\alpha_2(m_2) + \lambda_2(n_2)]}{P_2(m_3)},$$
$$m_3 = 0, \ldots, C-1. \tag{7.87}$$

The above rate of disappearance combines the rate at which servers vanish at the preceding level with the rate of customer arrivals at that level. By analogous reasoning, we can express the rate of server reappearance at level 3 as

$$\beta_3(m_3) = \frac{\sum_{m_2=0}^{m_3} \sum_{i_2=-b_2}^{b_2} p(n_2 = m_3 - m_2, m_2, i_2)[\beta_2(m_2) + n_2\xi_2(n_2, m_2)]}{P_2(m_3)},$$
$$m_3 = 1, \ldots, C. \tag{7.88}$$

Here, we are simply expressing the fact that a server will become available to level 3 if it becomes available to the immediately preceding level or if a customer completes at the immediately preceding level while no customers are waiting for service at that level. The rates are divided by $P_2(m_3)$ because these rates are conditioned on the fact that m_3 servers are unavailable at level 3, occupied serving customers at higher-priority levels.

We can express the rates of completion by service positions other than the selected one $\nu_3(n_3, m_3, i_3)$ using formulas analogous to (7.84) and (7.85), and we are then ready to solve level 3, just like level 2. We proceed in this way level by level. The rates of server disappearance and server reappearance are evaluated at each priority level – except the lowest level, of course – for the solution of the level immediately below it using formulas analogous to formulas (7.87) and (7.88).

It is clear that the analysis of priority level ℓ ($\ell = 1, \ldots, L$) gives us $p_\ell(n_\ell)$ the steady-state probability that there are n_ℓ customers at this level. From here, we get several customary performance indices. The mean number of customers at level ℓ is given by $E[N_\ell] = \sum_{n_\ell=0}^{K_\ell} n_\ell p_\ell(n_\ell)$. The attained throughput of customers at this level is simply $\theta_\ell = \sum_{n_\ell=0}^{K_\ell-1} \lambda_\ell(n_\ell)p_\ell(n_\ell)$, and the loss probability can be expressed as $\zeta_\ell = \lambda_\ell(K_\ell)p_\ell(K_\ell)/\sum_{n_\ell=0}^{K_\ell} \lambda_\ell(n_\ell)p_\ell(n_\ell)$. Armed with the mean number of customers and the throughput, we have no problem computing the mean sojourn time and the server utilization for the preemptive-resume discipline using Little's formula as $E[N_\ell]/\theta_\ell$ and $\theta_\ell T_\ell$, respectively, where T_ℓ denotes the mean service time, not counting service interruptions. For the preemptive-restart discipline, the computation of the server utilization is a bit more involved since, by the very nature of this discipline, service that was interrupted is lost although it needs to be counted toward the utilization of the servers. We can use $\sum_{n_\ell=1}^{K_\ell} \sum_{m_\ell=0}^{C-1} \sum_{i_\ell=1}^{b_\ell} p(n_\ell, m_\ell, i_\ell)$ to assess the server utilization in this case.

Our approximate solution, summarized in Algorithm 7.3, replaces the solution of a single system of balance equations, whose complexity grows combinatorially with the number of priority levels L and the number of servers C, by the solution of L systems of equations whose complexity in terms of the number of equations grows only linearly in C.

Algorithm 7.3.

Step 1. Consider level 1

- Solve the top level to obtain $p(n_1, i_1)$, $n_1 = 0, \ldots, K_1$; $i_1 = 0, \ldots, b_1$, and $p_1(n_1)$.
- Evaluate performance indices of interest pertaining to level 1.

- Compute $\alpha_2(m_2)$, $m_2 = 0, \ldots, C-1$, and $\beta_2(m_2)$, $m_2 = 1, \ldots, C$ for use in the solution of level 2 (formulas (7.82) and (7.83)).

Step 2. Consider levels $\ell = 2, \ldots, L$ in the order of decreasing priority. At level ℓ

- Solve the balance equations for the level considered using the approximation formula (analogous to) (7.85) in order to obtain $p(n_\ell, m_\ell, i_\ell)$ and $p_\ell(n_\ell)$.
- Evaluate performance indices of interest pertaining to level ℓ.
- If $\ell < L$, compute $\alpha_{\ell+1}(m_{\ell+1})$ and $\beta_{\ell+1}(m_{\ell+1})$ using formulas (analogous to) (7.87) and (7.88).

With memoryless arrivals, the approximation described carries two possible sources of error. First, even for exponentially distributed service times, inaccuracies are possible due to the assumption that the rates of server disappearance and reappearance at a lower-priority level depend only on the number of servers occupied at higher-priority levels. Second, even at the top priority level 1, where there are no service interruptions, the reduced-state description, which we use to account for non-exponential service times, introduces possible errors. A study of a fairly large set of numerical examples of preemptive-resume models indicates that, despite this, the accuracy of our level-by-level approximation is generally quite good. As an example, Table 7.13 shows the distribution of the relative errors for the mean number of customers in the system in the case of Poisson arrivals.

Table 7.13. Distribution of relative errors for the mean number of customers in the system with Poisson arrivals.

Class	Mean	Median	<1%	<5%	<10%	<15%	≥15%
1	0.07	0.05	100.00	100.00	100.00	100.00	0.00
2	0.09	0.05	99.80	100.00	100.00	100.00	0.00
3	0.17	0.08	97.78	100.00	100.00	100.00	0.00
4	0.37	0.10	88.51	99.80	100.00	100.00	0.00
5	0.89	0.15	76.41	95.56	99.19	100.00	0.00
All	0.32	0.07	92.50	99.07	99.84	100.00	0.00

In order to develop some intuition regarding how preemptive-resume priority systems behave at different priority levels, let's consider an example with $C = 16$ servers, $L = 4$ priority levels, and the maximum number of customers at each priority level limited to $K_\ell = 3C = 48$. The mean service times are $1, \frac{1}{2}, \frac{1}{4}$, and $\frac{1}{8}$ for priority levels $1, 2, 3$, and 4, respectively, while the coefficient of variation of the service time distribution is kept at 2 for all customer classes (priority levels). The rates of customer arrivals are $\lambda_1 = \lambda$, $\lambda_2 = \lambda/2$, $\lambda_3 = \lambda/4$, and $\lambda_4 = \lambda/8$ and we vary the factor λ to study the performance of customers at different priority levels as a function of the offered system load. In Figure 7.21, we show the results obtained for the attained customer throughput for each level. We include the results of discrete-event simulations for comparison.

It is quite striking to observe how the attained throughput collapses at lower-priority levels as a result of preemptions by higher-priority customers. Clearly, in this example, the agreement between simulation results and our level-by-level approximation is very good.

Keep in mind that quasi-Poisson arrivals, as considered so far in our discussion of the preemptive priority model, also include another

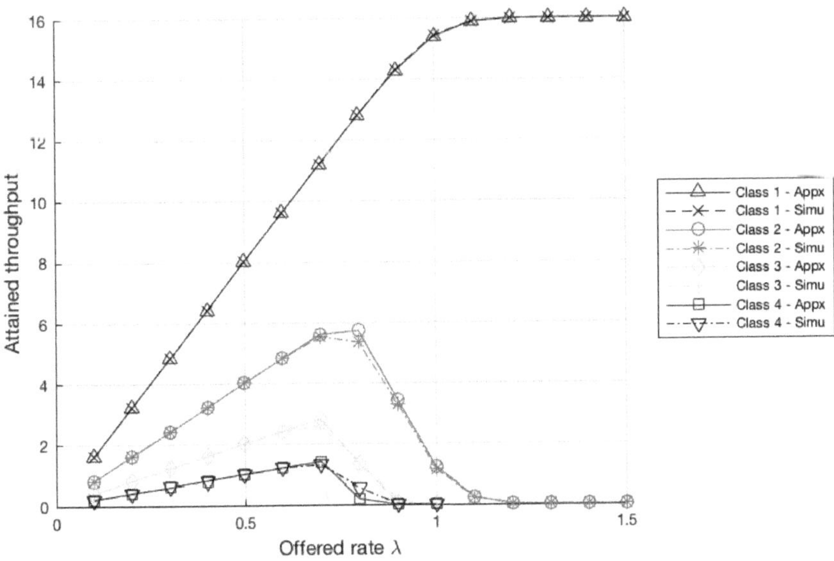

Figure 7.21. Attained throughput as a function of the offered rate with Poisson arrivals.

pattern of arrivals in which customers at each priority level come from a separate finite set of memoryless sources with K_ℓ sources for level ℓ. In such a model, a customer is either at the source or in the priority queue (waiting or at a service position being served or suspended). In this formulation, no customers are lost, and the rate of customer arrivals to level ℓ with n_ℓ customers already present is $(K_\ell - n_\ell)\phi_\ell$ where $1/\phi_\ell$ is the mean time a customer spends at the source on each pass through the system. This variant of the preemptive priority model is particularly well suited to model processor queues in systems with preemptive priority. The mean time at the source on each pass through the system then corresponds to the time a process spends at I/O devices or otherwise away from the CPU queues. This is illustrated in Figure 7.22.

For this type of quasi-Poisson arrivals, the accuracy of the level-by-level approximate solution also seems fairly good. Across all arrival patterns, the accuracy of the approximation, as measured by the mean relative errors, seems to degrade for lower-priority levels. The median relative errors seem to remain low (below 1%).

We are now ready to consider an extension of our priority model to the case where the times between consecutive customer arrivals at each priority level are distributed according to a phase-type distribution. This distribution may be different for each customer class. Thus, the times between customer arrivals at level ℓ are distributed

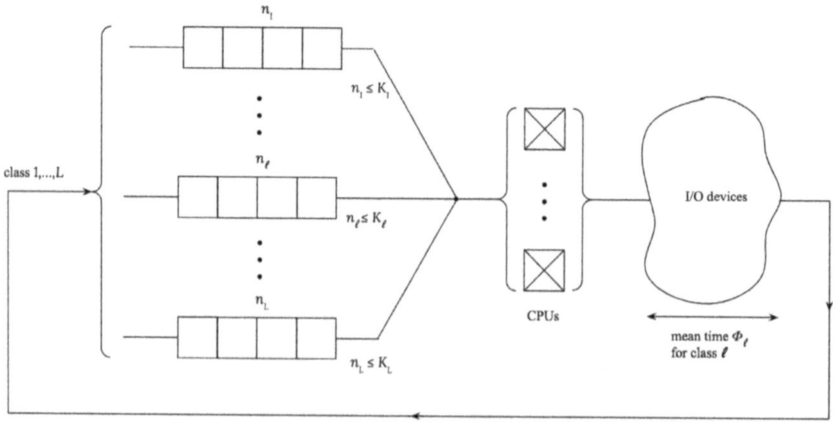

Figure 7.22. CPU priority model with discrete sources.

according to an acyclic phase-type distribution with a_ℓ phases, similar to the distribution used for the Ph/M/1/K or Ph/M/C/K models in previous sections. Referring to level ℓ, we denote by $\tau_{\ell,i}$ the probability that the time between arrival starts in phase i and by $\lambda_{\ell,i}$ the intensity of the corresponding phase, i.e., the mean duration of this phase is $1/\lambda_{\ell,i}$. The probability that the time between arrivals proceeds in phase j following the completion of phase i is given by $r_{\ell,ij}$ and the probability that an arrival happens with the completion of phase i is denoted by $\hat{r}_{\ell,i}$.

To extend our approximate solution to such phase-type distributions of times between arrivals, we can use the method described earlier for the Ph/Ph/C/K model, where we iterate between two simpler models. Indeed, we can view our system as a multi-server system in which, for lower-priority levels, the servers disappear and reappear with approximate rates $\alpha_\ell(m_\ell)$ and $\beta_\ell(m_\ell)$ where m_ℓ is the number of servers currently unavailable to the level considered. Such a system is shown in Figure 7.23.

Our fixed-point iteration is between a model with state-dependent memoryless arrival rates and phase-type service on the one hand and a model with phase-type times between arrivals and memoryless state-dependent service, i.e., Ph/M/C/K queue, on the other. As discussed in Section 2.7 regarding the Ph/M/1/K queue, a simple, numerically stable recurrent solution exists for this model, which

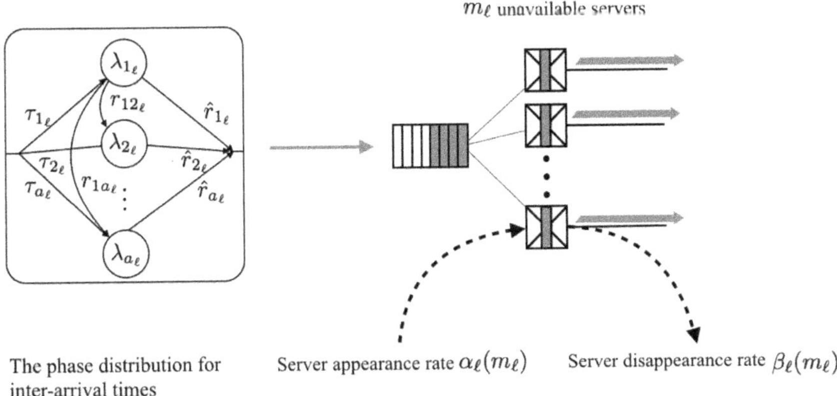

Figure 7.23. Model of a single level ℓ with phase-type times between arrivals and phase-type service times.

encompasses the Ph/M/C/K model. Algorithms 7.4 and 7.5 summarize the steps of this iteration between two simpler models.

Algorithm 7.4.

Step 1. Consider level 1 (highest priority)

- Use iteration described in Algorithm 7.5 to obtain approximate values for $p(n_1, i_1)$, $n_1 = 0, \ldots, K_1$, $i_1 = 0, \ldots, b_1$, and $p_1(n_1)$.
- Evaluate performance indices of interest pertaining to level 1.
- Compute $\alpha_2(m_2)$, $m_2 = 0, \ldots, C-1$, and $\beta_2(m_2)$, $m_2 = 1, \ldots, C$ for use in the solution of level 2 (formulas (7.82) and (7.83)).

Step 2. Consider levels $\ell = 2, \ldots, L$ in the order of decreasing priority. At level ℓ, perform the following:

- Use iteration described in Algorithm 7.5 to obtain approximate values for $p(n_\ell, m_\ell, i_\ell)$ and $p_\ell(n_\ell)$.
- Evaluate performance indices of interest pertaining to level ℓ.
- If $\ell < L$, compute $\alpha_{\ell+1}(m_{\ell+1})$ and $\beta_{\ell+1}(m_{\ell+1})$ using formulas (analogous to) (7.87) and (7.88).

Algorithm 7.5 summarizes the fixed-point iteration between models at each priority level. We denote by $\mathcal{M}$ the performance metric we use for the convergence test, e.g., the mean number of customers at the priority level considered.

Algorithm 7.5.

Step 1. Initialize the arrival rate values $\lambda_\ell(n_\ell)$ to the inverse of the mean time between customers' arrivals at level ℓ.

Step 2. Solve the model with state-dependent memoryless arrivals using the current values of $\lambda_\ell(n_\ell)$.

- Obtain current values for $p(n_\ell, m_\ell, i_\ell)$ and $p_\ell(n_\ell)$, as well as the equivalent service rate $u_\ell(n_\ell)$ as $u_\ell(n_\ell) = \lambda_\ell(n_\ell - 1)p_\ell(n_\ell - 1)$ $/p_\ell(n_\ell)$.
- Compute current values of the performance metric $\mathcal{M}$ from this model.

Step 3. Solve the Ph/M/C/K queue with the current values of $u_\ell(n_\ell)$ obtained in Step 2.

- Obtain current values for $p_\ell(n_\ell)$ and $\lambda_\ell(n_\ell)$.
- Compute the current value of $\mathcal{M}$ from this model.

Step 4. If the values of $\mathcal{M}$ from Steps 2 and 3 deviate by less than $\varepsilon > 0$, then stop the iteration; otherwise, return to Step 2.

Step 5. Use the values of $p(n_\ell, m_\ell, i_\ell)$ and $p_\ell(n_\ell)$ from the last execution of Step 2 as the solution of level ℓ.

As is often the case, there is no theoretical proof that this fixed-point iteration between models converges to a unique solution. In practice, however, the iteration between models tends to converge quite fast. We mentioned in our discussion of the Ph/Ph/C model that this type of approach introduces additional errors. Luckily, these errors seem generally small. A study of a system with four priority levels, $C = 16$, $K_\ell = 3C$ and service time with a coefficient of variation of 2 at all levels, and values of the coefficient of variation of the times between arrivals ranging from 2 to about 15 (a Pareto-like distribution with 16 phases) indicates that, overall, the relative errors for the mean number of customers in the system remain small even with a large coefficient of variation for the arrivals (Brandwajn & Begin, 2017).

We now have a tool to assess (at least approximately) the influence of the distribution of the times between arrivals on the performance of a system with preemptive priorities. Let's consider a simple example and compare the mean numbers of customers in a preemptive-resume priority system with just two classes of customers for Poisson versus non-Poisson arrivals. The mean service times are 1 and $1/2$ for priority levels 1 and 2, respectively, while the coefficient of variation of the service time is set to 2 for both customer classes. The respective rates of arrivals are λ and $\lambda/2$, and the coefficient of variation of the inter-arrival times is set to 4 for both priority levels, i.e., customer classes. The system in our example has $C = 8$ servers and the number of customers at each level is limited to $K_1 = K_2 = 24$.

The results, illustrated in Figure 7.24, confirm that our approximation closely matches discrete-event simulation results, including in the case of non-Poisson arrivals. The influence of the arrival pattern (Poisson vs. non-Poisson arrivals) in this example seems most apparent at medium system loads.

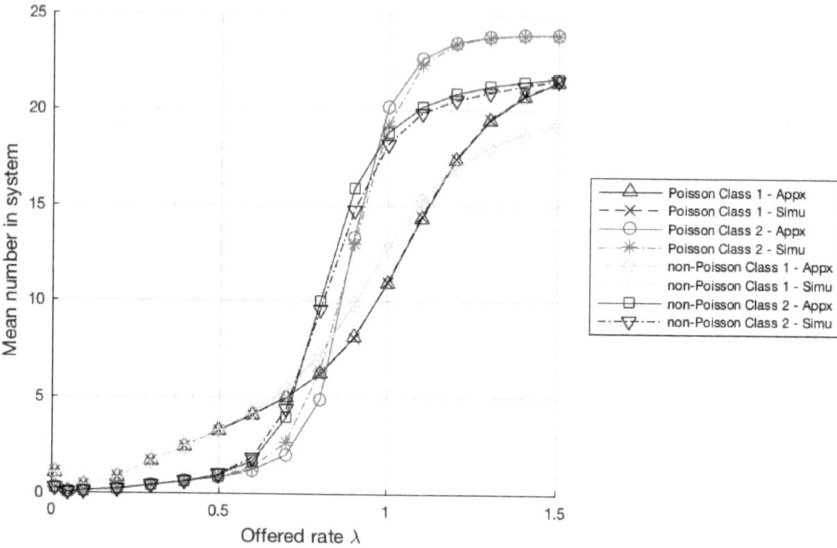

Figure 7.24. Mean number in a system with Poisson and non-Poisson arrivals as a function of offered load.

7.9 Model with multi-server jobs

The multi-server models we have discussed so far can be used as models of devices or computer subsystems with an inherent degree of parallel service or as very high-level models of whole systems. In particular, several authors (Khazaei *et al.*, 2011; Yang *et al.*, 2013; Atmaca *et al.*, 2015) considered the Ph/Ph/C or Ph/Ph/C/K model as a high-level model of a cloud computing system where the servers represent virtual or physical resources. All the models we have considered up to this point share a fundamental property: no server can be idle if there are customers queued waiting for service. This is the direct result of the fact that each customer (request) needs a single server to proceed with service. However, as pointed out by Mar Harchol-Balter (2021), in modern data centers, jobs of different classes, sharing the same FCFS queue, may require different numbers of servers to start service. Thus, it is possible for some servers to be idle while there is a queue of requests blocked behind a job needing more servers than currently available. As a consequence, it is not easy in general to determine even the processing capacity of such a system in the presence of a particular mix of customer classes with different requirements.

In reality, a variant of such a problem was considered in the past in the context of the sharing of the bandwidth of a bus in an I/O controller (Brandwajn & Sahai, 1993) under the condition that requests are subject to loss, i.e., requests that don't find enough remaining bandwidth are not queued but rejected and must be reissued at a later point. Specifically, the model studied includes a total service resource of B units, such as bus bandwidth of B GBytes/s, and set of L classes of customers. Each class of customers has its own set of sources of requests. We denote by N_i the number of sources of requests, i.e., customers, of class i, $i = 1, \ldots, L$. N_i is also the total number of class i customers since a customer of a given class is either in service using the resource or at its source preparing to generate a new request. The mean time for a source of class i to generate a new request is taken to be $1/\lambda_i$. In other words, λ_i is the idle (when not using the resource) rate of requests for this class. A customer of class i requires a "chunk" of size b_i (e.g., data rate) of the resource to proceed with service and uses it for an average time $1/\mu_i$ (e.g., data transfer time). The times for a source to issue a new request and the service times are assumed to be memoryless. If the remaining resource when a request of class i arrives is less than b_i, the request is rejected and the rejected customer returns to its source to generate a new request after a mean time of $1/\lambda_i$. The model described is represented in Figure 7.25.

Clearly, in the above system, there is no fixed number of servers. Rather, the service resource gets partitioned into variable numbers of servers according to the class makeup of the customers being served. Our goal is to obtain the expected number of customers of each class using the resource in steady state, denoted by $E[M_i]$, and the probability that an arriving request finds the necessary amount of resource, denoted by s_i. Over a long period of time of duration T, the number of class i requests processed is $TE[M_i]\mu_i$, and the total number of requests issued by the source is $T(N_i - E[M_i])\lambda_i$, which means that we must have $s_i = E[M_i]/[(N_i - E[M_i])\rho_i]$ where ρ_i denotes the ratio λ_i/μ_i. Thus, it suffices to compute the mean number of customers if each class using the resource, $E[M_i]$, to obtain both quantities.

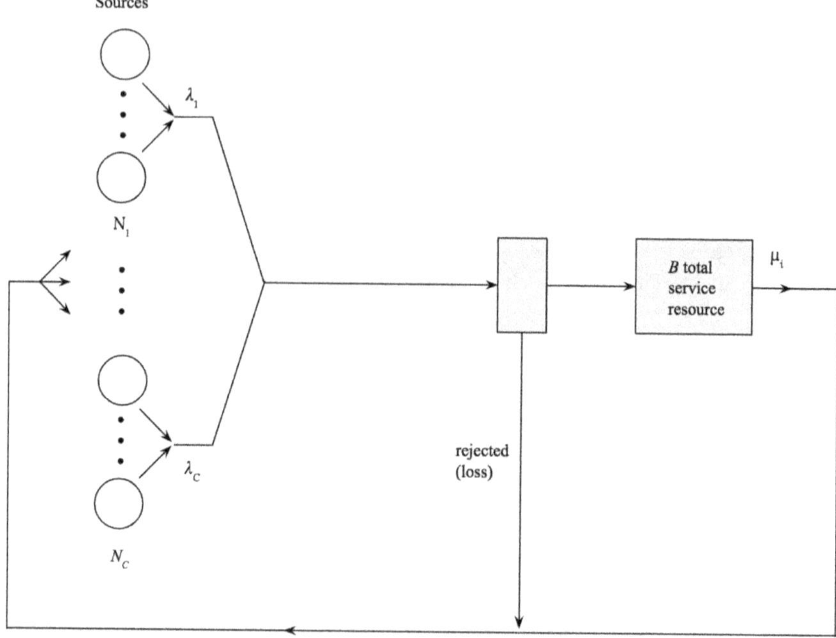

Figure 7.25. Generalized multiclass loss model.

Somewhat surprisingly, the model we just described possesses a simple product-form solution for the steady-state joint probability $p(m_1, \ldots, m_L)$, where $m_i, i = 1, \ldots, L$ is the number of class i customers being served. Indeed, for each class, local balance happens to hold between service activations and completions in a feasible state. This leads to

$$p(m_1, \ldots, m_L) = \frac{1}{\mathrm{G}} \prod_{i_1=1}^{m_1} \frac{(N_1 - i_1 + 1)\rho_1}{i_1} \cdots \prod_{i_L=1}^{m_L} \frac{(N_L - i_L + 1)\rho_L}{i_L}. \tag{7.89}$$

G is a normalizing constant, which corresponds to the sum of the product terms over all states compatible with the total resource and individual class requirements. Since the analytical solution is known, the issue becomes for us how to compute efficiently the probability that a customer is able to successfully acquire the needed resource for each class of requests. Unlike in a regular loss model with a fixed number of servers, the maximum number of customers of a given class in service doesn't necessarily increase when the total number of customers being served increases. Think, for instance, of a total resource of 10 and two classes of customers with resource requirements of 4 and 2, respectively. When there are 3 requests in service, there can be $0, 1$, or 2 customers of class 1 in service, which implies up to 3, 2, or 1 customers of class 2. With a total of 4 requests, there can be only 0 or 1 customers of class 1, since 2 such customers would use up 8 units of the total resource, leaving enough room for only a single class 2 customer.

It is possible to organize the enumeration of system states in an efficient way by repeatedly applying the equivalence and decomposition approach. We will illustrate this using an example with three classes of customers. The basis for the approach is the probability identity $p(m_1, m_2, m_3) = p(m_3|m_2, m_1)p(m_2|m_1)p(m_1)$, where $p(m_1)$ is the probability that there are m_1 class 1 customers in service, $p(m_2|m_1)$ is the conditional probability that there are m_2 class 2 customers given that there are m_1 class 1 customers in service and $p(m_3|m_2, m_1)$ denotes the probability that there are m_3 customers of class 3 given m_2 and m_1.

If we consider customers of class 1, the available resource allows for a maximum of $C_1 = \lfloor B/b_1 \rfloor$ customers in service, where $\lfloor . \rfloor$ denotes the "integer" function. For customers of this class, the system appears as a loss system with C_1 servers in which the interference of customers of other classes can be represented by an "activation" function $a_1(m_1)$ defined as follows:

$$a_1(m_1) = \sum_{m_2} \hat{a}_1(m_2, m_1)p(m_2|m_1), \tag{7.90}$$

where

$$\hat{a}_1(m_2, m_1) = \sum_{m_3: B - m_1 b_1 - m_2 b_2 - m_3 b_3 \geq b_1} p(m_3|m_2, m_1). \tag{7.91}$$

The activation function $a_1(m_1)$ is simply the probability that, given m_1, other customer classes leave enough resources for at least one additional class 1 customer. Similarly, $\hat{a}_1(m_2, m_1)$ is the probability that there is enough resource left for class 1 given that there are

m_2 and m_1 customers of classes 2 and 1, respectively. With known $a_1(m_1)$ the solution of such loss system is simply

$$p(m_1) = \frac{1}{\mathrm{G}_1}\prod_{i=1}^{m_1}\frac{(N_1 - i + 1)a_1(i-1)\rho_1}{i}, \qquad m_1 = 0, \ldots C_1. \tag{7.92}$$

Clearly, we have $E[M_1] = \sum_{m_1=1}^{C_1} m_1 p(m_1)$. To compute $a_1(m_1)$, we consider our system with m_1 customers of class 1 in service. The remaining resource is $B - m_1 b_1$, so that for class 2 customers given m_1 customers of class 1, the system can be viewed as a loss system with $C_2(m_1) = \lfloor (B - m_1 b_1)/b_2 \rfloor$ servers. The activation function to represent the interference of class 3 customers, denoted by $a_2(m_2, m_1)$, is defined as

$$a_2(m_2, m_1) = \sum_{m_3 : B - m_1 b_1 - m_2 b_2 - m_3 b_3 \geq b_2} p(m_3 | m_2, m_1). \tag{7.93}$$

(7.93) expresses the probability that there is enough resource left for at least one additional class 2 customer given the state (m_2, m_1). If we know the activation function $a_2(m_2, m_1)$, we can obtain the conditional probability $p(m_2|m_1)$ as

$$p(m_2|m_1) = \frac{1}{\mathrm{G}_2}\prod_{i=1}^{m_2}\frac{(N_2 - i + 1)a_2(i-1, m_1)\rho_2}{i},$$
$$m_2 = 0, \ldots C_2(m_1). \tag{7.94}$$

We can readily obtain the conditional expected number of class 2 customers in service given m_1 as $E[M_2|M_1] = \sum_{m_2=1}^{C_2(m_1)} m_2 p(m_2|m_1)$ and, hence, $E[M_2] = \sum_{m_1=0}^{C_1} E[M_2|M_1]p(m_1)$.

If we now consider the system with m_1 and m_2 customers of class 1 and 2 in service, respectively, the resource left is $B - m_1 b_1 - m_2 b_2$. Under these conditions, for class 3 users, we can view our system as a simple loss system with $C_3(m_2, m_1) = \lfloor (B - m_1 b_1 - m_2 b_2)/b_3 \rfloor$ servers. Since class 3 is the last in our chain of analysis, all the interference by other customer classes is already accounted for in the reduced amount of the service resource, and there is no need for an activation function at this level. Consequently, we can obtain the conditional probability $p(m_3|m_2, m_1)$ as

$$p(m_3|m_2, m_1) = \frac{1}{G_3} \prod_{i=1}^{m_3} \frac{(N_3 - i + 1)\rho_3}{i}, \quad m_3 = 0, \ldots, C_3(m_2, m_1). \tag{7.95}$$

Let $E[M_3|M_2, M_1] = \sum_{m_3=1}^{C_3(m_2,m_3)} m_3 p(m_3|m_2, m_1)$ and $E[M_3|M_1] = \sum_{m_2=0}^{C_2(m_1)} E[M_3|M_2, M_1]p(m_2|m_1)$.

The expected number of class 3 customers in service is then given by $E[M_3] = \sum_{m_1=0}^{C_1} E[M_3|M_1]p(m_1)$.

Clearly, from the solution of this last loss model, we also obtain the activation functions $a_2(m_2, m_1)$ for class 2 customers, as well as $\hat{a}_1(m_2, m_1)$ needed for the analysis of class 1. It can be shown that this equivalence and decomposition approach yields, in fact, the exact solution for the generalized loss (or Engset, as such models are often referred to) model described (Brandwajn & Sahai, 1993). In actual computation, we would enumerate states in the order $m_1 = 0, \ldots, C_1$ and for each value of m_1, in the order $m_2 = 0, \ldots, C_2(m_1)$. The solution starts with the "innermost" model for class 3 for each (m_2, m_1). The model for class 2 customers is then solved for each value of m_1, and finally, the loss model for class 1 is solved last.

This model has been originally applied in the context of requests subject to loss to provide an element of answer to the interesting question of whether it is better to have shorter transfers at faster speed (larger b_i, which may be more difficult to obtain but used for a shorter time) or longer transfers requiring a smaller amount of resources (smaller b_i, potentially easier to obtain but used for a longer time). More recently, the same general model has been used to examine at a high level the trade-offs in data center disaggregation (Begin *et al.*, 2022).

The generalized Engset model considered above allows no queueing, i.e., either the customer finds enough resources available or the incoming request is lost. In the problem formulation suggested by Harchol-Balter (2021) as a model applicable to modern data centers queueing for service is allowed. The specific model we consider (Brandwajn & Begin, 2022) is depicted in Figure 7.26.

As in the case of the generalized Engset model studied earlier, the system consists of a total resource of size B shared by L classes

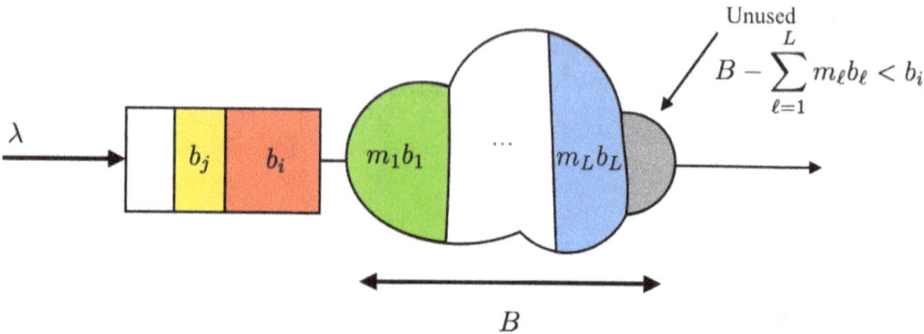

Figure 7.26. Resource of size B shared by L classes of customers.

of customers (jobs). We use the terms "customer" and "job" interchangeably. Each class of customers is characterized by the amount of resources it needs to proceed b_ℓ and the mean time the resource is used by the job during a single service period $t_\ell = 1/\mu_\ell$ for jobs of class $\ell = 1, \ldots, L$. In our model, jobs (customers) arrive to the system from a Poisson source with an overall rate λ and are served in the order of their arrival, i.e., the service discipline is FCFS. The probability that an arriving customer is of class ℓ is given by p_ℓ, and we have, of course, $\sum_{\ell=1}^{L} p_\ell = 1$. Our model is a generalization of the multi-server jobs in which jobs require multiple servers to proceed. It maps onto such a model if the values of B and b_ℓ are all integers.

Our first goal is to determine the processing capacity of such a system in the presence of a particular mix of customer classes with different requirements. As mentioned earlier, this is, in general, far from obvious because it is possible for some of the resources to remain idle, although there are jobs waiting in the queue. We start by selecting an appropriate state description for our system in the steady state if it exists. Let's use the number of jobs of each class currently in service (denoted by $m_\ell, \ell = 1, \ldots, L$), the class of the job at the head of the queue (if any), (denoted by i), and the total number of jobs in the system (denoted by n). We let $p(m_1, \ldots, m_L, i, n)$ be the corresponding steady-state probability, assuming it exists, and we denote by $m = \sum_{\ell=1}^{L} m_\ell$ the total current number of jobs in service. We use the value $i = 0$ when there are no jobs in the queue, i.e., when $n = m$.

We can enumerate the states of the system in a way similar to the one used for our generalized Engset model discussed earlier.

We start with jobs of class 1, and we note that we must have $m_1 = 0, \ldots, \lfloor B/b_1 \rfloor$, $m_2 = 0, \ldots, \lfloor (B - m_1 b_1)/b_2 \rfloor$, $m_3 = \lfloor (B - m_1 b_1 - m_2 b_2)/b_3 \rfloor$ and so on for consecutively increasing customer class numbers. Feasible states satisfy the condition that $\sum_{\ell=1}^{L} m_\ell b_\ell \leq B$, and, if there are jobs queued, i.e., when $n > m$, $B - \sum_{\ell=1}^{L} m_\ell b_\ell < b_i$, where $i = 1, \ldots, L$ is the class of the job at the head of the queue of jobs waiting for service. This latter condition simply states that there is not enough resource remaining to accommodate the customer at the head of the queue. Keeping these points in mind and assuming exponentially distributed service times for all job classes, we can generate the balance equations for the steady-state probabilities $p(m_1, \ldots, m_L, i, n)$ using our standard "rate of flow out = rate of flow in" approach.

With jobs waiting in the queue, a single departure in our system can trigger the start of service for a variable number of jobs, depending on how much resource is left following the departure and what the resource requirements are for the jobs waiting in the queue in FCFS order. Regardless of this fact, if we look at the total number of jobs in the system, n, it is clear that it increases with rate λ and decreases with rate $u(n)$ given by

$$u(n) = \sum_{S(n)} (m_1\mu_1 + \cdots + m_L\mu_L) p(m_1, \ldots, m_L, i|n), \tag{7.96}$$

where $S(n)$ denotes the set of states feasible for a given value of n jobs in the system, and $p(m_1, \ldots, m_L, i|n)$ is the conditional probability that the job mix in service is $(m_1, \ldots, m_L)$ and the class of the job at the head of the waiting line is i given the total number of jobs in the system n. It follows that the steady-state probability that there are n jobs in the system (in service and waiting for service), which we denote by $p(n)$, has a simple and familiar form:

$$p(n) = \frac{1}{\mathrm{G}} \prod_{j=1}^{n} \frac{\lambda}{u(n)}, \quad n = 0, 1, \ldots. \tag{7.97}$$

As usual, G is a normalizing constant that $\sum_n p(n) = 1$. From the definition of conditional probability, we have

$$p(m_1, \ldots, m_L, i, n) = p(n) p(m_1, \ldots, m_L, i|n) \tag{7.98}$$

If we use formulas (7.97) and (7.98) in the balance equations for our model, we readily obtain the set of equations for the conditional probabilities $p(m_1, \ldots, m_L, i|n)$. These equations can be solved using a form of our method of conditionals of Section 5.3, i.e., a fixed-point iteration together with formula (7.96) and the normalizing condition $\sum_{S(n)} p(m_1, \ldots, m_L, i|n) = 1$. For the moment being, however, we want to concentrate on the processing capacity of our system. For that, we need to consider what happens as the number of jobs in the system grows.

We let $\overline{\overline{p}}(m_1, \ldots, m_L, i) = \lim_{n\to\infty} p(m_1, \ldots, m_L, i|n)$ and, $\overline{\overline{u}} = \lim_{n\to\infty} u(n)$. It is clear from (7.97) that, for the steady state to exist, these limits must exist and we must have $\lambda < \overline{\overline{u}}$ since $p(n)$ is asymptotically geometric. We have seen similar asymptotic geometric behavior in the M/Ph/C model and it has been observed in a number of other queues (Takahashi, 1981). If we take the limit for $n \to \infty$ in the equations for the conditional probabilities $p(m_1, \ldots, m_L, i|n)$, we readily obtain a set of equations for the limiting probabilities $\overline{\overline{p}}(m_1, \ldots, m_L, i)$. To determine the maximum job processing rate of the system, it then suffices to solve this set of equations for a value of arrival rate $\lambda = \overline{\overline{u}}$. This is easily accomplished using a fixed-point iteration with $\overline{\overline{u}} = \sum_{S(\infty)} (m_1\mu_1 + \cdots + m_L\mu_L)\overline{\overline{p}}(m_1, \ldots, m_L, i)$ and the normalizing condition $\sum_{S(\infty)} \overline{\overline{p}}(m_1, \ldots, m_L, i) = 1$. All feasible states $S(\infty)$ are states $(m_1, \ldots, m_L, i)$ such that $\sum_{\ell=1}^{L} m_\ell b_\ell \leq B$ and $B - \sum_{\ell=1}^{L} m_\ell b_\ell < b_i$, $i = 1, \ldots, L$. Thus, we can determine the processing capacity of the system without having to obtain its steady-state solution.

A simple example will illustrate and clarify the above discussion. Let's consider a system with a total resource of size $B = 6$ and two job classes with respective resource requirements $b_1 = 2$ and $b_2 = 3$. Feasible states (m_1, m_2, i) when there are jobs waiting in the queue $(n > m)$ can be generated from $m_1 = 0, \ldots, [B/b_1]$ and $m_2 = [(B - m_1 b_1)/b_2]$: $(0,2,1)$, $(0,2,2)$, $(1,1,1)$, $(1,1,2)$, $(2,0,2)$, $(3,0,1)$, $(3,0,2)$. State $(2,0,1)$ is not feasible since $B - 2b_1 \geq b_1$, which violates our second condition for feasible states. As an example, for the states $(2,0,2)$ and (3,0,2), we have the following balance equations:

$$p(2,0,2,n)(2\mu_1 + \lambda) = p(2,0,2,n-1)\lambda(n-1) + p(1,1,1,n+1)\mu_2 p_2 + p(3,0,2,n+1)3\mu_1,$$

$$p(3,0,2,n)(3\mu_1+\lambda) = p(3,0,2,n-1)\lambda(n-1)$$
$$+p(1,1,1,n+1)\mu_2 p_1 p_2$$
$$+p(3,0,1,n+1)3\mu_1 p_2.$$

With the help of formulas (7.98) and (7.97), we transform these balance equations into the following equations for the conditional probabilities:

$$p(2,0,2|n)(2\mu_1+\lambda) = p(2,0,2|n-1)u(n) + [p(1,1,1|n+1)\mu_2 p_2$$
$$+p(3,0,2|n+1)3\mu_1]\lambda/u(n+1)$$
$$p(3,0,2|n)(3\mu_1+\lambda) = p(3,0,2|n-1)u(n) + [p(1,1,1|n+1)\mu_2 p_1 p_2$$
$$+p(3,0,1|n+1)3\mu_1 p_2]\lambda/u(n+1)$$

We now take the limit for $n \to \infty$ to obtain

$$\overline{\overline{p}}(2,0,2)(2\mu_1+\lambda) = \overline{\overline{p}}(2,0,2)\overline{\overline{u}} + [\overline{\overline{p}}(1,1,1)\mu_2 p_2 + \overline{\overline{p}}(3,0,2)3\mu_1]\lambda/\overline{\overline{u}}$$

and

$$\overline{\overline{p}}(3,0,2)(3\mu_1+\lambda) = \overline{\overline{p}}(3,0,2)\overline{\overline{u}} + [\overline{\overline{p}}(1,1,1)\mu_2 p_1 p_2$$
$$+\overline{\overline{p}}(3,0,1)3\mu_1 p_2]\lambda/\overline{\overline{u}}.$$

Let us now set $\lambda = \overline{\overline{u}}$ in the equations for $\overline{\overline{p}}(m_1, m_2, i)$, as needed to determine the limiting processing capacity of our system, and denote by $\tilde{p}(m_1, m_2, i)$ the resulting limiting conditional probabilities. The above equations become

$$\tilde{p}(2,0,2)(2\mu_1) = \tilde{p}(1,1,1)\mu_2 p_2 + \tilde{p}(3,0,2)3\mu_1$$

and

$$\tilde{p}(3,0,2)(3\mu_1) = \tilde{p}(1,1,1)\mu_2 p_1 p_2 + \tilde{p}(3,0,1)3\mu_1 p_2.$$

Proceeding similarly for other feasible states, we get

$$\begin{aligned}
\tilde{p}(3,0,1)(3\mu_1 p_2) &= \tilde{p}(1,1,1)\mu_2 p_1^2,\\
\tilde{p}(1,1,1)(\mu_1 p_2 + \mu_2) &= \tilde{p}(0,2,1)2\mu_2 p_1 + \tilde{p}(2,0,2)2\mu_1 p_1\\
&\quad + \tilde{p}(1,1,2)\mu_2 p_1,\\
\tilde{p}(1,1,2)(\mu_1 + \mu_2 p_1) &= \tilde{p}(0,2,1)2\mu_2 p_2 + \tilde{p}(1,1,1)\mu_1 p_2\\
&\quad + \tilde{p}(2,0,2)2\mu_1 p_2,\\
\tilde{p}(0,2,1)(2\mu_2) &= \tilde{p}(0,2,2)2\mu_2 p_1 + \tilde{p}(1,1,2)\mu_1 p_1,
\end{aligned}$$

and $\tilde{p}(0,2,2)(2\mu_2 p_1) = \tilde{p}(1,1,2)\mu_1 p_2$.

This is a rather simple system of linear equations for the limiting probabilities $\tilde{p}(m_1, m_2, i)$, which can be easily solved subject to the normalizing condition $\sum_{S(\infty)} \tilde{p}(m_1, m_2, i) = 1$. The asymptotic maximum job processing rate is then given by the formula $\tilde{u} = \sum_{S(\infty)} (m_1\mu_1 + \cdots + m_L\mu_L)\tilde{p}(m_1, \ldots, m_L, i)$. Figure 7.27

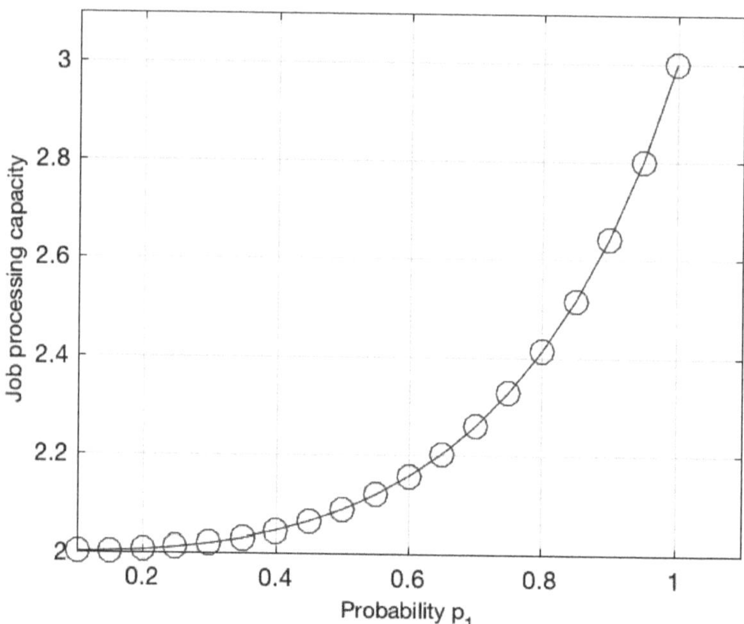

Figure 7.27. Maximum job processing capacity (asymptotic value $\tilde{u}$) as a function of job mix.

shows the values of this rate as a function of p_1, the probability that an arriving job is of class 1.

Obviously, we can easily determine the maximum job completion rate when all jobs belong to a single class (either class 1 or class 2 in our simple example). Outside these points, however, we see that the asymptotic maximum job completion rate is, in our example, a non-linear function of the probabilities p_ℓ and that even a small proportion of more resource-demanding jobs can have an important effect on the job processing capacity of a multi-server job queue. Thus, studying the limiting values of conditional probabilities of our state description allows us to determine the asymptotic maximum job processing capacity without having to obtain the full solution for the system.

7.10 Variance of the response time

Another important point raised in the paper by Mar Harchol-Balter (2021) is that, particularly in the context of Service Level Agreements in modern data centers, it is desirable for performance models to produce more than just the expected response times, viz., ideally one would like to be able to know the probabilities that the response time will, or will not, exceed a specified value. This is in general not an easy question to answer in the usual computer performance modeling analysis, as distributions of response time are known only for a limited number of models. Our bridge between the steady-state distributions of the numbers of customers in the system and the customer response times is the well-known Little's law, which links the expected response time to the expected number of customers and the mean rate of customer arrivals. As it turns out, there are generalizations of Little's law to higher moments of the response time, although, unfortunately, they have important limitations and are easy to apply only in much more restricted circumstances (Bertsimas & Nakazato, 1995).

Our second goal will be to determine the variance of the job response time in our model of Figure 7.26. To analyze our model beyond the maximum processing capacity, as explained above, we can determine the steady-state probability that there are n jobs in the system, $p(n)$, from formula (7.97) together with formula (7.96) and the solution of the equations for $p(m_1, \ldots, m_L, i|n)$. The mean number of jobs in the system, queued and in service, is then simply $E[N] = \sum_n np(n)$. It is, of course, straightforward to obtain the expected sojourn (i.e., response) time for a job $E[W]$ from Little's law as $E[W] = E[N]/\lambda$. Because at a service facility with multiple servers, it is possible for some jobs to overtake others, i.e., not finish in the order of arrival, extensions of Little's law that relate higher order moments of the number of customers in a system to higher order moments of the sojourn time cannot be used directly for the whole system. However, there is no overtaking in the FCFS waiting line, so these extensions can be applied to the queue of jobs waiting for service.

Let's denote by $n_q(m_1, \ldots, m_L, i, n)$ the number of jobs waiting for service in our system when the system state is $(m_1, \ldots, m_L, i, n)$. We also denote by $E[N_q^k]$ the k-th moment of the number of jobs in the FCFS waiting line. This quantity can be readily obtained as

$$E[N_q^k] = \sum_n p(n) \sum_{S(n)} [n_q(m_1, \ldots, m_L, i, n)]^k p(m_1, \ldots, m_L, |n). \tag{7.99}$$

Let's denote by $E[W_q^k]$ the k-th moment of the time a job spends queued waiting for service. From the generalization of Little's law to the second moment (Whitt, 1991), with Poisson arrivals, we have the following relationship:

$$E[W_q^2] = (E[N_q^2] - E[N_q])/\lambda^2. \tag{7.100}$$

As we can see, the second moment of the waiting time is expressed in terms of the first two moments of the number of jobs queued and the square of the arrival rate. Let's now move on to the service part of the sojourn of a job in the system. We denote by $E[S_\ell^k]$ the k-th moment of the service time for jobs of class ℓ and by $E[S^k]$ the k-th moment of the overall service time. We must have

$$E[S^k] = \sum_{\ell=1}^{L} p_\ell E[S_\ell^k]. \tag{7.101}$$

Formula (7.101) comes from the law of total moments (Allen, 1990; Trivedi, 2008). To obtain the second moment of the time a job spends in the system, which is the sum of the time a job has to wait for service and the job's service time, we assume that these two can be treated as independent. This allows us to express the second moment of the sojourn time in system as

$$E[W^2] = E[W_q^2] + E[S^2] + 2E[W_q]E[S]. \tag{7.102}$$

In formula (7.102), $E[W_q]$ is the expected time a job spends in the queue, and $E[S]$ is the mean service time of a job. With exponentially distributed service times, we have $E[S_\ell^k] = k!t_\ell^k$ (t_ℓ is the mean service time for a job of class ℓ). From Little's law, we readily get $E[W_q] = E[N_q]/\lambda$ where $E[N_q]$ is the mean number of jobs queued waiting for service, i.e., the first moment from formula (7.99).

Knowing $E[W^2]$ we get the variance of the sojourn time for a job as $\text{Var}[W] = E[W^2] - (E[W])^2$. The variance, by its very definition, is a measure of how much the job response time varies around its mean $E[W]$. We can also use it to obtain an upper bound on the probability that the sojourn (response) time exceeds some value t greater than the mean sojourn time $E[W]$. In particular, we can use the so-called one-sided inequality (Allen, 1990) which states

$$\text{Prob}\{W > t\} \leq \frac{\text{Var}[W]}{\text{Var}[W] + (t - E[W])^2}. \tag{7.103}$$

Of course, how useful this bound may be in practice will depend on the specific values of the quantities involved. In any case, we have the second moment of the job sojourn time, and its higher moments can be obtained from the generalized Little's law in an analogous manner.

To get a feel of how accurate our results are for the second moment of the job sojourn time, we consider again our simple example of a system with a resource of size $B = 6$ and two job classes with respective resource requirements $b_1 = 2$ and $b_2 = 3$. We assume that we have $p_1 = p_2 = 0.5$, but each job class has a different mean service time $t_1 = 1/\mu_1 = 0.4$ and $t_2 = 1/\mu_2 = 1.6$ so that the overall mean service time is 1. We show in Table 7.14 a few values of the first two moments of the job sojourn time, as well as the second moment of the time in queue obtained by the approach described above for different values of the arrival rate λ. As a "sanity check", we give

Table 7.14. Example of results for second moments of sojourn and queueing times.

Arrival rate λ	$E[W]$	$E[W^2]$	$E[W_q^2]$	$\text{Prob}\{W > t\}$
1.0	1.438 (1.438)	4.763 (4.843)	1.167 (1.245)	0.036 (0.003)
1.2	1.743 (1.741)	6.758 (6.814)	2.553 (2.610)	0.052 (0.006)
1.4	2.271 (2.273)	11.120 (11.216)	5.858 (5.946)	0.091 (0.016)
1.6	3.341 (3.347)	23.416 (23.513)	16.013 (16.090)	0.217 (0.056)
1.8	6.471 (6.467)	85.707 (86.367)	72.044 (72.702)	0.779 (0.216)

in parenthesis the corresponding moments estimated in a discrete-event simulation using 7 independent replications of 5,000,000 job completions each.

The results in Table 7.14 also include the values for the bound for the probability $\text{Prob}\{W > t = 10\}$ obtained using the one-sided inequality (7.103). The values in parenthesis in this column give the fraction of sojourn times that exceeded the given target value of $t = 10$ in the simulation runs. Clearly, the assessed values for the second moments are generally in good agreement with those estimated in the simulation, but the bounds for the response time tend to be overly conservative. In fact, for $\lambda = 1.8$, a better upper bound for $\text{Prob}\{W > 10\}$ can be obtained from the elementary Markov's inequality (Allen, 1990; Trivedi, 2008) $P\{W > t\} \leq E[W]/t = 0.6471$.

Before leaving the subject of multi-server jobs, let's note that although the state description used for our system $(m_1, \ldots, m_L, i, n)$ certainly allows us to escape the complexity of describing in detail the state of the waiting line, we are still faced with the enumeration of all possible combinations of jobs of different classes in service. Obviously, the number of states to consider may grow very fast with the size of the total resource and the number of customer classes.

References

Allen, A. O. (1990). *Probability, Statistics, and Queueing Theory.* Gulf Professional Publishing.

Andrew, S. T., & Herbert, B. (2015). *Modern Operating Systems.* Pearson Education.

Atmaca, T., Begin, T., Brandwajn, A., & Castel-Taleb, H. (2015). Performance evaluation of cloud computing centers with general arrivals and service. *IEEE Transactions on Parallel and Distributed Systems*, 27(8), 2341–2348.

Baskett, F., Chandy, K. M., Muntz, R. R., & Palacios, F. G. (1975). Open, closed, and mixed networks of queues with different classes of customers. *Journal of the ACM (JACM)*, 22(2), 248–260.

Begin, T., & Brandwajn, A. (2009). Considerations in workload characterization for parallel access volumes. In *Int. CMG Conference*, 6–11 December 2009, Dallas, Texas, USA.

Begin, T., & Brandwajn, A. (2013, July). A note on the accuracy of several existing approximations for M/Ph/m queues. In *2013 IEEE 37th Annual Computer Software and Applications Conference Workshops* (pp. 730–735). IEEE.

Begin, T., Brandwajn, A., & Tchana, A. (2022, August). Data center disaggregation: When and how much? In *2022 9th International Conference on Future Internet of Things and Cloud (FiCloud)* (pp. 62–66). IEEE.

Bertsimas, D. J., & Nakazato, D. (1995). The general distributional little's law and its applications. *Operations Research*, 43(2), 298–310.

Bini, D. A., Latouche, G., & Meini, B. (2005). *Numerical Methods for Structured Markov Chains*. OUP Oxford.

Björklund, M., & Elldin, A. (1964). A practical method of calculation for certain types of complex common control systems. *Ericsson Technics*, 20, 3–75.

Bobbio, A., Horváth, A., & Telek, M. (2005). Matching three moments with minimal acyclic phase type distributions. *Stochastic Models*, 21(2–3), 303–326.

Bolch, G., Greiner, S., De Meer, H., & Trivedi, K. S. (2006). *Queueing Networks and Markov Chains: Modeling and Performance Evaluation with Computer Science Applications*. John Wiley & Sons.

Boxma, O. J., Cohen, J. W., & Huffels, N. (1979). Approximations of the mean waiting time in an M/G/s queueing system. *Operations Research*, 27(6), 1115–1127.

Brandwajn, A., & Sahai, A. K. (1993). Aspects of the solution of some multiclass loss systems. *Performance Evaluation*, 17(2), 141–154.

Brandwajn, A., & Begin, T. (2009). Higher-order distributional properties in closed queueing networks. *Performance Evaluation*, 66(11), 607–620.

Brandwajn, A., & Begin, T. (2014). Reduced complexity in M/Ph/c/N queues. *Performance Evaluation*, 78, 42–54.

Brandwajn, A., & Begin, T. (2016). Breaking the dimensionality curse in multi-server queues. *Computers & Operations Research*, 73, 141–149.

Brandwajn, A., & Begin, T. (2017). Multi-server preemptive priority queue with general arrivals and service times. *Performance Evaluation*, 115, 150–164.

Brandwajn, A., & Begin, T. (2019). First-come-first-served queues with multiple servers and customer classes. *Performance Evaluation*, 130, 51–63.

Brandwajn, A., & Begin, T. (2022, December). A note on the determination of the processing capacity in a multiserver job system as a model of cloud datacenters. In *2022 IEEE International Conference on Cloud Computing Technology and Science (CloudCom)* (pp. 49–52). IEEE.

Cooper, R. B. (1984). *Introduction to Queueing Theory.* Second Edition. North-Holland.

Cosmetatos, G. P. (1974). Approximate equilibrium results for the multi-server queue (GI/M/r). *Journal of the Operational Research Society*, 25, 625–634.

Crovella, M. E., & Bestavros, A. (1997). Self-similarity in world wide web traffic: Evidence and possible causes. *IEEE/ACM Transactions on Networking*, 5(6), 835–846.

Ellens, W., Akkerboom, J., Litjens, R., & Van Den Berg, H. (2012, June). Performance of cloud computing centers with multiple priority classes. In *2012 IEEE Fifth International Conference on Cloud Computing* (pp. 245–252). IEEE.

Feldmann, A., & Whitt, W. (1998). Fitting mixtures of exponentials to long-tail distributions to analyze network performance models. *Performance Evaluation*, 31(3–4), 245–279.

Gupta, V., Dai, J., Harchol-Balter, M., & Zwart, B. (2007). The effect of higher moments of job size distribution on the performance of an M/G/s queueing system. *ACM SIGMETRICS Performance Evaluation Review*, 35(2), 12–14.

Harchol-Balter, M. (2021). Open problems in queueing theory inspired by datacenter computing. *Queueing Systems*, 97(1–2), 3–37.

Horváth, A., & Telek, M. (2002, April). PhFit: A general phase-type fitting tool. In *International Conference on Modelling Techniques and Tools for Computer Performance Evaluation* (pp. 82–91). Berlin, Heidelberg: Springer.

Hsu, W. W., & Smith, A. J. (2003). Characteristics of I/O traffic in personal computer and server workloads. *IBM Systems Journal*, 42(2), 347–372.

Jiang, H., & Dovrolis, C. (2005, June). Why is the internet traffic bursty in short time scales? In *Proceedings of the 2005 ACM SIGMETRICS International Conference on Measurement and Modeling of Computer Systems* (pp. 241–252). Association for Computing Machinery, New York, NY, USA.

Khayari, R. E. A., Sadre, R., & Haverkort, B. R. (2003). Fitting world-wide web request traces with the EM-algorithm. *Performance Evaluation*, 52(2–3), 175–191.

Khazaei, H., Misic, J., & Misic, V. B. (2011). Performance analysis of cloud computing centers using m/g/m/m+ r queuing systems. *IEEE Transactions on Parallel and Distributed Systems*, 23(5), 936–943.

Kimura, T. (1986). A two-moment approximation for the mean waiting time in the GI/G/s queue. *Management Science*, 32(6), 751–763.

Kurinckx, A., & Pujolle, G. (1979, February). Analytic methods for multiprocessor system modelling. In *Proceedings of the Third International Symposium on Modelling and Performance Evaluation of Computer Systems: Performance of Computer Systems* (pp. 305–318). Vienna, Austria, February 6–8, 1979. Amsterdam: North Holland Publishing Co.

Kuzler, C., Norman, P., Pate, A., & Wolf, R. (1999). IBM Enterprise Storage Server. San Jose, California, IBM Corp., Int. Tech. Support Org.

Latouche, G., & Ramaswami, V. (1993). A logarithmic reduction algorithm for quasi-birth-death processes. *Journal of Applied Probability*, 30(3), 650–674.

Lee, A. M., & Longton, P. A. (1959). Queueing processes associated with airline passenger check-in. *Journal of the Operational Research Society*, 10(1), 56–71.

Martin, J. (1978). *Systems Analysis for Data Transmission*. Prentice Hall PTR.

McNutt, B. (2005). *The Fractal Structure of Data Reference: Applications to the Memory Hierarchy* (Vol. 22). Springer Science & Business Media.

Neuts, M. F. (1994). *Matrix-geometric Solutions in Stochastic Models: An Algorithmic Approach*. Courier Corporation.

O'Cinneide, C. A. (1990). Characterization of phase-type distributions. *Stochastic Models*, 6(1), 1–57.

Osogami, T., & Harchol-Balter, M. (2006). Closed form solutions for mapping general distributions to quasi-minimal PH distributions. *Performance Evaluation*, 63(6), 524–552.

Puigjaner, R., & Potier, D. (2012). *Modeling Techniques and Tools for Computer Performance Evaluation*. Springer Science & Business Media.

Ramaswami, V., & Lucantoni, D. M. (1985). Algorithms for the multi-server queue with phase type service. *Stochastic Models*, 1(3), 393–417.

Sasaki, Y., Imai, H., Tsunoyama, M., & Ishii, I. (2004). Approximation of probability distribution functions by Coxian distribution to evaluate multimedia systems. *Systems and Computers in Japan*, 35(2), 16–24.

Seelen, L. P. (1986). An algorithm for Ph/Ph/c queues. *European Journal of Operational Research*, 23(1), 118–127.

Stallings, W. (1998). *Operating Systems Internals and Design Principles*. Prentice-Hall, Inc.

Stewart, W. J. (1995). *Introduction to the Numerical Solution of Markov Chains*. Princeton University Press.

Takagi, H., & Takahashi, Y. (1991). Priority queues with batch Poisson arrivals. *Operations Research Letters*, 10(4), 225–232.

Takahashi, Y. (1981). Asymptotic exponentiality of the tail of the waiting-time distribution in a PH/PH/c queue. *Advances in Applied Probability*, 13(3), 619–630.

Thummler, A., Buchholz, P., & Telek, M. (2006). A novel approach for phase-type fitting with the EM algorithm. *IEEE Transactions on Dependable and Secure Computing*, 3(3), 245–258.

Trivedi, K. S. (2008). *Probability & Statistics with Reliability, Queuing and Computer Science Applications*. John Wiley & Sons.

Tijms, H. C. (1986). *Stochastic Modelling and Analysis: A Computational Approach*. John Wiley & Sons, Inc.

Whitt, W. (1980). The effect of variability in the GI/G/s queue. *Journal of Applied Probability*, 17(4), 1062–1071.

Whitt, W. (1991). A review of $L = \lambda W$ and extensions. *Queueing Systems*, 9, 235–268.

Willinger, W., Taqqu, M. S., Sherman, R., & Wilson, D. V. (1997). Self-similarity through high-variability: Statistical analysis of Ethernet LAN traffic at the source level. *IEEE/ACM Transactions on Networking*, 5(1), 71–86.

Willinger, W., Alderson, D., & Li, L. (2004, October). A pragmatic approach to dealing with high-variability in network measurements. In *Proceedings of the 4th ACM SIGCOMM Conference on Internet Measurement* (pp. 88–100). Association for Computing Machinery, New York, NY, USA.

Wolff, R. W. (1977). The effect of service time regularity on system performance. In *Computer Performance*, (pp. 297–304). K.M. Chandy and M. Relser (Eds.), Elsevier North-Holland, Inc, New York, USA.

Xiong, K., & Perros, H. (2009, July). Service performance and analysis in cloud computing. In *2009 Congress on Services-I* (pp. 693–700). IEEE.

Yang, B., Tan, F., & Dai, Y. S. (2013). Performance evaluation of cloud service considering fault recovery. *The Journal of Supercomputing*, 65, 426–444.

Chapter 8

Simple Techniques

8.1 Operational analysis

In the preceding chapters, we studied a number of computer performance models and different ways to solve them, i.e., to obtain meaningful performance metrics, whether exact or approximate. You certainly noticed that words such as "probability" and "exponential distribution" kept coming back and, even with our efforts to keep things simple, a certain amount of knowledge of probabilities was needed to understand the solution. In particular, one of the problems is that our models incorporate assumptions difficult or impossible to either relate directly to measurable quantities or to verify.

Sure, we can elect to think of steady-state probabilities in our balance equations as being simply fractions of time during which a specific condition (such as the buffer full or a specific number of customers in the queue) prevailed. Strictly speaking, the overall measurement period would have to be infinite, but let's say that for the common of mortals a period "long enough" would do. After all, that's how state probabilities are estimated in discrete-event simulations. However, distributional assumptions are more problematic. Distributions, such as the exponential distribution that appears many times in the preceding chapters of this text, are theoretical constructs whose validity or applicability in a specific real-life situation can't be verified within a finite interval of time. Although statisticians might take issue with this statement, this was the feeling of several influential computer modeling authors and practitioners. The result of their reflections on the matter is the so-called operational analysis

(Denning & Buzen, 1978; Dallery and Cao, 1992; Denning, 2006) whose main premise is that it deals with measurable quantities over finite intervals of time.

The goals of the operational analysis are to

- provide precise mathematical expressions to relate measurable quantities,
- obtain relationships which can then be used to verify the internal consistency of measured data,
- derive formulas to predict the effects of system and/or workload modifications.

In its very principle, the operational analysis tries to stay close to system measurements. Therefore, it is based on the notion of observation interval and operational variables. The observation interval is simply the interval during which we decide (or happen) to observe and measure the behavior of the system being considered, and the operational variables refer to measurable quantities we wish to manipulate. Following the original designers' description (Denning & Buzen, 1978), the operational method is to proceed in the following three steps:

- Define a set of operational variables that correspond in a natural way to intuitive quantities of interest.
- Derive mathematical relationships between the operational variables to characterize the behavior of the system during the observation interval.
- Apply as desired and consistent with the goals of the approach.

To illustrate the operational analysis, let's first consider a system with s servers. Here, the term "server" refers in fact to a service station. All we are saying about the system is that it processes jobs of some nature (e.g., transactions or queries). We denote by T the length of the observation interval on which we define our operational variables. The first such variable is the number of jobs completed by the system during the interval T, denoted by J. We now define additional operational variables: B_i is the time during which server i was busy and C_i is the number of service requests completed by server i, $i = 1, \ldots, s$, during our observation interval. A job can result in a number of requests directed to the servers during its lifetime in the system. Clearly, the operational variables defined so far correspond quite naturally and directly to intuitive, measurable concepts of interest.

We now define the throughput of our system, denoted by X, as

$$X = J/T. \tag{8.1}$$

The utilization of server i is defined as

$$U_i = B_i/T. \tag{8.2}$$

The average service time for server i is defined as

$$S_i = B_i/C_i. \tag{8.3}$$

We note that the server utilization is defined very naturally as the fraction of the observation interval during which the server was busy. Similarly, the average service time is simply defined as the ratio of the time the given server was busy to the number of service requests completed by the server (during the observation interval), all measurable quantities. We define the average number of requests for server i, $i = 1, \ldots, s$, per job as

$$D_i = C_i/J. \tag{8.4}$$

We are now ready to derive our first operational relationships: throughput law and utilization equality. The throughput law links the system throughput to the utilization of a server, its average service time, and the average number of serviced requests for the server per job:

$$X = \frac{U_i}{S_i D_i}, \quad i = 1, \ldots, s. \tag{8.5}$$

The proof of this law is straightforward. Indeed, simply using the respective definitions of the operational quantities involved, we have

$$\frac{U_i}{S_i D_i} = \frac{B_i/T}{(B_i/C_i)(C_i/J)} = \frac{J}{T} = X, \quad i = 1, \ldots, s. \tag{8.6}$$

It is important to stress that the operational "throughput law" is independent of considerations such as service time distributions or statistical equilibrium, since it is, in a sense, an accounting identity.

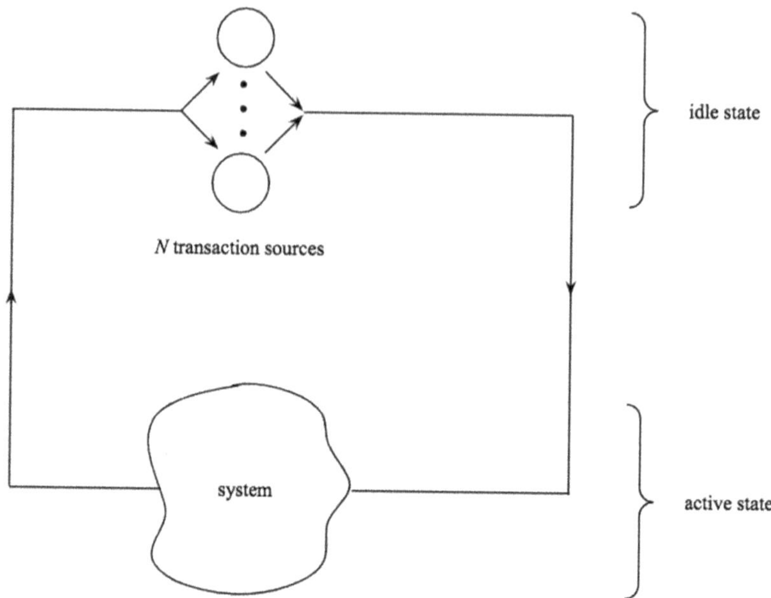

Figure 8.1. A model of an interactive system.

A direct consequence of this law is the utilization equality, which states that we must have

$$\frac{U_i}{S_i D_i} = \frac{U_j}{S_j D_j}, \quad i = 1, \ldots, s; \quad j = 1, \ldots, s. \tag{8.7}$$

Next, to derive operational response time laws, we consider the model of an interactive system with a set of N transaction (job) sources shown in Figure 8.1. We use the terms transaction and job interchangeably.

We do make some important assumptions here, viz., we assume that the interactive process represented by our model consists of two states: the process is idle (preparing a new transaction) or the process is active (transaction is in the system). In this system, a source cannot generate a new transaction until the preceding transaction from this source has been completed by the system. In this context, an interaction is completed each time a process leaves the active state, and the response time for a transaction (interaction) is the time between the moment a given process enters and exits the active state.

With these considerations in mind, we denote by T the duration of our observation interval and by J the number of interactions completed during this period. The average response time is denoted by R and is the time spent in the active state per interaction. Let's denote by r_k the total time process number k, $k = 1, \ldots, N$, spends in the active state during our observation period. The average response time can then be written as

$$R = \frac{1}{J} \sum_{k=1}^{N} r_k. \tag{8.8}$$

In a similar way, if we denote by z_k the total time process k is in the idle state and by $\hat{J}$ the number of transitions from idle to active state, we can express the average time in idle state as

$$Z = \frac{1}{\hat{J}} \sum_{k=1}^{N} z_k. \tag{8.9}$$

We can now derive the operational response time law which states

$$R = N \frac{S_i D_i}{U_i} - \frac{\hat{J}}{J} Z, \quad i = 1, \ldots, s. \tag{8.10}$$

The proof of the above law is not complicated. Since a process can only be idle or active, we must have $z_k + r_k = T$ for all $k = 1, \ldots, s$. This means that $\sum_{k=1}^{N} (z_k + r_k) = NT$. If we divide this relationship by J, we get

$$\frac{NT}{J} = \frac{1}{J} \sum_{k=1}^{N} z_k + \frac{1}{J} \sum_{k=1}^{N} r_k = \frac{\hat{J}}{J} Z + R. \tag{8.11}$$

Using the fact that $\frac{T}{J} = \frac{1}{X}$ and the previously derived throughput law, we get our response time law (8.10), which we can also write as

$$R = N \frac{1}{X} - \frac{\hat{J}}{J} Z. \tag{8.12}$$

As before, this law was derived using only intuitive, measurable quantities. If our observation period is so long that $\frac{\hat{J}}{J} \to 1$, we will have

$$R \to N \frac{1}{X} - Z. \tag{8.13}$$

Now let's consider the average number of jobs present in the system during our observation period T. This operational quantity can be defined as

$$L = \sum_{j=1}^{N} jT(j)/T. \tag{8.14}$$

$T(j)$ is the time during which there were j jobs in the system, so that $T(j)/T$ is the fraction of the observation period with j jobs in the system. Consider formula (8.12) and multiply both sides by the throughput X. We have

$$RX = N - \frac{\hat{J}}{J}ZX = N - \frac{\hat{J}}{J}Z\frac{J}{T} = N - \frac{\hat{J}Z}{T} = N - \frac{\sum_{k=1}^{N} z_k}{T}. \tag{8.15}$$

The term $\sum_{k=1}^{N} z_k/T$ is in fact the average number of processes in idle state during our observation period. Indeed, summing over all processes effectively multiplies the time during which there were n processes $(n = 0, 1, \ldots, N)$ by n. Consequently, we must have $N - \sum_{k=1}^{N} z_k/T = \sum_{j=1}^{N} jt(j)/T = L$ and thus

$$RX = L. \tag{8.16}$$

Relationship (8.16) is the operational equivalent of Little's law for our interactive system. Again, we note that its derivation does not rely on statistical equilibrium or particular distributional assumptions. It is in effect a simple accounting relationship among operational variables.

A large number of classical results obtained for networks of queues in steady-state, such as Jacksonian networks, can be re-derived within the framework of operational analysis. However, to achieve this, specific assumptions, such as that only one-step transitions are possible, or that routing frequencies and service times don't depend on global system state, referred to as homogeneity assumptions, are needed (Denning & Buzen, 1978). These assumptions have the effect to mimic the properties of networks of queues with memoryless service or inter-arrival times and only locally state-dependent service times. This allows one to obtain product-form solutions for such models in the operational analysis domain. These product-form solutions are for the fractions of time that various system states prevail and are very much similar to the product-form solutions for steady-state probabilities. Clearly, unlike distributional assumptions in the probability domain, one can, at least in theory, verify whether a given homogeneity assumption holds over the observation interval. One can also re-derive equivalence and decomposition approaches in the framework of operational analysis.

How well does operational analysis live up to its stated goals? Obviously, it manipulates measurable quantities of interest (as opposed to theoretical probability notions). There is no doubt that it provides an excellent means to check the consistency of measurement data. Indeed, relationships such as the operational throughput law (8.5), operational response time law (8.10), or Little's law equivalent (8.16) must be satisfied by measurement data over any observation period, provided that the system satisfies the assumptions under which the laws were derived. If the latter condition is true but the relationships are violated, there must be an error in the system instrumentation.

Things are less clear-cut when it comes to using operational analysis for predictive purposes. To predict the effects of system or workload changes, additional assumptions must be made that certain operational quantities will remain invariant. Operational analysis refers to such assumptions as invariance assumptions. As an example, if the average service time of some device (server) was obtained during some observation period under a light system load, one invariance assumption might be that the average service time will remain unchanged as the system workload increases. In some situations, the nature of the system being modeled may indicate that this must indeed be the case. But, in general, this will be only an assumption in a predictive study.

Things are even less clear with state-dependent service times. Let's switch back to our usual probability domain and consider the simple single-server system shown in Figure 8.2.

Times between arrivals are memoryless with mean time between arrivals $1/\lambda$. The service time of the single server is the result of operations that sometimes proceed using a very fast internal cache (probability p_1 and mean service time $1/\mu_1$) and sometimes using a much slower device with a different technology (probability $p_2 = 1 - p_1$ and mean service time $1/\mu_2$). For simplicity of solution, we assume that

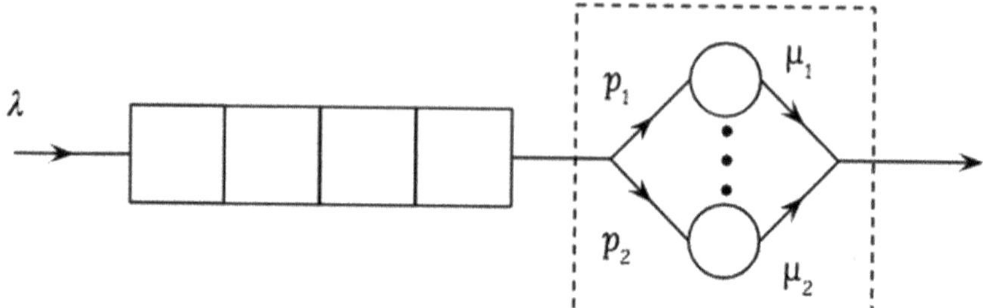

Figure 8.2. A simple model of a system with internal cache.

the service times at both devices are exponentially distributed with respective rates μ_1 and μ_2. We denote by n the current number of customers in the system, and we assume that this number is limited to K.

We are interested in the long-run probability that there are n customers in the system, which we denote by $p(n)$. We also denote by $p(n,i)$ the probability that there are n customers and that the service is in phase i, $i = 1, 2$. We have $p(n,i) = p(n)p(i|n)$, where $p(i|n)$ is the conditional probability of the service phase given n. If we let

$$u(n) = \sum_{j=1}^{2} p(j|n)\mu_j \tag{8.17}$$

be the equivalent conditional service rate of the server, we know from analogous derivations in previous chapters that the solution of this M/Ph/1/K model can be written as

$$p(n) = \frac{1}{G}\prod_{k=1}^{n}\frac{\lambda}{u(k)}, \quad n = 0, 1, \ldots, K. \tag{8.18}$$

From the balance equations for $p(n,i)$ we obtain the following equations for $p(i|n)$ $(i = 1, 2)$:

$$\begin{aligned} p(i|1)(\lambda + \mu_i) &= p_i u(1) + p_i\lambda, \\ p(i|n)(\lambda + \mu_i) &= p(i|n-1)u(n) + p_i\lambda, \quad n = 2, \ldots, K-1, \\ p(i|K)\mu_i &= p(i|K-1)u(K). \end{aligned} \tag{8.19}$$

We note that we must have $p(1|n) + p(2|n) = 1$ for $n = 1, \ldots, K$. With this condition, it is quite straightforward to solve the system of equations (8.19) recurrently for increasing values of $n = 1, \ldots, K$ and thus obtain the conditional rate of service $u(n)$. A numerical example will help us visualize things. Let's set $p_1 = 0.8$, $\mu_1 = 4$ and $p_2 = 0.2$, $\mu_2 = 0.25$. The resulting mean service time is $0.8 \times 0.25 + 0.2 \times 4$, i.e., equal to 1. We set the maximum number of customers in the system to $K = 10$. We show in Figure 8.3 the equivalent conditional service rate $u(n)$ for two values of the customer rate of arrivals: $\lambda = 0.4$ and $\lambda = 0.8$.

We observe that the equivalent conditional rate of service changes considerably as the system workload changes. Imagine now that we use operational analysis to study our system, and we are unaware of the internal server organization or just choose to ignore it. If we describe the state of the system by the number of customers (jobs) in the system n, through stratified measurements for different values of n, we will obtain a set of average service times conditioned on n, say $S(n)$. If the observation period is long enough, these measured values

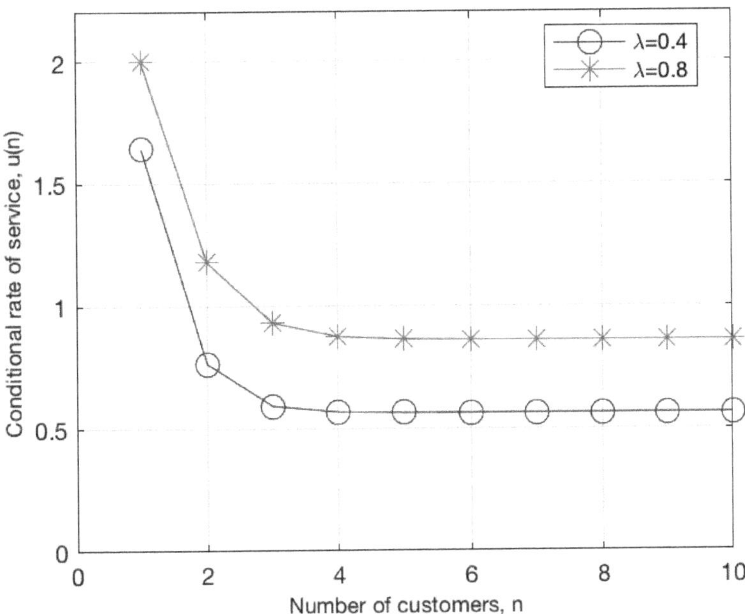

Figure 8.3. Conditional service rate $u(n)$ for different workload levels.

will likely be close to $1/u(n)$. However, as we just saw in the probability domain, these values cannot be assumed invariant, and if we were to measure the conditional average service times for another level of system load, we would get a markedly different set of values. In other words, to use state-dependent values obtained from measurements in a predictive study, we need additional assumptions, viz., that these values don't change with system load. Such assumptions may be impossible to validate. Similarly, although we can check on the measurement data the validity of a specific homogeneity assumption for a given observation period, predictive studies require a leap of faith that the homogeneity property in question remains valid over all other measurement periods, which certainly can be assumed but cannot be verified.

8.2 High-level modeling

The majority of the performance models considered so far are constructive in nature, in the sense that they attempt to model the structure of the system considered, and, to whatever degree, the nature of its workload. In complex modern systems, it is not always easy to know the internal structure of the system. In some cases, performance analysts may face a computer system whose inner structure and operation are mostly unknown but in which they have access to performance measurements. We assume that this knowledge of the system performance is represented by a set of measurement points such as, for instance, the mean sojourn time of customers at different levels of system throughput (see Figure 8.4). The black-box nature of the system clearly precludes any constructive approach that aims to mimic the system's behavior into an analytical model. However, one can wonder whether, in these circumstances, making predictions about the system's behavior remains possible. Said differently, can we do more with the measurement points than a simple linear or polynomial regression to forecast the system's behavior at unrecorded levels of workload?

In 2010, we proposed an automatic modeling method (Begin *et al.*, 2010b) to make predictions on a system behavior if it was to be running at another level of workload under the assumption that the knowledge of the system is limited to a set of measurement points.

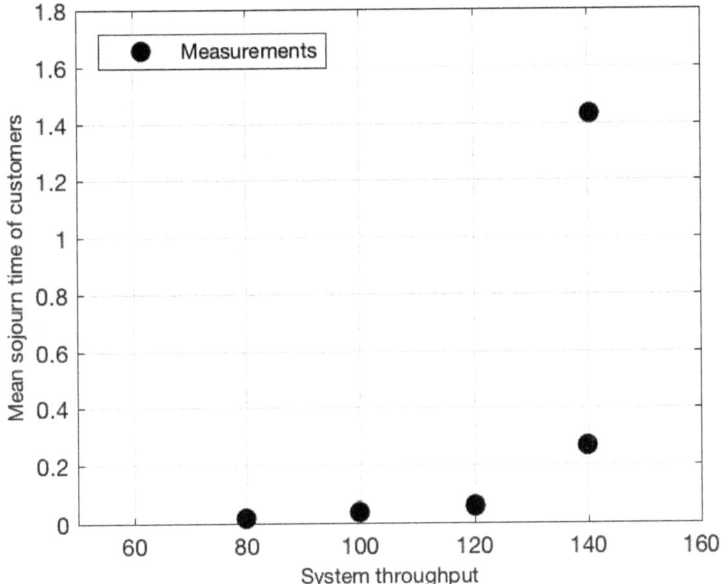

Figure 8.4. Example of measurement points.

We provide here a rapid outline of this method. The proposed method proceeds as follows.

Denote by M the set of measurement points taken from the system at different workload levels (e.g., rate of customer sources and number of customer sources). In its simplest form, M can relate the values of the mean sojourn time of customers to different levels of system throughput. Typically, M might include 5 to 20 measurement points. Note that other performance indices than the mean sojourn time can be considered (e.g., mean queue length and loss probability).

Our automatic modeling method includes a set of building blocks issued from queueing theory. Building blocks can be classical queues (e.g., M/M/C, M/M/C/K, M/G/1, and M/G/C) or more advanced models. An example of a more advanced building block could be an M/M/C model in which the service time derives from the sojourn time of another M/M/C queue; such a building block targets systems whose mean service times are not constant but instead grow as the workload increases. Each building block is characterized by a set of parameters whose values are not known in advance. These parameters can be discrete (e.g., the number of servers) or continuous (e.g., the mean service time of servers). As we shall see shortly, the modeling procedure involves an optimization task, and in order to facilitate this task, we treat discrete parameters as if they were continuous. When the analytical solution to a building block cannot be easily adapted to a continuous value for one of its parameters (e.g., a non-discrete number of servers), we simply use a straightforward linear interpolation between the two closest integer values to compute the building block performance. Finally, we extend every building block to include an additional continuous parameter, referred to as "offset", which represents a constant delay extending the mean sojourn time of the building block. One can think of this parameter as the incompressible minimum overhead associated with the given building block.

We define an objective function f to quantify the difference between the performance indices returned by a building block with specified numerical values for its parameters (we refer to such a building block as "specified") and the measurement points of M. This is illustrated in Figure 8.5. In general, f does not possess a closed-form expression and does not verify any specific properties (e.g., f is not continuously differentiable nor continuous) and f may exhibit multiple local optima, as well as large flat valleys. Additionally,

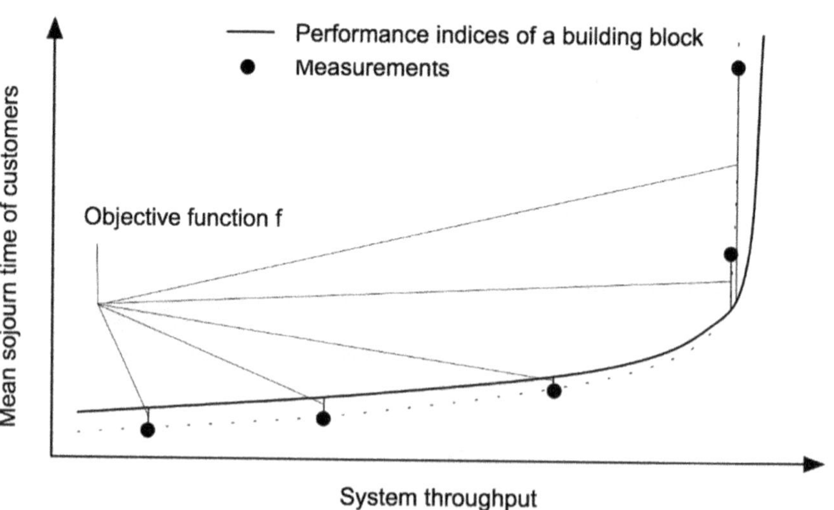

Figure 8.5. Illustration of the computation of the objective function f.

we need f to satisfy "sanity" constraints (e.g., the sum of the mean service time of a model and the offset must be less than the minimum sojourn time recorded in the measurement set M) that may be nonlinear (e.g., the product of the number of servers by the rate of service must be larger than the maximum level of workload reported in M). Minimizing f turns to be a difficult matter and clearly any gradient-based approaches (Polak, 2012) are inapplicable here.

For each building block, our goal is to find the numerical values for its parameters that minimize f while satisfying the constraints outlined above. Given the complexity of this task, we rely on an iterative search that makes use of a Derivatives-Free Optimization (DFO) algorithm (Powell, 2007; Conn *et al.*, 2009). Starting with an initial feasible parameter configuration, our algorithm aims at improving this configuration. The improved configuration then becomes the next current parameter configuration, and so on until convergence is found (local optima). To help find promising configuration candidates in the huge space of configurations, our DFO algorithm relies on a quadratic surrogate model whose parameters are found at each iteration so as to fit as closely as possible to the neighborhood of the current configuration. The minimum of the surrogate model is much easier to compute, and we use this minimum to explore configurations that can improve the current configuration. The use of this surrogate model speeds up the exploration and is thought to reduce the risks of being stuck too early in a local optimum. Note that in practice, we gradually decrease the size of the surrogate model to ensure convergence. For more information about the optimization of f we refer the reader to Begin *et al.* (2010a).

We repeat the search for the best set of parameters for the given set of measurement points for every building block. Then, we compare the score, i.e., the value of the function f, found for each building block, and we retain the one with the best score (i.e., lowest value of f) as the "laureate" solution selected by our automatic modeling method. Note that in case of a tie (or close scores), we favor the simplest building block, that is, the one with the lowest number of parameters (following the principle of Occam's razor (MacKay, 2003), which favors simplicity).

To illustrate the application of this automatic performance modeling for black-box systems, we consider two sets of measurements, referred to as Set 1 and Set 2, obtained for disk controllers in

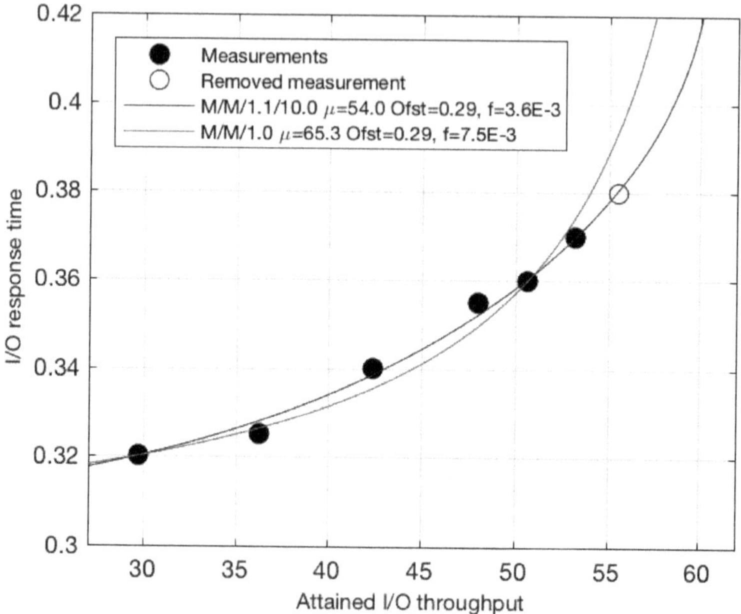

Figure 8.6. Results with measurement Set 1 for enterprise disk controllers.

mainframe environments. The measurement points in question give the expected I/O response time as a function of the attained I/O throughput (I/O requests completed per time unit).

Figures 8.6 and 8.7 illustrate the results obtained for measurement Sets 1 and 2, respectively. We used our procedure to obtain the parameter values for the building blocks with one or more measurement points removed. In other words, we on purpose removed one or more measurement points from each set and applied our procedure to the remaining measurement points. This allows us to compare the values predicted by the laureate model with the values of the removed points, i.e., actual measured values. We observe in Figure 8.6 that the "best" model among the building blocks considered in this case is the M/M/C/K queue, while for the measurement set of Figure 8.7, it is the M/G/1 queue.

In both cases, the laureate models not only closely reproduce the data points used in the calibration process but are also able to adequately predict the performance for the removed data points (not used in the search for the parameter values of the building block).

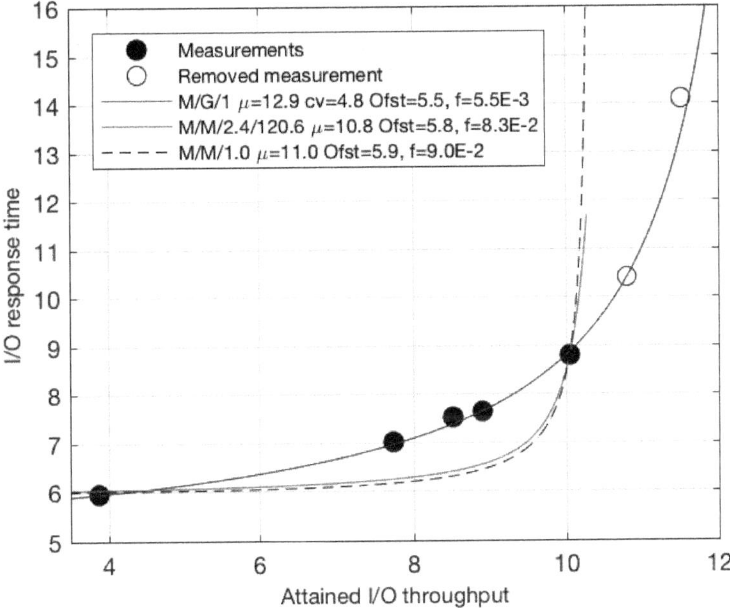

Figure 8.7. Results with measurement Set 2 for enterprise disk controllers.

The relative prediction errors for the removed data points are below 1% both for the expected throughput and the mean I/O time in Figure 8.6 and below 5% in Figure 8.7. It is interesting to note that, in this study, removing different points during the calibration process lead to similarly successful predictions by the laureate models. Clearly, this might not necessarily be the case for other measurement sets.

It may be of interest in building a constructive model of I/O performance for the controller considered in the examples above that for Set 2 neither the M/M/C nor the M/M/C/K models appear adequate. Our results show that even with the best possible combination of parameters these two building blocks fall short from matching the observed performance. In addition, as mentioned before, the laureate model may be of help in capacity planning. For example, the analyst can use the laureate model to forecast the system behavior in the face of projected growth in the I/O workload. Consider, for instance, measurement Set 1 (shown in Figure 8.6). With the help of the laureate model, we forecast that in the vicinity of the last removed measurement point (I/O rate of around 55.5), an increase in the I/O rate of, say, 5% will only cause a 5% degradation in the system average response time. By contrast, in the case of measurement Set 2 (shown in Figure 8.7) an increase of 5% in the I/O rate in the vicinity of the last removed point (I/O rate 11.5) will result in a 37% surge of the mean response time.

The high-level modeling approach can be extended to include more advanced building blocks and to handle situations where we wish to forecast the system behavior for changing numbers of customers (Begin *et al.*, 2010b). Contrary to the classical modeling approach, in which we aim to rely on understanding and expertise of the system under study to build a model, and we may use the measurement points to calibrate the parameters of our constructive model, the high-level performance modeling relies solely on measurement points to discover a model with some prediction capability. This allows essentially automated model building. Of course, the model being non-constructive, it cannot in general be used to predict the performance effects of a radical redesign of system internals. One can easily draw a parallel between this automatic performance modeling and machine learning solutions wherein sets of data are used to determine a predictive model. Finally, it is interesting to note that Scherr's model of a time-sharing system (discussed in Sections 2.4 and 3.4) may be viewed as an example of a high-level only non-constructive model (or minimally constructive) since in this model the whole system, with its internal queues of processes and various devices, is represented as a single server whose mean service time is known from measurements.

8.3 Simple performance bounds: Bottleneck analysis

The solution of a properly constructed model allows us to make predictions for selected performance measures. In some situations, we may be able to obtain meaningful bounds on important performance indices without obtaining a full solution of the model. As an example, let's consider the Central Server model of Figure 3.4. We denote by θ the job throughput in our system, i.e., the expected number of jobs going through the "new job" loop per time unit. When we used this system as an instance of a Jacksonian queueing network, we derived the frequencies of visits as $e_0 = 1$ for the central server and $e_i = p_i$ for the peripheral servers numbered $i = 1, \ldots, L$. Here we consider frequencies of visits relative to the "new job" loop, which we denote by ϕ_i, $i = 0, 1, \ldots, L$. Since at every passage by the central server the probability of a job going through the "new job" loop is p_o, we have $\phi_i = e_i/p_0$. We let $t_i = 1/u_i$ denote the mean service time at server i, $i = 0, 1, \ldots, L$. Our goal is to find bounds on the job throughput θ in this system. If we denote by $E[W_i]$ the mean time a customer spends at server i, by Little's law we must have

$$\theta = \frac{n}{\sum_{i=0}^{L} \phi_i E[W_i]} \tag{8.20}$$

if n is the total number of jobs in the system. Clearly, the mean service times t_i are the smallest possible values for the respective $E[W_i]$ since they imply that no jobs experience queueing delays at any devices. This is the case when $n = 1$, and the corresponding throughput is given by

$$\theta_a = \frac{1}{\sum_{i=0}^{L} \phi_i t_i}. \tag{8.21}$$

It is clear from formula (8.20) that we must have the following upper bound:

$$\frac{n}{\sum_{i=0}^{L} \phi_i t_i} \geq \theta. \tag{8.22}$$

This means that, if we graph the system throughput as a function of the number of jobs in the system, a straight line with the slope θ_a bounds the throughput for smaller values of n.

As the number of customers in the system increases, at some point at least one of the servers in the system will saturate, i.e., its utilization will become close to 1. From Little's law, the utilization of server i is given by $U_i = \theta\phi_i t_i$ so that we must have $\theta \leq 1/\phi_i t_i$ since $U_i \leq 1$. Let's use the subscript b to refer to any server capable of saturating as the number of jobs n increases. It is common to call such servers bottlenecks as they limit the overall throughput of the system. For a server that is a bottleneck, we must have $\phi_b t_b = \max_i (\phi_i t_i)$. Let's denote by θ_b the corresponding limiting throughput. Clearly, θ_b is an upper limit on system throughput, which means that we must have

$$\theta \leq \theta_b = 1/(\phi_b t_b) = 1/\max_i(\phi_i t_i). \tag{8.23}$$

This is illustrated in Figure 8.8(a). We note that the throughput of the system θ is upper bounded by two straight lines. One passing through the origin with a slope θ_a (this value is attained for $n = 1$), and the second horizontal with value θ_b. The point at which these two straight lines intersect, corresponding to the number of customers $\tilde{n} = \sum_{i=0}^{L} \phi_i t_i/(\phi_b t_b)$, is sometimes referred to as the saturation point of the system. Indeed, for $n > \tilde{n}$, we expect jobs to queue somewhere in the system. Similar results apply to other closed Jacksonian models.

If we now consider a model of a transaction processing system of the type shown in Figure 8.1 with a total of N_t sources of requests (transactions and jobs), in which the system part is a Central Server network, we can derive analogous results for the mean response time of the system $E[W]$. We assume that the time for an idle source (the corresponding process is not using the system) is exponentially distributed with mean $1/\alpha$. From Little's law, we must have $E[W] = N_t/\theta - 1/\alpha$. The throughput θ is the mean number of transactions (requests) processed per time unit. When $N_t = 1$, we get $E[W] = E[W]_a = 1/\theta_a - 1/\alpha$. θ_a denotes the request throughput with a single source. As the number of request sources N_t becomes large, system throughput is limited by $\theta_b = 1/\phi_b t_b$. As before, the subscript b refers to a bottleneck server in the system part of our model, i.e., a server whose utilization reaches 1 as the number of jobs in the system increases. The frequencies of visits for the servers in the system, ϕ_i, are taken relative to the system part of the model. This implies that we must have

$$E[W] \geq E[W]_b = N_t \phi_b t_b - 1/\alpha \geq N_t \phi_i t_i - 1/\alpha, \quad \forall i. \tag{8.24}$$

The index i refers to server centers inside the system part of our model, and single servers are assumed at each service center. Figure 8.8(b) shows an example of response time in such a system.

Overall, the response time in our transaction processing system is lower bounded by two straight lines: a horizontal line with value

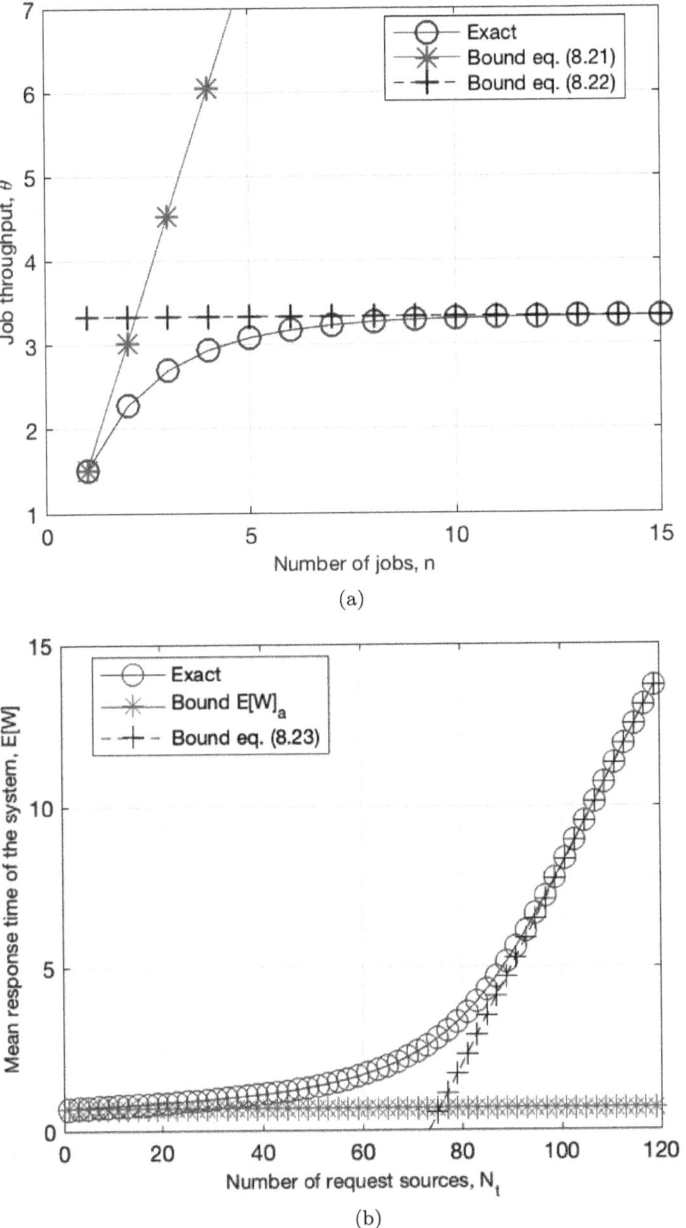

Figure 8.8. (a) Job throughput as a function of the number of jobs in a Central Server model. (b) System response time as a function of the number of request sources N_t.

$E[W]_a$ and the line defined by $E[W]_b$. $E[W]_b$ provides an asymptote for the response time as the number of request sources N_t becomes large. This response time asymptote intersects with the minimum mean response time $E[W]_a$ at $\tilde{N}_t = (E[W]_a + 1/\alpha)/(\phi_b t_b)$. For $N_t > \tilde{N}_t$, we expect queueing to occur in the system. It is interesting to note that the asymptote $E[W]_b$ intersects the horizontal axis at the point $N_b = (1/\alpha)\theta_b = (1/\alpha)/(\phi_b t_b)$. Since θ_b is the limiting throughput for a bottlenecked system, N_b is the mean number of request sources in idle state (generating new requests) in a saturated system.

Analogous results can be derived in the operational domain under the assumption that the visit ratios for servers and the mean service times of the servers remain invariant, in addition to the assumption of job flow balance, which leads to a product-form solution for the state occupancies in the system (Denning & Buzen, 1978).

The type of bounding analysis discussed above is referred to as bottleneck analysis. It can be quite valuable in understanding, without having to obtain a complete and detailed solution of the model, the overall effects of certain configuration changes. In particular, it shows clearly that replacing a non-bottleneck device with a faster one will not change the maximum processing capacity of the system and will likely have minimal effect on the mean response time of the system.

8.4 Sampling bounds with equivalence and decomposition

There has been a relatively large number of performance bounds published in the literature, e.g., (Eager & Sevcik, 1983; Kurose, 1992; Koukopoulos *et al.*, 2004), and it is not our objective to discuss them all. As mentioned before, the goal of bounds is most often to provide a reasonably good estimate of performance measures in a system without a full solution of the model. As we already know from our discussion of approximation methods, a system like the one shown in Figure 8.1, in which a total of N_t customers circulate alternating between a delay server (the request source) and a subnetwork, is a good candidate for the application of the equivalence and decomposition approach. Recall that in this approach we analyze the subnetwork (the inner model) in isolation to produce the conditional

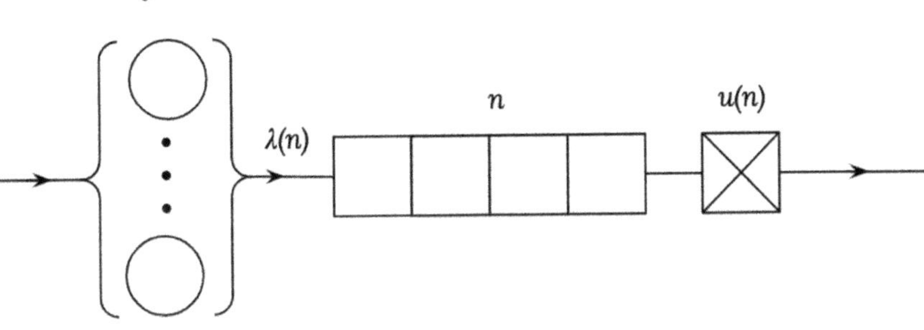

Figure 8.9. The equivalent (outer model).

throughput $u(n)$ for $n = 1, \ldots, N_t$, which is the rate with which customers leave the subnetwork given that the current number of customers in the subnetwork is n. The values of $u(n)$ are then used in the simpler equivalent model (the outer model), like the one shown in Figure 8.9 (or Figure 6.2, if you prefer).

As discussed earlier, the equivalence and decomposition approach requires that we solve the inner model, i.e., the subnetwork, for all values of the state vector used in the outer model. In our case, this means the subnetwork would have to be solved for all population levels $n = 1, \ldots, N_t$. Clearly, for a large value of N_t, this can represent a significant computational effort. As pointed out in the literature (Brandwajn, 1998), it seems reasonable to assume that in a large system with very many customers only a relatively small subset of all possible system states would constitute the "operating point(s)" of the system, i.e., the region(s) of the state space where the system is most likely to be.

Let's take the example of a transaction processing system represented in Figure 8.10.

In this model, in addition to a set of N_t request sources (delay servers) and a processing part of the system, assumed to consist of a set of servers with memoryless service times, there is an admission control limiting the number of customers (transactions) in the processing part of the system, which we denote by m, to a total of M. An FCFS admission queue holds customers awaiting admission

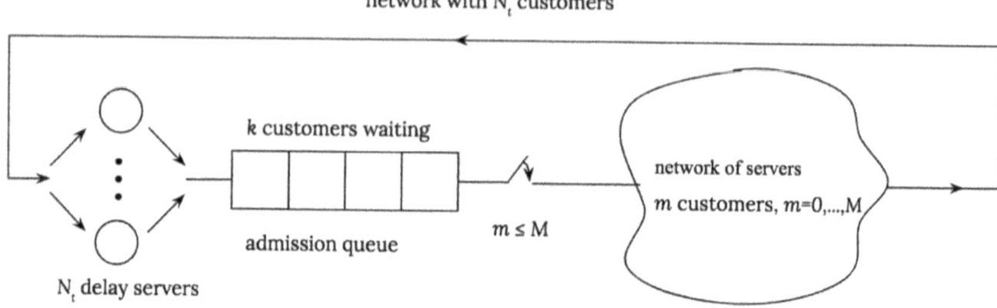

Figure 8.10. A transaction processing system with admission control.

if the number of customers already admitted into processing by the system is M. It is clear that with this type of admission control, if we know the total number of customers outside the delay servers, we also know the number of customers admitted into processing. Indeed, we have

$$m(n) = \begin{cases} n, & \text{if } n < M \\ M, & \text{otherwise.} \end{cases} \tag{8.25}$$

We can easily check that, without global dependencies, this model possesses a product-form solution and the equivalence and decomposition method produces the exact solution of this queueing network. By global dependencies we mean, for example, service times and routing probabilities that would depend on the number of jobs in the processing part of the system. We can verify the fact that the equivalence and decomposition approach produces the exact solution by directly substituting the solution produced by this approach into the actual balance equations of the model. Let's denote by $p(n)$ the steady-state probability that there are n ($n = 0, \ldots, N_t$) customers outside the delay servers. We show in Figure 8.11 an example of the shape of $p(n)$ as a function of n, as well as the corresponding conditional throughput $u(.)$ as a function of m, the number of customers admitted into the subnetwork of queues. These results were obtained using classical equivalence and decomposition, i.e., the subnetwork representing the processing part of the system was evaluated in isolation for all population levels $m = 1, \ldots, M$ with $M \leq N_t$.

We note in Figure 8.11 that the probability $p(n)$ tends to concentrate within a relatively small subset of possible values of the number of customers n. This subset clearly corresponds to the most

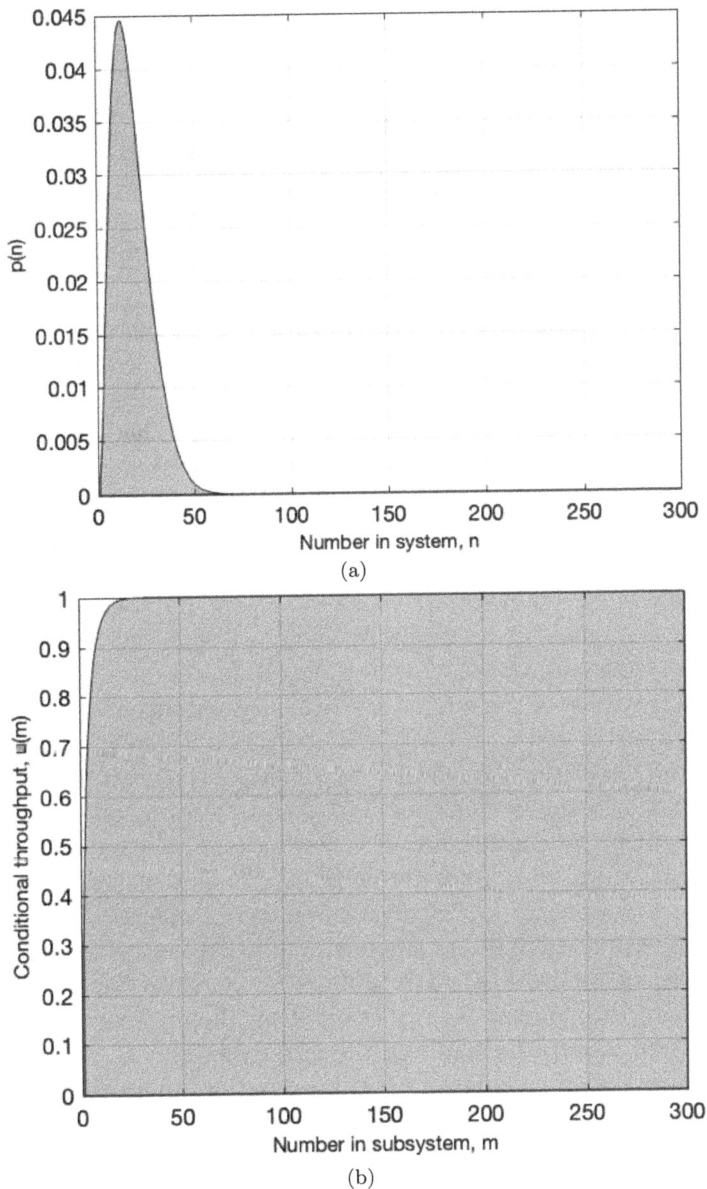

Figure 8.11. Example of the probabilities (a) $p(n)$ and the corresponding (b) conditional throughput $u(m)$.

probable "operating region(s)" of our system. In fact, in our example, the state probabilities $p(n)$ are close to 0 outside of the well-visible concentrated area. This observation suggests that the precise evaluation of the conditional throughputs $u(.)$ should be important close to the operating points of the system, but outside these points, just a reasonable approximation for $u(.)$ could be sufficient. Of course, a "reasonable" approximation should not alter excessively the shape of the probability distribution $p(n)$, and, in particular, the location of the system operation regions.

The catch is that, when dealing with a queueing model, we generally don't know where its most likely points of operation are. As it turns out, we can design a simple fixed-point procedure (you guessed it) that quickly finds the most important points. The inner model (in our case, the subnetwork representing the processing part of the system) is solved mainly for these points. Outside these points we can use an upper and lower bound for the conditional throughput $u(.)$, giving us an optimistic and a pessimistic bound for performance measures such as system response time or expected numbers of customers in admission queue or in the processing part of the system. Since the operating region(s) of the system typically represent a small fraction of the state space of the model, we avoid expending energy on regions of low importance.

For our derivation of bounds, we concentrate our attention on systems for which the equivalence and decomposition solution is exact and in which the conditional throughput $u(.)$ has the general form shown in Figure 8.11. As mentioned before, this includes models in which the processing part of the system is a Jacksonian subnetwork and there are no global dependencies. This is the case for our example model of Figure 8.10 with admission control if we consider the conditional throughput as a function of m, the number of customers in the system proper, excluding the admission queue.

Assume that the values of $u(m)$ have been evaluated for the edges of the domain of interest, i.e., for the points $m = 1$ and $m = M$, as well as a few intermediate points. Let k be the number of intervals defined by the known points. In our discussion, we assume that $k \geq 3$. Interval i is defined by points x_i and x_{i+1}. Let's denote by c_i the chord through these two points. Given the shape of the function $u(m)$, it is clear that c_i defines a lower bound for $u(x)$ over this interval, i.e., for any $x \in [x_i, x_{i+1}]$. From the convexity of $u(.)$, it is also clear that each of the two chords c_{i-1} and c_{i+1} defines an upper bound for the conditional throughput $u(.)$ in the same interval. To obtain the tightest bound, we select the lowest of the two values for a given number of customers x. For the first interval, we can use c_2 as the upper limit. For the last interval, we use the smaller of the values on the chord through the next to last interval, c_{k-1}, and $u(M)$. Denoting by $u_{inf}(.)$ and $u_{sup}(.)$ the lower and upper bounds for $u(.)$, respectively, we can summarize our very simple bounding as follows:

$$\begin{aligned}
&\text{For } x \in (x_i, x_{i+1}): && u_{inf}(x) = c_i(x), u_{sup}(x) = \min(c_{i-1}(x), c_{i+1}(x)), \\
& && i = 2, \ldots, k-1 \\
&\text{For } x \in (1, x_2): && u_{inf}(x) = c_i(x), u_{sup}(x) = c_2(x), \\
&\text{For } x \in (x_k, M): && u_{inf}(x) = c_i(x), u_{sup}(x) = \min(c_{k-1}(x), u(M)).
\end{aligned}$$

If we solve the outer model using the lower bound $u_{inf}(.)$ for all points for which the true value of $u(.)$ has not been obtained from the solution of the inner model, we should obtain pessimistic bounds for overall system performance. This is because $u_{inf}(.)$ corresponds in essence to a system with a slower processing speed. Conversely, if we use the upper bound $u_{sup}(.)$ for the conditional throughput, we should obtain optimistic values for system performance measures. It is important to keep in mind that in real-life situations, either bound may be sufficient to answer an essential question. For example, if during a design or planning phase, a pessimistic bound for the mean response time shows that our system will meet the expectations (or Service Level Agreement), this may be enough to proceed with the design. By the same token, if even with an optimistic bound the system performance falls short of objectives, we know for sure that a redesign or overhaul of the system is in order.

Let's now outline a possible fixed-point procedure to take advantage of the bounds on conditional customer throughput discussed above. In order to keep the computational effort low, we should attempt to select the evaluation points (i.e., points at which the inner model is solved fully) so that they contribute most to the accuracy of the results. Intuitively, these points would be in the vicinity of the most likely operating regions of the system. Since these regions are not known *a priori*, we start by computing the values of $u(m)$ for $m = 1$, $m = M$, and a small number of intermediate points. As an example, we use four evenly spaced intermediate points.

This allows us to create upper and lower bounds for all missing values of $u(m)$ and gives us two probability distributions for the number of customers in the system. We denote by $q(n)$ and $r(n)$ the distributions obtained using the upper and lower bound for $u(n)$, respectively. We use $u(n)$ as shorthand for $u(m(n))$, where $m(n)$ is given by 8.24.

We select the additional evaluation point for $u(.)$ as being the point for which the difference between the upper and the lower bounds weighted by the probability of occurrence of the given state is the largest in absolute value, i.e., which maximizes the quantity $|u_{sup}(n)q(n) - u_{inf}(n)r(n)|$. Assume j is the corresponding population level. In an attempt to reduce the number of iterations needed, we can additionally evaluate the conditional throughput for the points $j - \Delta$ and $j + \Delta$, where Δ is some relatively small number, for instance, $\Delta = 3$. If a population level selected in this way corresponds to a point already evaluated, we may select a somewhat lower or somewhat higher point instead.

We repeat the procedure outlined above until the consecutive values of a given performance measure (e.g., the mean response time) have stabilized within a specified accuracy or no new points are selected for the evaluation of the conditional throughput $u(.)$. Interestingly, this procedure tends to converge within just a few iterations. How many evaluation points for $u(.)$ are actually necessary depends on the values of model parameters, but, typically, the total number of points at which $u(.)$ is evaluated tends to be a small fraction of the total number of population levels $M + 1$. We illustrate this with the results of a few examples in Table 8.1.

Table 8.1. Example of results obtained for the mean number of customers in the system.

Total number of customers (N_t)	Maximum number admitted (M)	Mean number in system		Number of points evaluated	Number of iterations
		Optimistic	Pessimistic		
300	300	16.968	17.765	33	11
300	300	85.714	86.056	9	3
300	100	27.658	30.740	15	5
300	100	69.231	69.244	9	3

Here, we use the mean number of customers in the system as the selected performance metric. We note that the optimistic and pessimistic values of the mean number in the system differ by less than 10%, while the number of evaluation points used is about 10% to 15% of the total number of population levels. Of course, in addition to the computation of the values of $u(.)$ for the selected evaluation points, there is a small computational cost to obtain the values $u_{inf}(.)$ and $u_{sup}(.)$. We can think of this method as a form of "guided sampling" since we sample the exact values of the conditional throughput guided by a function of the probability of occurrence of the state.

Clearly, with the bounding method described above, we could get arbitrarily tight bounds. In general, the bounds shown in Table 8.1 are tight enough to be used as a viable approximation. It is notable that the number of points needed to get relatively tight bounds tends to grow quite slowly at the total population size N_t (or M) grows.

For systems in which the equivalence and decomposition method doesn't produce an exact solution, for instance, those with global dependencies where the conditional throughput exhibits an inflection point, the bounding described above doesn't work. This is because the conditional throughput $u(.)$ obtained by evaluating the subnetwork (inner model) in isolation for such models is not exact so that an upper or lower bound on approximate throughput is not necessarily a bound on the actual throughput. The sampling approach can, however, be used as an additional layer of approximation to significantly reduce the computational cost.

References

Begin, T., Baynat, B., Sourd, F., & Brandwajn, A. (2010a). A DFO technique to calibrate queueing models. *Computers & Operations Research*, 37(2), 273–281.

Begin, T., Brandwajn, A., Baynat, B., Wolfinger, B. E., & Fdida, S. (2010b). High-level approach to modeling of observed system behavior. *Performance Evaluation*, 67(5), 386–405.

Brandwajn, A. (1998, October). Fast decomposition in large stochastic models. In *SMC'98 Conference Proceedings. 1998 IEEE International Conference on Systems, Man, and Cybernetics (Cat. No. 98CH36218)* (Vol. 4, pp. 3073–3078). IEEE.

Conn, A. R., Scheinberg, K., & Vicente, L. N. (2009). *Introduction to derivative-free optimization*. Society for Industrial and Applied Mathematics.

Dallery, Y., & Cao, X. R. (1992). Operational analysis of stochastic closed queueing networks. *Performance Evaluation*, 14(1), 43–61.

Denning, P. J., & Buzen, J. P. (1978). The operational analysis of queueing network models. *ACM Computing Surveys (CSUR)*, 10(3), 225–261.

Denning, P. J. (2006). Operational analysis. In *Computer System Performance Modeling In Perspective: A Tribute to the Work of Prof Kenneth C Sevcik* (pp. 21–33).

Eager, D. L., & Sevcik, K. C. (1983). Performance bound hierarchies for queueing networks. *ACM Transactions on Computer Systems (TOCS)*, 1(2), 99–115.

Koukopoulos, D., Mavronicolas, M., & Spirakis, P. (2004, May). Performance and stability bounds for dynamic networks. In *7th International Symposium on Parallel Architectures, Algorithms and Networks, 2004. Proceedings.* (pp. 239–246). IEEE.

Kurose, J. (1992). On computing per-session performance bounds in high-speed multi-hop computer networks. *ACM SIGMETRICS Performance Evaluation Review*, 20(1), 128–139.

MacKay, D. J. (2003). *Information Theory, Inference and Learning Algorithms*. Cambridge University Press.

Polak, E. (Ed.). (2012). *Optimization: Algorithms and Consistent Approximations* (Vol. 124). Springer Science & Business Media.

Powell, M. J. (2007). A view of algorithms for optimization without derivatives. *Mathematics Today-Bulletin of the Institute of Mathematics and its Applications*, 43(5), 170–174.

Chapter 9

Model Validation and Robustness

9.1 Basic notions

Assume we have built our model and were able to select and implement a suitable approach to its solution, whether exact or approximate. As we stated several times in this text, the goal of modeling, besides intellectual satisfaction that comes from improved understanding, is to be able to use our model to make performance predictions. Clearly, before we start relying on the results of our model, we need to gain some level of confidence that the model and its solution are valid. This is the objective of model validation. In particular, if the model represents an existing system, we could feed measured parameter values into the model and compare the model results with actual measurements.

In many, if not most, situations, certain model parameters may not be known directly. Good examples are the values of system overheads in a cloud system (e.g., the average time to activate a VM and average time to migrate a VM) or the average seek times of disk storage devices in a particular environment. This means that we will need to select appropriate values for the unknown parameters in order to calibrate the model so as to bring its results sufficiently close to measured values. Thus, the calibration of the model is closely intertwined with model validation.

It is possible that regardless of the values we select (within reasonable bounds) for the unknown model parameters, the model results fall too far from the expected values, and the system behavior predicted by the model is significantly different from the observed behavior. If there are no major inaccuracies in the solution approach and no errors in the implementation of the solution, this usually means that our model misses one or more major performance aspects of the system. This may be because we don't fully understand the internals of the system being modeled or because we incorrectly represent the performance impact of important aspects of the system. Whatever the cause, the missing or misrepresented features will have to be incorporated, and the calibration and validation process reattempted. Clearly, with bad luck, we might need to repeat this cycle more than once. Thus, the development of a calibrated and validated model may be an iterative process.

9.2 Validation

The objective of model validation is to gain confidence that our model is valid. If the model represents an existing system for which we have (or can have) access to measurement results, the most straightforward approach is to select what we would consider a representative measurement period and set the model parameters based on observed values. (If not all model parameters can be found directly in the measured data, some model calibration, discussed in more detail in the next section, might be needed.) We would then execute the model with these parameter values and compare the values of selected performance indices, such as, e.g., the mean response times or device utilizations, predicted by our model with those actually observed, i.e., measured during the selected observation period. If the two sets of values are close enough, we would be comforted in our feeling that our model is valid. How close the two sets of values should be depends on the system being modeled and our personal tolerance. In most cases, a deviation of less than 10–15% would be considered a reasonable goal for model validation, but in some environments, a 20% discrepancy might be quite acceptable. In environments in which the workload is very well understood and characterized, a deviation of less than 5% might be required for the performance analysts to feel good about their model.

Ideally, if possible, we would like to validate our model using several sets of measurement points. With that respect, a potentially important pitfall to avoid is the use of measurement points that are very similar, not allowing us to check the model behavior for a range of workloads.

It is, of course, much more difficult to validate a model if we have no actual measurements against which to compare our model results. This is the case, in particular, when the model represents a system under development or a study of design alternatives for a hypothetical system. A minimal level of validation may be to ensure that our model results pass the "sanity check". By this we mean that the model behaves as could be reasonably expected in specific scenarios. As an example, we certainly would expect the mean response time of a transaction processing system to be greater than the weighted sum of the mean service times, or we would expect a lower priority class to exhibit a reasonable level of interference from higher priority classes, etc. Another level of validation may be provided by comparing the results of our model with those of a more detailed discrete-event simulation if our model is solved by a different method. Clearly, this validation approach really makes sense only if the more detailed discrete-event simulation and our model don't use the same simplifying assumptions. As an extreme example, if we model a transaction processing facility as an M/M/1 queue and compare our model with a discrete-event simulation of an M/M/1 queue, we are not validating our model beyond verifying that both solution methods produce matching results.

9.3 Calibration

Closely related to model validation is model calibration, i.e., the selection of values for model input parameters so that model results match known measurement results. Model calibration requires the availability of actual measurement results or of the results of a more detailed simulation model. In an ideal world, we would have at our disposal the correct values for all model input parameters, including, for instance, all system overheads, internal data rates within an I/O controller as well as disk device mean seek times. In reality, this is not always the case.

We will use the example of a disk I/O model in a mainframe system to illustrate some of the issues in model calibration.

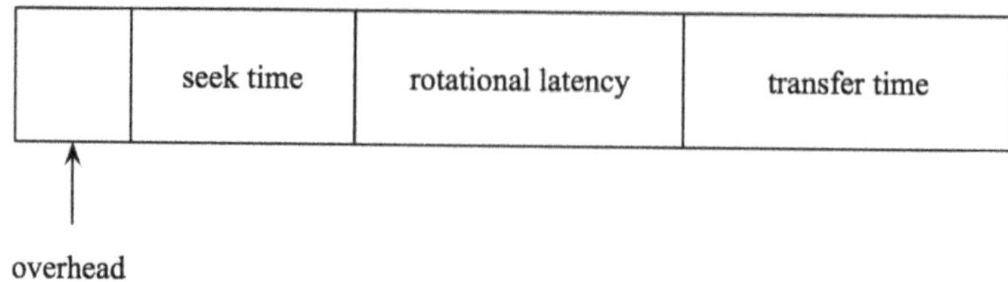

Figure 9.1. Service time components of un-cached I/O.

For simplicity, let's consider a device that is not using cached I/Os. As shown in Figure 9.1, at a high level, the service time of such a device could be viewed as a sum of overhead, disk seek time, rotational latency, and transfer time. We assume in our discussion that there is a sufficiently high number of available I/O paths, as well as a sufficient amount of buffer space along the I/O path, so that any additional waits can be neglected and the device service time represented by the main components shown in Figure 9.1. Of course, at the system level, there is a queue of requests waiting for access to the device, and the mean I/O time includes the mean time an I/O request is queued in addition to the mean service time.

We wish to use our model to predict the effect of changing the device on the expected I/O time. Contemplated changes include switching to a new generation device with a faster seek time, higher RPM (revolutions per minute), and a faster data transfer rate. This means that we need to model the effect of such changes on each component of the I/O service time and the resulting mean queueing time.

Let's assume that the capacity of the software queue is large enough so that we don't have to worry about exceeding it in a typical environment. Let's also assume, for starters, that the I/O arrival process for the device considered is relatively well behaved and can be represented as a Poisson process. With these assumptions, we model the queueing time for our disk device as the queueing time in an M/G/1 queue. Recall that the expected queueing time in the M/G/1 queue can be obtained from the Pollaczek–Khintchine formula (2.35) together with Little's law (2.16).

Given our objective, the parameters of our model include the I/O overhead time, the mean and the coefficient of variation of the transfer length (number of bytes transferred per I/0), the device characteristics such as the data transfer rate, the RPM, which gives us the device revolution time (rotational latency), nominal average and maximum seek times, as well as the percentage of zero seeks, i.e., the fraction of I/Os for which no seek phase was actually required. The latter quantity is a characteristic of the workload and data layout on the device. For our purposes here, we use a very simple model for the seek part of the I/O service time. We assume that the seek component is a combination of zero seek time for sure with a probability equal to the zero seek percentage and is otherwise given by the device average seek time.

Our task at hand now is to calibrate our model using the available measurement data. We have at our disposal measurement data provided by the standard system instrumentation (IBM OS/390; IBM OS/390; z/OS, 2001; z/OS, 2003), which include (after some manipulation) the measured average queueing time for an I/O request at this device, the average I/O service time during which the device was not using the I/O channel ("disconnect" time) and the average time during which the I/O channel was used ("connect" time).

For our model to serve the intended purpose, we need to be able to determine the mean values for each of the components of the I/O service time, as well as their respective variances. This is because, according to the Pollaczek–Khintchine formula (2.35), the mean queueing time in our model depends on the mean and the variance (or coefficient of variation) of the service time. Let's analyze the components of the I/O service time one by one. We assume, of course, that we know the device type for which the measurements were obtained, i.e., we know the manufacturer-specified average seek time, the device RPM, and the data transfer rate. We also assume that we know reasonable values for the I/O overheads from published data or benchmark measurements. We have two major difficulties in trying to calibrate our model.

First, the system instrumentation doesn't give us directly the average number of bytes transferred per I/O. Instead, we are given the average time the I/O channel was busy per I/O on behalf of the given device. So, we need to subtract from this "connect" time the average I/O channel overhead to obtain the average transfer time and divide the result by the data transfer rate of the current device to obtain the average number of bytes transferred per I/O. Here, we need a reasonable value for the channel overhead per I/O for our approach to work. The standard system instrumentation has no information regarding the variance of the transfer length, so this becomes an "adjustment knob" in our calibration of the average queueing time. In some particular situations, the nature of the workload may give us the needed information about the number of bytes transferred per I/O, e.g., if records of known constant length are used by specific applications. In this case, we can use this knowledge to confirm or sharpen our estimate of the I/O channel overhead.

Our second difficulty is that the system instrumentation does not measure the percentage of I/O operations for which there was no actual seek time. In fact, we are only given the sum of the average seek time and the average latency as the "disconnect" time. Under normal conditions, the latency time is uniformly distributed between zero and a full device revolution time. Consequently, the nominal mean latency time is equal to one-half revolution time. The seek times can vary from zero to the maximum seek time for the device, but, as mentioned above, a possibly large percentage of I/O requests result in no seek time. By backing out the average latency time from the measured "disconnect" time, we get a measure of the effective average seek time. As mentioned above, we use a very simple model in which we assume that a certain percentage of I/O requests require for sure no seek time, and the remainder experience the nominal average seek time for the device. We can then determine the hypothetical "percentage of zero seeks". Of course, it could happen that this calculation yields a negative result (i.e., the estimated average seek time is higher than the manufacturer-specified nominal device average seek time). Assuming that the system instrumentation is correct, this would imply that either the latency was not equal to the nominal half revolution or that the I/O pattern was such that the device had to seek on (almost) all operations. This latter possibility is fine as long as the resulting seek time does not exceed the maximum seek time for the device.

Once we have determined the average number of bytes transferred per I/O and the percentage of zero seeks, we have calibrated our model for the mean service time. It remains now to determine reasonable values for the variance of the number of bytes per I/O and the variance of the seek component. (The variance of the rotational latency is given if we assume a uniform distribution between zero and a full device revolution.) We need to select these two variances so that the resulting overall variance of the I/O service time, when used in the Pollaczek–Khintchine formula, gives us a mean queueing time close enough to the measured average queueing time. Thus, we have two "knobs" to turn to try to calibrate this part of our model: the variance of the transfer length and the variance of the seek time. Of course, there are limits within which we can adjust the variance of the transfer length. The smallest value is 0, and although the variance could theoretically be infinite, in practice, there is a limit on the amount of data transferred per I/O. Similarly, the variance of the seek time must be within some reasonable limits. Assuming that the different components of the I/O service time are independent, we can then compute the variance of the service time as the sum of the variances of its components.

If we can find reasonable values for these two parameters (the variance of the transfer length and the variance of the seek time) that allow the results of our M/G/1 model to match the measured mean queueing time, we have a calibrated model. If, however, we cannot find reasonable values for our "adjustment knobs", we may need a different model. In particular, we might want to consider a model that would allow us to account for the variability of the times between I/O requests. Among the models discussed in this text, the Ph/Ph/1 model would give us this additional flexibility. Of course, this new model will need to be calibrated before we decide to use it for predictive purposes.

Both in model validation and calibration we need a metric of distance between available measurement points and corresponding model results. The commonly used absolute values of the difference or of the relative difference between these quantities each have their advantages and drawbacks. A 1 ms difference can be viewed as quite small if the measured value is on the order of 50 ms or very important if we are talking about a measured value of 2 ms. On the other hand, when dealing with small quantities, large relative errors arise easily but may not be very meaningful. Of course, one can devise more sophisticated metrics of distance, see, for example, Begin *et al.* (2010).

A related issue is that not all measurement points are equally trustworthy. As an example, imagine that we are trying to validate a model of a large I/O controller supporting thousands of logical I/O devices. In most environments, only a subset of these devices will be active during a given measurement period. Consequently, there will be a significant number of devices for which there is very little activity in the measured data. Intuitively, the corresponding measurement points are less trustworthy than data points for high activity devices.

We think that it is reasonable, and in fact advisable, to "prune" from our model such low-activity devices, and disregard them in the validation and calibration of the model. Clearly, if very many devices with a low (but non-zero) activity get pruned, the utilization of controller elements by these devices will be missing in our model but not in the measurement data. Ideally, it would be great if there was a simple way to introduce "catch all" dummy devices to account for the missing workload, but depending on the nature of the system being modeled, this may be more work than it's worth.

Based on the idea that measurement points with more activity are more trustworthy, we think that a reasonable measure of overall "distance" between measurements and model results is a weighted average of distances for each measurement point, where the weights are proportional to the frequency of use. In the case of our example I/O controller, this would mean an average of, for example, relative differences in I/O times between the model and measurement data, weighted by the relative I/O rates of each device.

9.4 Iterative model development

We stated at the beginning of this text that, in our view, a good model is the simplest model that correctly reproduces the system features or behaviors we are aiming to model. Of course, as we have been stressing all along, the ultimate goal of modeling is prediction, so our models must possess predictive power. With these points in mind, it is not surprising that model development can, and oftentimes is, an iterative process in which the model may be refined or changed altogether in favor of a better-adapted model. An example in point comes from our preceding discussion of the calibration of a simple model of non-cached I/O. If adjusting model input parameters within reasonable limits in the M/G/1 model doesn't allow us to attain an acceptable calibration, we need to revise our model to include I/O arrival variability and try again.

Sometimes, the need to revise a model stems from incomplete knowledge of the system being modeled. A good example comes from the development by one of the authors some time ago of a model of an I/O controller with a high level of parallelism within the controller. Following the description of the internals of the controller and discussions with design engineers, we developed a model whose goal was to assess the expected overheads and service time elongations caused by I/O path contention within the controller. When the model results were compared with available benchmark measurements, there was not only a significant systematic difference but, importantly, a fundamentally different pattern of increase as the I/O rates were increasing. Instead of the slow increase followed by an abrupt growth one would expect with multiple servers, the pattern exhibited by the measured values was strikingly similar to the saturation pattern in a single-server queue. As a reminder, we discussed saturation patterns in Section 7.1. This behavior was quite baffling in a controller with many internal paths at each end of the controller until another round of discussions with the design engineers disclosed that every I/O operation had to go several times through a single shared element. As a result, it was this shared component that was effectively driving the performance of the controller. Consequently, the model had to be revised to include an explicit model of the shared element, which then allowed the model to correctly calibrate to measurement results.

Before we leave the subject of iterative model development, we would like to note that some model designs lend themselves more easily to modifications and refinements than others. This is the case, in particular, of decomposition approaches, in which changes can be made to parts of the model without necessarily requiring a total model redesign.

9.5 Robustness

By now, we have studied in some level of detail a number of models of computer or computer-based systems. We have seen when learning about BCMP queueing networks that in many cases the steady-state behavior in such networks was insensitive to distributional assumptions. In particular, this was the case for service/queueing disciplines considered in the BCMP theorem other than the "old and trusted" FCFS. In that sense, BCMP-based models are robust vis-à-vis distributional assumptions. In other words, within the constraints of the BCMP networks, with the exception of FCFS service stations, the results would not change if we changed assumptions on service time distributions.

In a similar vein, models in which no queueing is allowed, i.e., loss systems, tend to be insensitive to service time distributions and, under some circumstances, to the distribution of times between arrivals. If the system being modeled falls into this lucky category, we can think of our model as being robust.

On the other hand, as we have seen, systems with FCFS queueing, most priority disciplines other than LCFS-PR (considered in the BCPM theorem), as well as most realistic types of blocking, are sensitive to distributional assumptions. As we have seen in our simple derivation of the Pollaczek–Khintchine formula (2.35), this is related to the fact that residual times exhibit intrinsic distributional dependence. Therefore, a solution that relies on memoryless assumptions in such models is not robust.

Regarding certain fundamental results, Little's formula is considerably more robust than its generalizations to higher moments. The example of the use of the latter in formula (7.100) depends in a fundamental way on the assumption of Poisson arrivals, and the results become significantly more involved if this assumption is relaxed.

If we consider different solution methods, it is pretty clear that a properly designed discrete-event simulation will tend to be more robust, i.e., resilient to changes in assumptions, than a corresponding analytical solution. By "properly designed", we mean a simulation that does not exploit specific properties of a given distribution. For example, if a simulation of a preemptive-resume priority system with exponentially distributed service times uses the memoryless property of the exponential distribution to simplify the simulation design, such a simulation becomes much less robust than a simulation that explicitly keeps track of residual service times at each service interruption. At the risk of repeating ourselves, in our view, in general, models (including those solved using discrete-event simulation) that use decomposition approaches tend to be more robust than models solved as a monolithic entity.

9.6 Art of computer modeling

Throughout this text, we have explored a selected set of approaches that can be used in designing and solving models of computer systems. The particular selection of approaches presented reflects personal preferences and, to some extent, biases informed by many years of successfully practicing computer modeling. As stated in the introduction, at its core, a model is an abstraction of reality. The art of computer modeling, if we elect to call it that, is to find the right abstraction with the right set of assumptions to capture the relevant features of the system being modeled. By assumptions we mean any simplifying assumptions regarding the system internals and operation, including assumptions on the nature of system workload. With that respect, less is more, so to speak, in the sense that we should avoid including in our model minutia that adds little to our understanding of the system being modeled and may actually detract from it. To be useful, our model must represent the particular system behavior or behaviors we wish to examine and, importantly, it must be able to have adequate predictive power. Of course, our model must be tractable, i.e., we must be able to solve it by some method at our disposal. Like in many areas of life, it is all about striking the right balance, and practice is important in guiding our intuition in that endeavor.

References

Begin, T., Brandwajn, A., Baynat, B., Wolfinger, B. E., & Fdida, S. (2010). High-level approach to modeling of observed system behavior. *Performance Evaluation*, 67(5), 386–405.

Bolch, G., Greiner, S., De Meer, H., & Trivedi, K. S. (2006). *Queueing Networks and Markov Chains: Modeling and Performance Evaluation with Computer Science Applications*. John Wiley & Sons.

Gelenbe, E., & Mitrani, I. (2010). *Analysis and Synthesis of Computer Systems* (Vol. 4). World Scientific.

Harchol-Balter, M. (2013). *Performance Modeling and Design of Computer Systems: Queueing Theory in Action*. Cambridge University Press.

IBM OS/390 — MVS System Management Facilities (SMF) — GC28-1783-05. International Business Machines Corporation, Armonk, NY, USA.

IBM OS/390 — Resource Measurement Facility (RMF) User's Guide — SC28-1949-03. International Business Machines Corporation, Armonk, NY, USA.

Jain, R. (1991). *The Art of Computer Systems Performance Analysis: Techniques for Experimental Design, Measurement, Simulation, and Modeling* (Vol. 1). Wiley.

Kobayashi, H., & Mark, B. L. (2009). *System Modeling and Analysis: Foundations of System Performance Evaluation.* Pearson Education India.

Lavenberg, S. (Ed.). (1983). *Computer Performance Modeling Handbook.* Elsevier.

Lazowska, E. D., Zahorjan, J., Graham, G. S., & Sevcik, K. C. (1984). *Quantitative System Performance: Computer System Analysis Using Queueing Network Models.* Prentice-Hall, Inc.

Menasce, D. A., Almeida, V. A., Dowdy, L. W., & Dowdy, L. (2004). *Performance by Design: Computer Capacity Planning by Example.* Prentice Hall Professional.

Sauer, C.H., & Chandy, K.M (1981). *Computer Systems Performance Modeling.* Prentice-Hall.

z/OS Resource Measurement Facility: Performance Management Guide (2001). Manual SC33-7992-00, IBM Corporation. International Business Machines Corporation, Armonk, USA.

z/OS Resource Measurement Facility: Report Analysis (2003). Manual SC34-2665-00, IBM Corporation. International Business Machines Corporation, Armonk, NY, USA.

Index

A

abstraction of reality, 2, 335
analytical solution, 3–4, 126, 132
approximate solution, 4
approximation, 24, 126, 129–131, 134, 137–142, 144, 146, 149–150, 153–158, 161, 163, 165–166, 175, 179–180, 196–199, 207, 212–213
arrival theorem, 202–205, 207, 242
asymptotic convergence, 232–233, 241
asymptotic probabilities, 230
automatic modeling, 306–307, 309

B

balance equations, 16, 22, 24, 28–29, 36, 39–40, 101–102, 107, 109, 193, 203, 210, 212, 215, 221, 238, 240, 245, 264, 266–267, 269, 285, 287, 297, 304, 318
batch means, 88–89, 91, 146
BCMP network, 69–70, 72, 74–75
bottleneck analysis, 316
bound, 291–292, 312–313, 316, 320–322
building block, 307–310, 312

C

central limit theorem, 81, 92
central server, 52–54, 60, 66
class aggregation, 151–152, 256, 259, 261
classes of customers, 68, 70, 75
cloud computing, 97, 181, 261, 278
completion rate, 153, 161, 186
computer networks, 97
computer system, 47, 54, 61, 68, 75
conditional probability, 29, 33, 37, 111, 117, 236, 281–283, 285
confidence interval, 86–89, 91–93, 96
constructive model, 311–312
convergence, 103–105, 110, 114, 116–118, 219–220, 230
Coxian distribution, 27, 200, 234
CPU utilization, 49–50, 52, 73, 75
customer classes, 242–243, 250, 252, 254, 259, 262, 273, 277, 279, 281, 284, 292

D

decomposition, 125–127, 129, 131–133, 141, 163, 165–167, 189
degree of multiprogramming, 47, 49–50
design alternatives, 327
deterministic models, 2
dimensionality curse, 209, 221
discrete-event simulation, 79–80, 82, 84, 93–94

distributional assumptions, 334
distributional dependence, 196

E

Embedded Markov Chain, 25
Engset model, 283–284
equivalence, 125, 131–132, 141, 161, 166
equivalent model, 132, 164, 166
equivalent queueing network, 125
evaluation points, 321, 323

F

First-Come-First-Served (FCFS), 8–9, 21, 52, 55, 58, 68–69, 71–73, 76, 242–244, 247, 250, 252, 259, 261, 263, 278, 284–285, 290
finite capacity, 17, 19
fixed-point iteration, 121, 135–137, 139, 141, 148, 155–156, 158, 164–165, 217, 220, 230, 249, 275–276, 286
fixed-priority, 147
floating-point computation, 101–102
floating-point overflow, 119
full state description, 137, 140, 156, 208, 213

G

generalization of Little's law, 290
global balance, 41

H

heavy-tail, 35
high variability, 36, 94, 97
HOL priority, 146
homogeneity assumption, 303, 306
hybrid solution, 181, 186, 189

I

I/O model, 327
I/O rates, 147, 152
I/O subsystem, 161
ill conditioned, 3
independent replications, 88–89, 93, 146, 152, 221, 292
infinite server, 68
inner model, 132, 163, 166–167
iterative scheme, 103, 106, 109

J

Jacksonian network, 54, 61–62, 64, 68, 75

K

Kendall's notation, 9
kurtosis, 201

L

LCFS-PR, 68, 70–72, 76
Little's law, 14–15, 17–18, 197, 250–252, 289–291, 302, 314, 328
live migration, 181, 185, 189
local balance, 40–41, 53, 62, 70, 75, 280
long-run behavior, 12
loss model, 280, 283
loss probability, 18, 41–43
loss system, 39, 42–43

M

γ-method, 102, 110, 117, 138
M/G/1, 22–23, 25, 196, 328, 331–332
M/G/C, 199, 201, 208
M/G/C/K, 208
M/M/1, 9, 12–13, 17, 19, 58, 62, 75, 83, 85, 89–90, 93
M/M/1/K, 17–20, 29, 33, 41
M/M/C, 193–194, 196–197
M/M/C/K, 193
M/Ph/1/K, 31, 37, 39, 155
M/Ph/C/K, 208–209, 213, 215, 220–221, 223, 233, 236, 238, 240, 245, 264
machine repair, 21
marginal probability, 29, 210, 245
mean value analysis, 64, 66–67, 75

measurable quantities, 297–298, 301, 303
measurements, 4, 35, 298, 305–306, 309, 312, 325, 327, 329, 332–333
memoryless, 9–10, 15–17, 22, 25, 28, 31, 36
method of conditionals, 110, 113–114, 117–118, 125, 131, 188, 198, 220, 247, 267, 269, 286
model calibration, 326–327
model validation, 325–327, 331
multiple classes, 146
multiple servers, 193, 196, 200, 242, 259, 261–262, 284

N

numerical method, 188
numerical solution, 3–4

O

objective function, 308
operating points, 320
operating region, 166
operational variables, 298–299
optical packet switching, 167
outer model, 132, 163
output analysis, 83, 85, 89
overflow, 106, 120

P

PASTA property, 23, 224–225
performance indices, 14, 18, 21
Ph/M/1/K, 31, 33, 154–155
Ph/Ph/1, 331
Ph/Ph/1/K, 153, 155–156
phase-type distribution, 26–27, 31, 35, 208, 221, 235–236, 243, 252, 263, 275
physical machines, 181–182
Poisson arrivals, 23, 25, 31, 35, 41, 75, 169
Pollaczek–Khintchine formula, 23, 196, 201, 328–329, 331, 334
practical implementation, 106
preemptive-resume, 261, 263–265, 271–272, 277
PRI priority, 169, 180
priority, 146, 149–151, 160, 169–170, 173, 175
probabilistic model, 2
processing capacity, 279, 284, 286–287, 289
processor sharing, 68
product-form, 62, 64, 68, 70, 75, 302, 316, 318
pseudo-random number, 80–82
pure delay, 68, 70, 72–73, 75

Q

quasi-Poisson, 16

R

recurrence, 29, 31, 34–35, 38
recurrent solution, 155
reduced-state, 156–157, 209–210, 213, 221–222, 226–227, 235–236, 241, 244, 261, 264, 272
response time law, 300–301, 303
routing chain, 69–70, 72, 74
routing probabilities, 55, 58

S

sampling, 165, 167, 323
Scherr's model, 21, 63
semi-analytical, 110, 186
service center, 55–56, 58, 60, 68–71
service discipline, 22
simulation, 3–4, 125, 141, 144–146, 152, 181, 184, 186–187, 189, 297
simulation time, 79–80, 83
skewness, 198, 200–201, 221, 223
stability condition, 13
state-dependent arrival, 20, 29
state-dependent service, 29
steady-state probability, 22, 28, 32–33, 37
stopping criterion, 105

Student's t-distribution, 86, 87
subnetwork, 126–128, 167
supplementary variable, 25–26
system instrumentation, 329–330

T

tandem network, 133, 137, 139
throughput law, 299
time scale, 125, 131, 141, 181
transaction processing, 15, 166, 314, 317
transient solution, 12

U

utilization equality, 299–300

V

variate, 81, 93
virtual machines, 181–182
virtual memory, 47–48, 72

W

waiting line, 8
warm-up, 85, 88–89, 93
weighted average, 332

www.ingramcontent.com/pod-product-compliance
Lightning Source LLC
LaVergne TN
LVHW020504100826
845148LV00003B/698

* 9 7 8 9 8 1 1 2 9 2 5 2 1 *